Comparative Pathobiology of Viral Diseases

T0204059

Volume I

Editors
Richard G. Olsen, Ph.D.
Professor
Department of Veterinary Pathobiology
College of Veterinary Medicine
Ohio State University
Columbus, Ohio

Steven Krakowka, D.V.M., Ph.D.
Professor
Department of Veterinary Pathobiology
College of Veterinary Medicine
Ohio State University
Columbus, Ohio

James R. Blakeslee, Jr., Ph.D.
Associate Professor
Department of Veterinary Pathobiology
College of Veterinary Medicine
Ohio State University
Columbus, Ohio

CRC Press
Taylor & Francis Group
Boca Raton London New York

CRC Press is an imprint of the
Taylor & Francis Group, an informa business

CRC Press
Taylor & Francis Group
6000 Broken Sound Parkway NW, Suite 300
Boca Raton, FL 33487-2742

Reissued 2019 by CRC Press

A Library of Congress record exists under LC control number:

ISBN 13: 978-0-367-25215-1 (hbk)
ISBN 13: 978-0-367-25221-2 (pbk)
ISBN 13: 978-0-429-28662-9 (ebk)

Visit the Taylor & Francis Web site at http://www.taylorandfrancis.com and the CRC Press Web site at http://www.crcpress.com

FOREWORD

Pathobiology has been introduced as a new expression in medical terminology within the last 3 decades. It reflects the expansion of pathology into related basic sciences (microbiology, immunology, biochemistry, molecular genetics etc.) in order to fulfill one of its traditional missions of redefining the ever changing concepts and principles of disease. Virchow, introducing the concepts of cellular pathology in the last century, already remarked in his keynote address to the Pathology Congress in Berlin that, unless methods are developed to reveal the function of the cellular structures discovered on the microscope, all our work will be in vain. These methods are now available. Modern pathologists (pathobiologists) employ a multitude of methods, ranging from gross observation to molecular genetics, to investigate disease-related problems.

A group of scientists from the Department of Veterinary Pathobiology, Ohio State University (a department combining pathology, microbiology, parasitology, and immunology) presents in 21 chapters of these volumes the current state of knowledge of a selected group of viral diseases in animals to which these authors made substantial contributions over the past decade. This publication reflects several years of cooperative research of a successful team of investigators trained in different disciplines but sharing common research interests. The book does not attempt to cover the whole field of virology but discusses selected viral diseases of broad interest to both veterinary and human medicine. It will fulfill an important need of all investigators involved in the study of viral diseases by providing them (in two volumes) with valuable information presently scattered in many national and international journals.

The title of the volumes, *Comparative Pathobiology of Viral Diseases* appropriately reflects the interdisciplinary team work of the authors. The range of interest is broader than the title indicates since most of the diseases treated in the volumes are excellent models of comparable human diseases. This two-volume set is greatly welcomed at this time.

Adalbert Koestner, D.V.M., Ph.D.
Michigan State University
East Lansing, Michigan
August, 1984

PREFACE

The pathobiology of viral diseases encompasses areas of knowledge of virology and the physiology of the host. These various parameters culminate in the discipline of "biology of viral disease". It was our intent in this treatise to select a few viral diseases of animal species to identify the various virus-host parameters that may be the basis of viral pathobiology. We feel that this level of understanding of viral diseases will be the basis of more effective control of viral diseases by prophylaxis and preventive medicine. Moreover, this approach may be the basis of understanding and developing innate disease resistance in animals of economic importance.

Richard G. Olsen
Steven Krakowka
James R. Blakeslee

ACKNOWLEDGMENTS

The senior editor wishes to acknowledge especially Dr. Eldon Davis, Dr. Lois Baumgartner, and Dr. William Beckenhauser of the Norden Company for their professional and personal friendship and for their many outstanding contributions to animal health research.

Dr. Blakeslee and Dr. Olsen had the distinct pleasure of working with Dr. Herold Cox during his tenure as chairman of the Department of Viral Oncology at Roswell Park Memorial Institute. They are indeed grateful to have been associated with this outstanding virologist, outdoorsman, and gentleman.

THE EDITORS

Richard G. Olsen, Ph.D., is currently Professor of Virology and Immunology in the Department of Veterinary Pathobiology, College of Veterinary Medicine, Microbiology, Department of Biological Sciences, and the Comprehensive Cancer Center, The Ohio State University. Professor Olsen is a native of Independence, Missouri and graduated with a B.A. from the University of Missouri and Kansas City. He obtained a M.S. degree from Atlanta University and a Ph.D. in Virology from the State University of New York (Roswell Park Memorial Cancer Institute Division), Buffalo, New York.

Professor Olsen joined the faculty at Ohio State University in 1969 and since then has developed a graduate program in pathobiology of viral diseases. He holds grants from the National Institutes of Health and the Department of Defense. He has published with his colleagues and graduate students over 200 papers in the fields of virology, immunology, immunopharmacology, and immunopathology. Professor Olsen has patented a unique procedure for the production of a feline leukemia vaccine. This patent reflects the pathobiologic approach of the immunotoxic effects of feline leukemia disease and indentifies the essential viral factor that cats immunologically recognize to resist disease. He is a member of many national and international associations. His current research interests are delineation of the mechanism of the retrovirus-induced acquired immune deficiency in cats, characterization of the preneoplastic events of retroviral disease, delineation of the biochemical mechanism of hydrazine-induced suppressor cell defects, and characterization of lichen planus dermatopathy as a presquamous cell carcinoma.

Steven Krakowka, D.V.M., Ph.D., received his D.V.M. degree from Washington State University in 1971 and a Ph.D. from the Ohio State University in 1974. He is currently a professor in the Department of Veterinary Pathobiology, College of Veterinary Medicine, The Ohio State University. His research interests are neuropathology, virology, and immunology. He is a co-author with Dr. Richard G. Olsen of a previous book entitled *Immunology and Immunopathology of Domestic Animals.*

James R. Blakeslee, Jr., Ph.D., received his B.S. in bacteriology in 1962 from the University of Pittsburgh and his M.S. and Ph.D. in Microbiology from the Roswell Park Memorial Institute Division of Microbiology of the State University of New York at Buffalo in 1971.

Dr. Blakeslee joined the faculty of the Department of Veterinary Pathobiology at the Ohio State University in 1973 and is an Associate Professor of virology and immunology in that department, the Department of Microbiology, and the Ohio State University Comprehensive Cancer Research Center. His research interests are mainly concerned with the interactions and effects of environmental chemicals on virus-induced neoplasias. Current research efforts are directed towards investigation of human and nonhuman primate T-cell lymphotropic viruses and co-factors that may interact with the infected host resulting in frank T-cell leukemias.

Dr. Blakeslee was a National Cancer Institute pre-doctoral fellow and, recently, an International Fellow of the Japan Society for the Promotion of Science while a Visiting Professor at the Institute for Virus Research at Kyoto University, Japan. He is a member of several national and international associations and has been co-editor of several books on comparative leukemia research and a contributor to a text on feline leukemia. He has over 50 publications on the subject of oncogenic viruses, modulators of virus expression, and host immunity.

CONTRIBUTORS

Jennifer Alexander, Ph.D.
Department of Microbiology
Medical University of Southern Africa
Pretoria, Republic of South Africa

S. Aspinall, Ph.D.
Department of Microbiology
Medical University of Southern Africa
Pretoria, Republic of South Africa

James R. Blakeslee, Jr., Ph.D.
Department of Veterinary Pathobiology
Ohio State University
Columbus, Ohio

Anthony E. Castro, D.V.M., Ph.D.
Associate Professor
Animal Disease Diagnostic Laboratory
Oklahoma State University
Stillwater, Oklahoma

Linda Eskra
Department of Veterinary Science
University of Wisconsin
Madison, Wisconsin

Werner P. Heuschele, D.V.M., Ph.D.
Research Department
Zoological Society of San Diego
San Diego, California

David J. Jasko, D.V.M.
Department of Veterinary Clinical Sciences
Cornell University
Ithaca, New York

Steven Krakowka, D.V.M., Ph.D.
Department of Veterinary Pathobiology
Ohio State University
Columbus, Ohio

John F. Long, D.V.M., Ph.D.
Department of Veterinary Pathobiology
Ohio State University
Columbus, Ohio

Michele Miller-Edge, Ph.D.
Department of Veterinary Science
University of Wisconsin
Madison, Wisconsin

Peter L. Nara, D.V.M.
Postdoctoral Staff Fellow
Virus Control Section
National Cancer Institute
Frederick, Maryland

Richard G. Olsen, Ph.D.
Department of Veterinary Pathobiology
Ohio State University
Columbus, Ohio

Colin R. Parrish, Ph.D.
Assistant Professor of Microbiology
Baker Institute for Animal Health
Cornell University
Ithaca, New York

Roy V. H. Pollock, D.V.M., Ph.D.
Assistant Professor of Microbiology
Baker Institute for Animal Health
Cornell University
Ithaca, New York

Gary A. Splitter, D.V.M., Ph.D.
Associate Professor
Department of Veterinary Science
University of Wisconsin
Madison, Wisconsin

Jackie L. Splitter, Ph.D.
Department of Veterinary Science
University of Wisconsin
Madison, Wisconsin

Mary Stiff, Ph.D.
Department of Veterinary Pathobiology
Ohio State University
Columbus, Ohio

Melinda J. Tarr, D.V.M., Ph.D.
Department of Veterinary Pathobiology
Ohio State University
Columbus, Ohio

Darrell Tuomari, D.V.M., Ph.D.
Department of Veterinary Pathobiology
Ohio State University
Columbus, Ohio

TABLE OF CONTENTS

Volume I

Chapter 1*

PATHOLOGY OF VIRAL DISEASES: GENERAL CONCEPTS

Steven Krakowka

TABLE OF CONTENTS

* Supported in part by Grant No. NS14827-NIH, PHS.

I. INTRODUCTION

Recognition and delineation of the virus-specific etiologies of disease is a science that has now come of age. Lesions formerly attributable to bacterial infection, metabolic disorders, and/or inherited defects are now known to be due to viruses. However, there are very few lesions, either grossly or microsopically that can be definitely attributed to a single viral agent. Rather, general patterns of injury are recognizable. These patterns of injury, with study, have been delineated for certain groups of viruses and certain types of host/virus interactions. For descriptive purposes, a viral infection can be viewed as occurring in one of two general forms: an acute form in which the course of disease is limited in time and space, and a chronic form in which the viral disease process and host/virus interaction tend to occur over a long period of time.

A. General Features of Acute Viral Diseases

Though not invariably true, acute viral diseases tend to be accompanied by florid but short term (2 to 3 weeks) virus replication cycles in tissues whereas the chronic or persistent viral infections tend to have low (or difficult to detect) amounts of infectious virus in tissues. Further, chronic forms of viral disease tend to be dominated by antiviral immune responses and/or host tissue proliferation.

The accompanying clinical signs and/or structural lesions characteristic of the acute interaction may follow one of several different patterns. For many viral diseases, clinical signs are coincident with the appearance of virus in the tissues and thus the lesions and/or signs of disease are directly attributable to the virus. An example of this type of acute, selflimiting infection would be canine parvovirus infection or feline panleukopenia virus infection in intestinal epithelium and lymphoreticular tissues of the respective host species.

A second variant of this acute, selflimiting disease process is that seen in typical latent viral infections. Of these, a classic example is Herpes simplex virus infection of man. Throughout the course of the animal's life or person's life, a series of acute clinical episodes occurs which directly coincides with and occurs consistently with shedding of virus or infectious viral material. In this type of host/virus interaction virus can regularly be recovered during the acute clinical episodes but is not readily detectable during the clinically normal episodes.

A third type of acute infection process is that which occurs with many of the noncytopathogenic viruses of both animals and man; that is the host experiences a one time only acute disease in the usual instance. Clinical signs, if detectable, are limited to that single occurrence although the virus may exist as a subclinical or inapparent infection for the life of the animal. There is minimal evidence of clinical disease associated with this and one would predict that disease manifested by these agents is very unusual. An example of this sort of host/virus interaction is equine herpesvirus 2 infection, a leukocyte-associated herpesvirus infection.

B. General Features of Chronic Viral Diseases

The second major type of host/virus disease syndrome is that of chronic virus-associated disease. Operationally, it is distinguished from the acute forms by the fact that manifestations of disease persist for long periods of time after the initiating infectious event. In these chronic viral diseases it is important to distinguish the effects of a permanently damaged organ vs. damage mediated by the presence of viral material in the tissue. It can be readily appreciated that in *in utero* viral infections, the virus need no longer be present to have produced a permanent deficit in the postnatal animal or human. In fact, such is the case with feline panleukopenia virus infection *in utero*. In this instance, the virus destroys rapidly dividing cells of the external granule cell layer

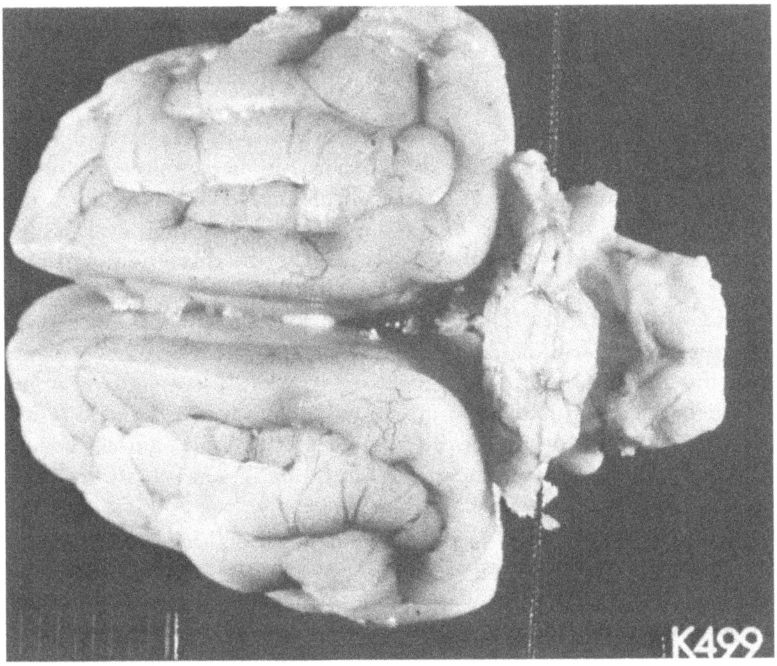

FIGURE 1. Cerebellar hypoplasia in a cat infected *in utero* with feline panleukopenia virus.

of the cerebellum. The net result of the host/virus interaction is varying degrees of postnatal cerebellar hypoplasia (Figure 1). Thus, virus need no longer be present to produce a static and permanent neurological deficit. Similarly, it is known that certain parainfluenza viruses, as well as certain measles virus variants can produce hydrocephalus in infants infected *in utero*. The lesion seems to occur secondary to virus-mediated destruction of the ependymal lining cells of the ventricles. The clinical signs of disease are not related to the viral infection but rather to the progressive dilation of the ventricular system and subsequent loss of cerebrocortical nervous tissue. In both of the instances cited above, the static sequelae (i.e., the cerebellar hypoplasia) or the progressive sequelae (i.e., the hydrocephalus) to acute, selflimiting disease, viral antigen and/or infectious virus may no longer be present during the clinical expression of the disease process.

Occasionally, one can identify both the virus and the presence of chronic progressive lesions attributable to virus infection. In these clinical conditions, the course of disease is slow and, of course, progressive, occurring over a period of months to years and, for the individual, is frequently fatal. Examples of host/virus interactions in which progressive disease associated with progressive infection occur, are old dog encephalitis, a canine distemper virus-associated neurologic disease, scrapie and related spongiform encephalopathies including transmissible mink encephalopathy, and the nononcogenic lentivirus infections of animals, namely visna, maedi, and equine infectious anemia.

The final form of chronic host/viral interaction that may or may not have expressed virus associated with it, is that of virus-induced neoplastic transformation, or cancer. It is now well recognized that a number of different viral agents of the oncornavirus group and also the herpesvirus group are capable of inducing malignant transformation in susceptible host cells. This transformational event can occur in the presence of the etiologic agent as occurs with most cases of feline leukemia virus infection, or it can occur in the absence of demonstrable complete viral material, as occurs with Marek's disease, a herpesvirus infection of chickens, certain of the herpesvirus-induced lymphomas in monkeys, and in Epstein-Barr virus-associated Burkitt's lymphoma of man.

Given this generalized overview of possible host/viral interactions, it would not be surprising to realize that the types and kinds of virus-induced lesions within the respective hosts will vary tremendously. As a general rule, acute viral infections tend to produce specific cytopathic or cellular damaging events in an organ and tissues. This is frequently accompanied by specific and nonspecific inflammation. If one is lucky, direct evidence of viral infection, namely the presence of inclusion bodies or other structural materials can be demonstrated. In contrast, chronic viral infections may or may not have direct virus-associated lesions accompanying the disease. However, many chronic viral infections include either local or systemic tissue and/or organ related structural deficits attributable to viral infection. Thus, virus-associated lesions within tissues can range in complexity to those of the viral effect only, namely cellular degeneration, cellular cytolysis, syncytium formation, or malignant transformation. Lesions can include the virus-specific lesions noted above plus the specific and nonspecific manifestations of both the humoral and cellular inflammatory components of the inflammatory response. Within a tissue, this inflammatory response may well include both vascular components and cellular components, and like any inflammatory response, may be both beneficial and harmful to the host tissue or organ. Finally, if the viral infection is in the repair stages when observed, all of the above lesions, plus evidence of host tissue repair may be observed. In this instance, structural changes may well include parenchymal cell mitosis and proliferation. If the tissue, however, is incapable of producing this response, scar tissue or healing by fibrosis may be observed.

For diagnostic pathologists and virologists of today, the gross and histopathologic diagnosis of a viral infection can be rendered with confidence if the lesion detected is characteristic. These pathognomic interactions are uncommon or transient. Fortunately, however, many viral infections are accompanied by virus-associated structural changes within a tissue. These include the presence of intracytoplasmic and intranuclear inclusion bodies and syncytial giant cell formation. Occasionally, cytopathic changes are diagnostic as is the appearance of scrapie in sheep neurons (Figure 2). Of course, one would prefer to recover the infectious agent from the tissue of the suspect animal thereby permitting identification of infectious agent by conventional means. Unfortunately, of course, not all of these procedures are available or necessary for identifying viral infections of animals and, in fact, very infrequently is the diagnosis dependent upon virus recovery in tissue culture. One way of circumventing this diagnostic problem is to apply modern techniques of viral antigen tracing to tissues. This, of course, is not a new idea and in fact, frozen or unfixed tissues have for years been examined for viral antigen by antibody tagging techniques, such as the immunofluorescence and immunoperoxidase procedures. Recent technical advances have permitted the application of these same technologies to formalin-and/or glutaraldehyde-fixed tissues. Thus, identification of viral structural proteins within tissues, I believe, is the most convenient, most readily accessible, and the most dependable way of ascribing a specific lesion to a specific viral infection or viral infectious event, and it is this area of diagnostic histopathology that will experience a great deal of growth and sophistication within the next several years.

II. VIRUS-INDUCED HOST CELL DAMAGE AT THE CELLULAR LEVEL

A. Lytic Infection

It is beyond the scope of this chapter to discuss in detail the range of virus/host cell

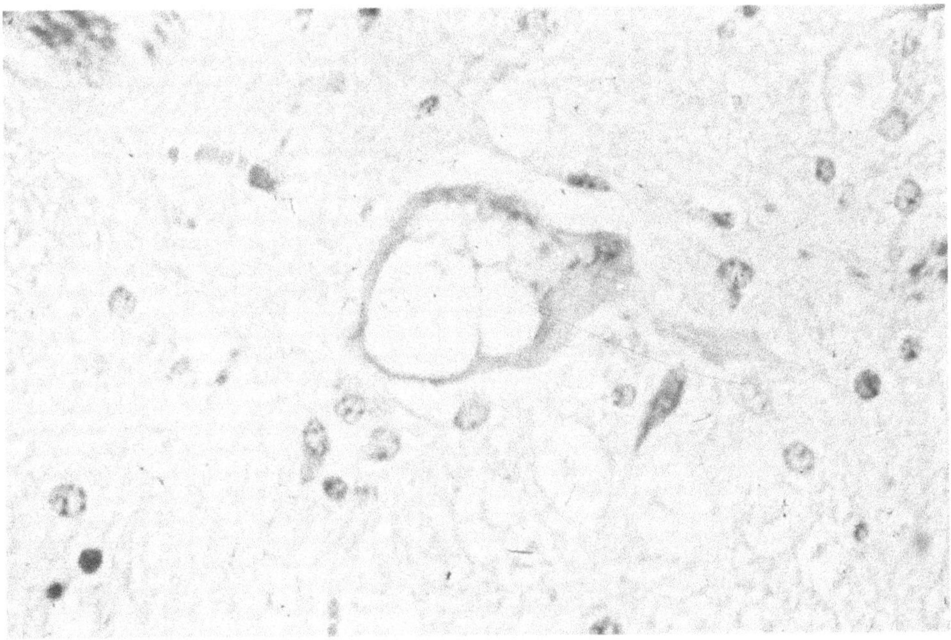

FIGURE 2. Neuronal vacuolar change in a neuron from a sheep affected with scrapie. (Courtesy of D. Morton, D.V.M., The Department of Veterinary Pathobiology, The Ohio State University.)

interactions that can be delineated by today's biochemists and cellular biologists. Rather, it is the purpose of this segment of the chapter to outline briefly the major types of host/viral interactions of pathologic importance. Perhaps the most fundamental morphologically recognizable structural alteration induced by a typical virus on a host cell is that of progressive cellular swelling. This change in the size and shape of an infected cell occurs with time and is thought to be due to viral protein-mediated inhibition or interference with the membrane-bound sodium pump mechanisms in the infected cell. This can occur directly by mechanical interference of viral proteins within the host cell membrane, or indirectly by virus-associated subversion of host cellular protein synthetic mechanisms. By whatever mechanism, the virus-infected cell will experience progressive cellular swelling as water enters into the cytoplasm. This cellular swelling has the effect of disrupting cellular organelles, dilating rough and smooth endoplasmic reticulum, and swelling mitochondria. In tissues this acute swelling phenomenon can be observed as progressive cloudy swelling and hyropic degeneration of virus-infected cells (Figure 3). A late manifestation of this cellular swelling and degeneration phenomenon in vivo is mineralization of cellular organelles, chiefly mitochondria. This tends to occur in tissues which have a high intrinsic metabolic rate related to calcium ion exchanges across cellular membranes such as skeletal or cardiac muscle (Figure 4). Of course cell and tissues do not swell indefinitely. One of two choices is presented to this cell: it may either rupture, discharging its contents including both complete and incomplete virus for subsequent infection of adjacent cells or the cell may experience some sort of fusion phenomenon with some or several uninfected adjacent cells. This fusion phenomenon is invariably mediated by virus-coded envelope glycoproteins. In both instances, however, the infected cell is permanently damaged and rendered nonfunctional by the virus.

B. Persistent Infection

The second major type of viral host cell interaction is that of the steady state or

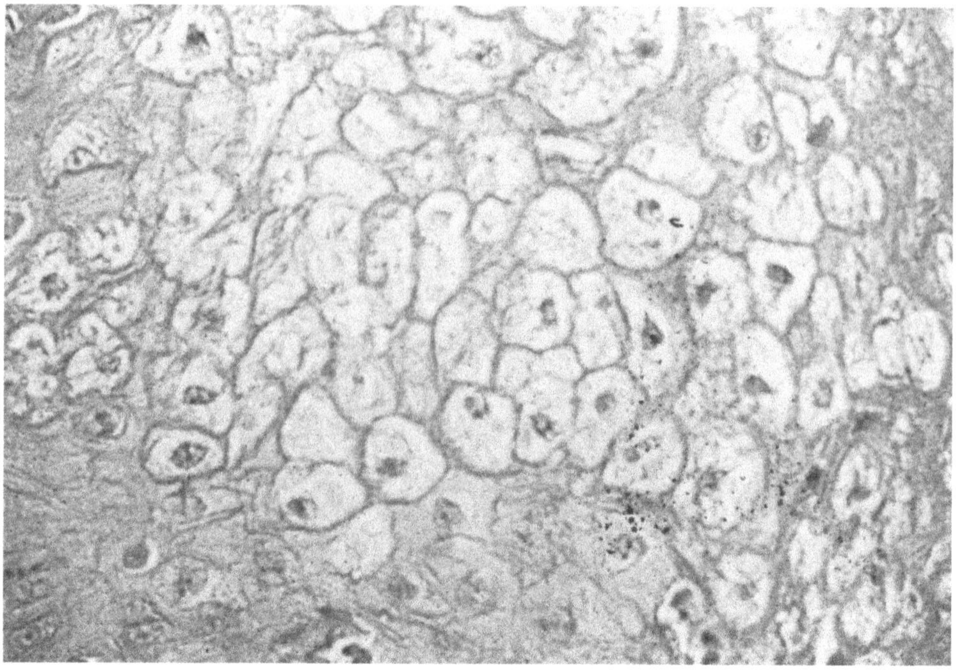

FIGURE 3. Bovine papillary stomatitis virus infection of the hard palate which produces progressive acute cellular swelling and hydropic degeneration.

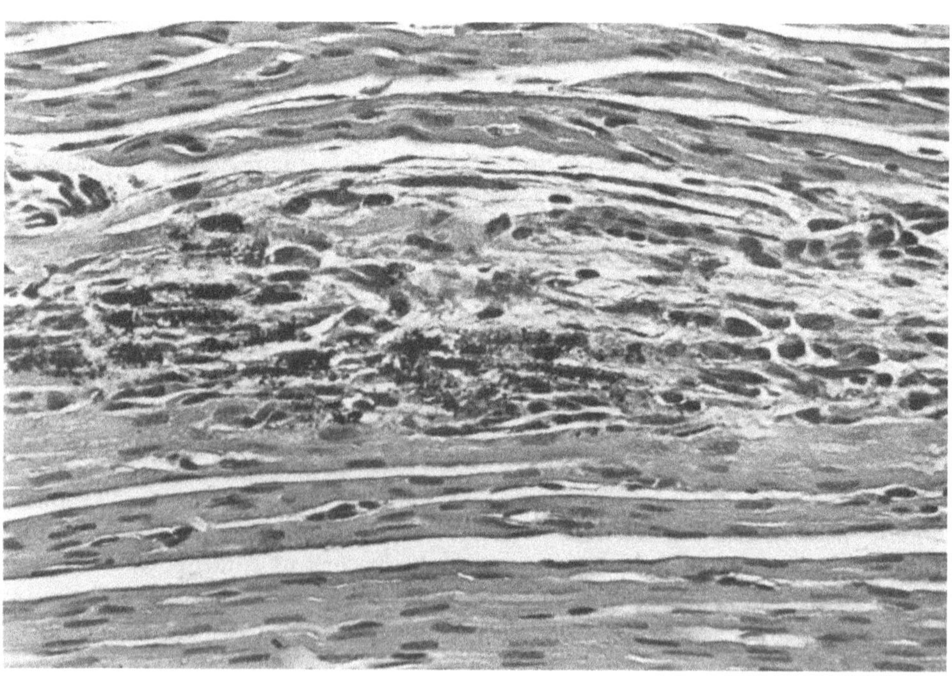

FIGURE 4. Mineralization of the myocardium in a young dog fatally affected with canine distemper virus.

persistent infection. In these conditions, a state of persistence is established in which cellular lysis is not an obvious or major component of the host/virus relationship. Rather, the host cell will shed viral material, whole infectious virus, or viral polypeptides either intermittently or constantly throughout the life of the cell. There is little doubt that these cells experience degenerative or retrogressive changes vs. their uninfected counterparts, but this type of degenerative change is subtle and difficult to delineate. The steady state or persistent virus infections have a number of ramifications, not only for the virus but also for the host tissues.

C. Transformation Infection

The third fundamental type of host cell/virus interaction is that of malignant transformation. It is clearly beyond the scope of this chapter to provide a detailed discussion of the mechanisms of transformation associated with the oncogenic viruses. One of the tenets of modern biology that has emerged from the virtual explosion in molecular genetics is the discovery that normal host cells contain genes whose expression is associated with malignant transformation. These oncogenes are normally very strongly repressed in an untransformed cell. These oncogenes hybridize with, and thus, are identical to similar gene sequences found in retrovial nucleic acid. It is thought that these oncogenes are activated during the process of transcription and translation of either proviral DNA from the retrovirus group, or secondary to and associated with insertion of oncogenic herpesvirus DNA into cellular DNA. The product(s) encoded by these oncogenes has not been determined, but they are most likely protein kinases or similar regulatory polypeptides. In any event, expression of this oncogene results in an irreversible transformation of this normal cell to a cell with an immortal lifespan, which is no longer responsive to exogenous or endogenous controls on proliferation. The net effect of this is development of malignant tumors in the host.

Given these general categories of changes, that is, a standard lytic infection, a steady state or persistent infection, and a transformation infection, it is easy to appreciate that the tissue response to each of these kinds of host/virus interactions will vary.

III. TYPES OF TISSUE RESPONSES TO VIRAL DAMAGE AND INJURY

A. Structural Changes in Cells Directly Attributable to Virus Replication

As indicated above, the simplest and most fundamental of the cellular responses to injury in viral infection is that of progressive hydropic degeneration and eventual cytolysis (Figure 3). This is a frequently detected component of many viral diseases. Generally speaking, this cytolytic event coincides with peak appearance of clinical signs and, of course, is responsible for and facilitates recovery of viral infectious material from these tissues. In the animal these cytolytic events are most obvious on epithelial surfaces of the skin, oronasal mucosa, and gastrointestinal tract. In these areas, the lesions progress in a centripetal fashion from the foci of infection. A gradation of lesions from peripheral unaffected areas to central cytolysis is seen. The net result of this type of interaction is the formation of interepithelial vesicles which eventually rupture to produce either erosions or ulcerations (Figure 5). This is an important diagnostic feature for the vesicular diseases of animals, namely foot and mouth disease, vesicular stomatitis, and vesicular exanthema. The lesion is also seen with various pox virus infections, local herpesvirus infections, and viral infectious diseases such as bovine virus diarrhea and mucosal disease complex.

Acute cellular necrosis and progressive cytolysis is, of course, not limited to epithelial surfaces. In systemic viral diseases, foci of infection can be found virtually anywhere. These expanding foci of coagulation necrosis may involve many parenchymal cells as occurs with murine hepatitis virus infection (Figure 6) or only single cells as occurs in young dogs infected with canine distemper virus (CDV) in which CDV-asso-

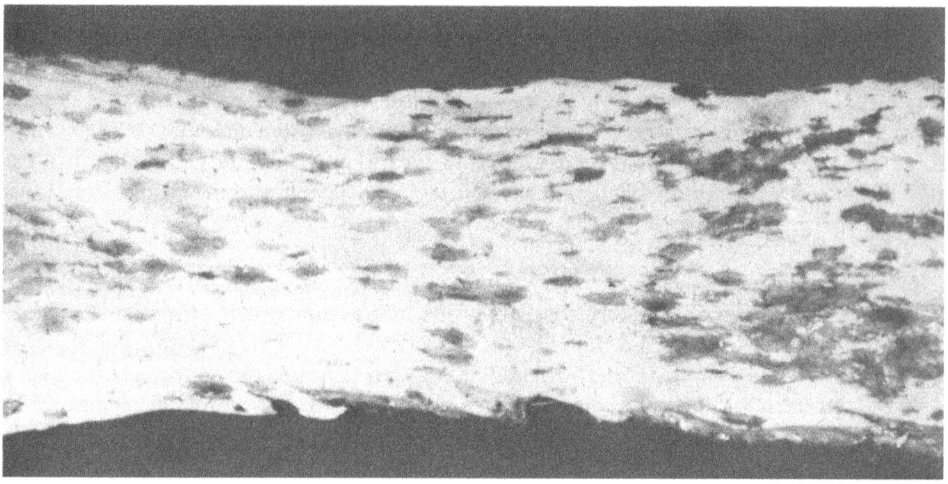

FIGURE 5. Mucosal disease (bovine virus diarrhea virus)-induced esophageal lesions and ulcerations.

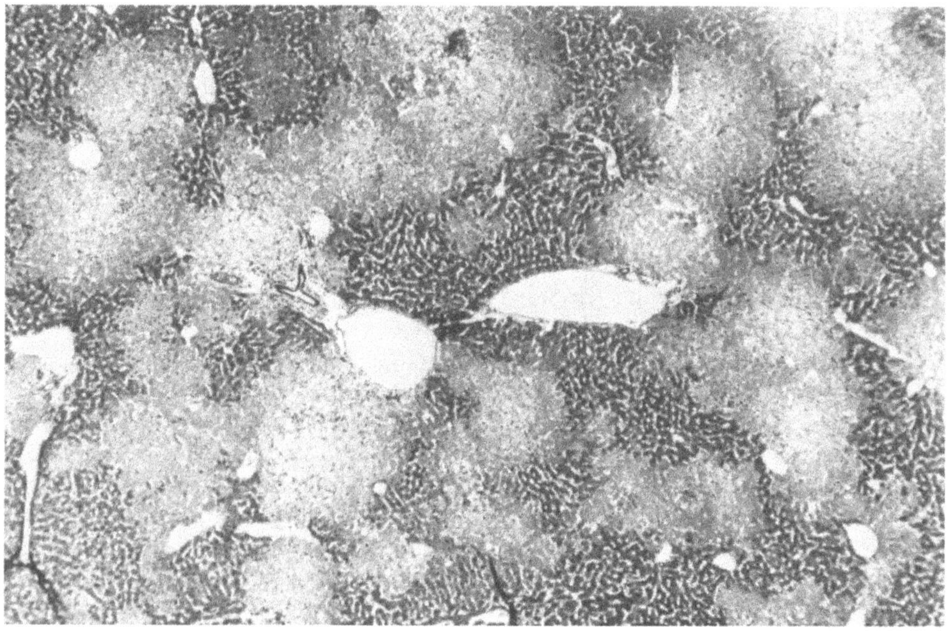

FIGURE 6. Multifocal hepatocellular coagulation necrosis in a mouse fatally affected with mouse hepatitis virus.

ciated single cell necrosis in the brain is recognized histologically as a "dark" neuron (Figure 7). In considering the overall significance of cellular death, it is important to determine if the cytolytic event occurred in a labile, that is, normally rapidly replicating cellular population such as epithelium, or if it occurred in a cell type in which postnatal replication does not occur such as neurons. Thus, depending on the extent of involvement, infection of labile cellular populations can be resolved without permanent injury. However, infection and cytolysis of permanent cells of the brain, optic nerve and

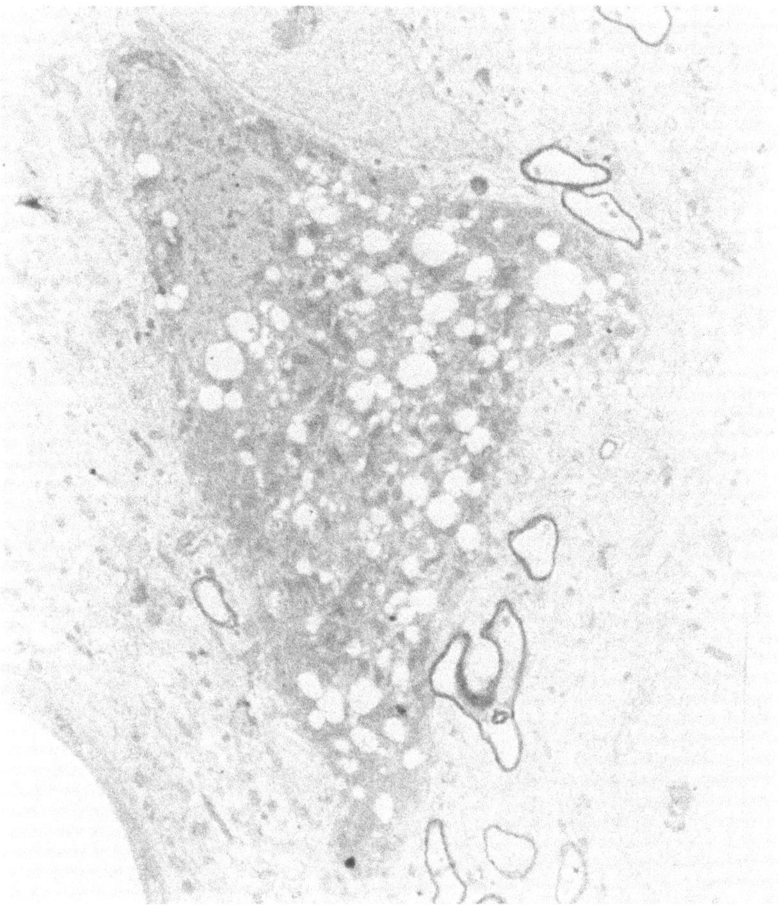

FIGURE 7. Acute neuronal necrosis induced by canine distemper virus infection in a young dog.

tract, and adrenal medulla, will produce permanent damage and is, in effect, a permanent organ or tissue deficit.

A second tissue change directly attributable to viral replication is the formation of multinucleated syncytial giant cells. In this form of tissue damage, infected cells fuse with one another, having thus a common cytoplasm, a common cytoplasmic membrane, but multiple nuclei (Figure 8). This type of cytopathology is common and is characteristic of those viral diseases with lipid envelopes, namely the paramyxovirus-orthomyxovirus group which includes morbilliviruses and related viruses. For DNA viruses, syncytial formation is notable at least with the herpesvirus group. Syncytium formation is mediated by virus-specific envelope glycoproteins, principally the fusion (F) polypeptide and hemagglutinin (HA)-equivalent glyprotein. Cell fusion is thought to occur by bridge formation between viral glycoproteins integrated into infected cell membranes and the adjacent normal cellular membrane. Syncytial giant cell formation is known to occur during the acute phase of infection. Although these giant cells are not necessarily dead there can be little doubt that these cells eventually will suffer from either direct viral cytolysis or from cell death secondary to immune and nonimmune mediated clearance of the viral infection. It is important to point out that cellular fusion phenomena are thought responsible for the subsequent cell to cell transmission of viral infections in the face of protective immune response, and thus serve as one

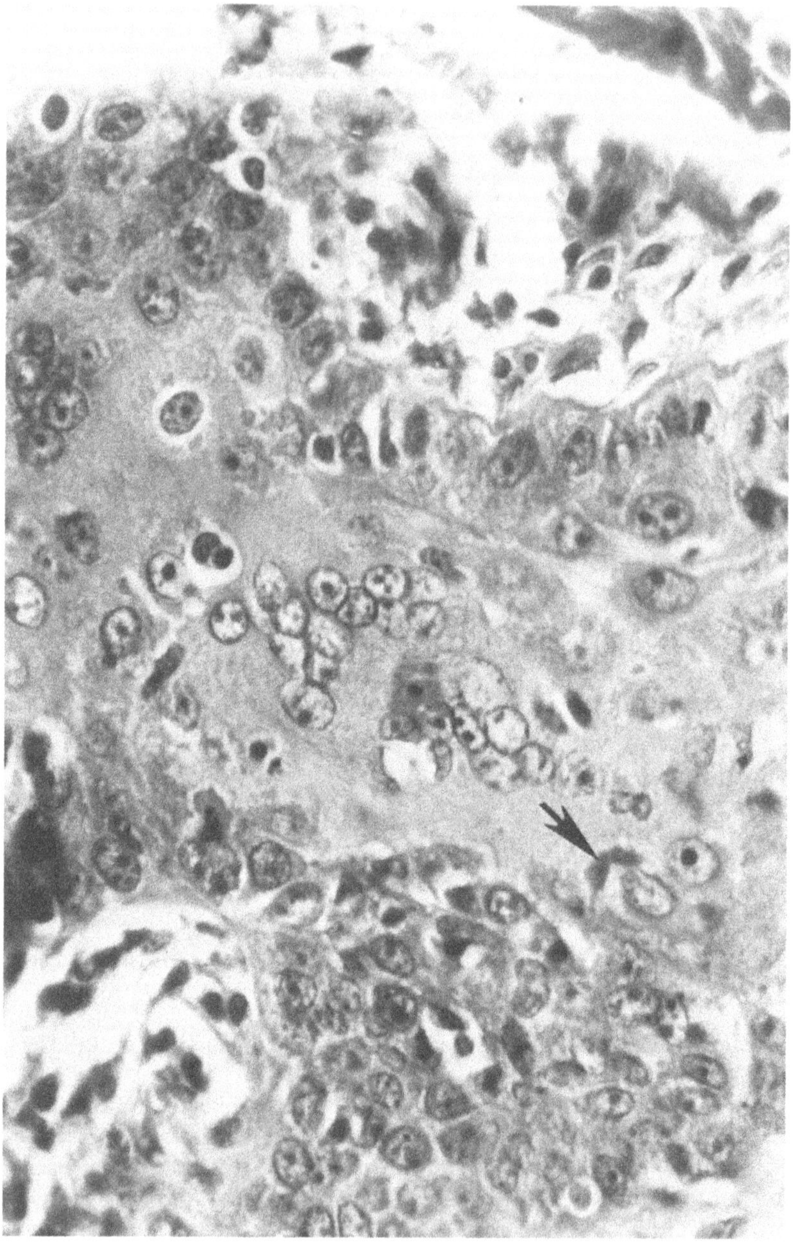

FIGURE 8. Syncytial giant cell formation in the epithelium of the foot pad of a ferret infected with canine distemper virus. Note the eosinophilic cytoplasmic viral inclusion bodies (arrow).

mechanism whereby a viral infection is maintained within the tissues of an actively immune individual.

Of course, one of the most diagnostic features of viral infection in tissues is the development of either intracytoplasmic or intranuclear virus-associated inclusion bodies. Depending upon the virus under consideration, the inclusion body can consist of soluble structural viral protein only, a complex of viral protein and viral nucleic acid or, in fact, complete infectious virions. These structures are recognized by light mi-

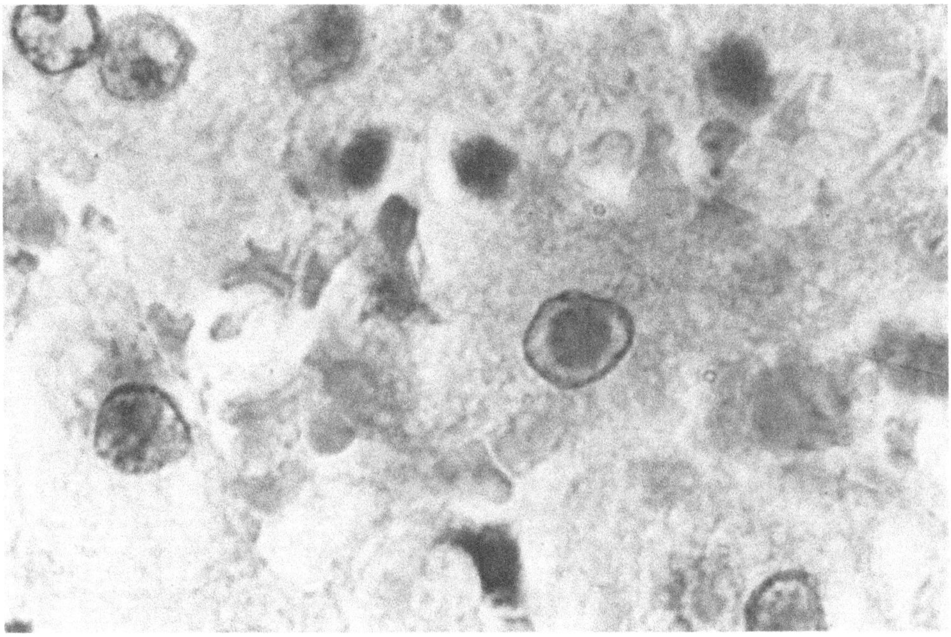

FIGURE 9. A lightly basophilic intranuclear inclusion body in a renal tubular epithelial cell of a dog with infectious canine hepatitis.

croscopy as homogeneous aggregates of dense eosinophilic, or sometimes slightly basophilic intracytoplasmic or intranuclear structures. Generally speaking, there is rarefaction of either the cytoplasm or nuclear contents around these inclusion bodies, producing a halo-like appearance (Figures 8 and 9). With practice, the diagnostician can readily assign a type of specific viral disease to the presence of viral inclusion body material. Of course, it is important, in fact imperative, to point out that the presence of these visible aggregates of viral infection within tissues greatly underestimates the extent of viral involvement within this tissue.

A third virus-encoded specific cytopathic change noted in tissues is that of local cellular proliferation. This local proliferative response can occur in the recovery phases of many viral infections. It is characteristic of viral infections by certain of the pox (Figure 10) and papillomavirus (Figure 11) subgroups. There is, in fact, some controversy as to whether the latter local host cell (epithelial or underlying mesenchymal origin) proliferative responses represent infection-associated hyperplasia, or whether it represents true virus-induced cellular transformation. Viral material, either antigen or nucleic acid can usually be demonstrated within these proliferative tissues, and while this controversy cannot be resolved with information presently available, this author still feels that these local forms represent local benign neoplastic transformation rather than just hyperplasia.

The final type of cellular response to viral damage, is that of malignant transformation and subsequent proliferation of these transformed cells (Figures 12 and 13). As indicated above, the viral agents may or may not be demonstrable in the tissues at the time the tumor becomes histologically or grossly evident. Nonetheless, the characteristic feature of this type of virus-induced cytopathic effect is the anointing of the transformed cell with the capacity for uncontrolled and/or uninterrupted cellular proliferation. Accompanying this change is alteration in cellular membrane characteristics, including expression of novel tumor-associated antigens. In general the morphological

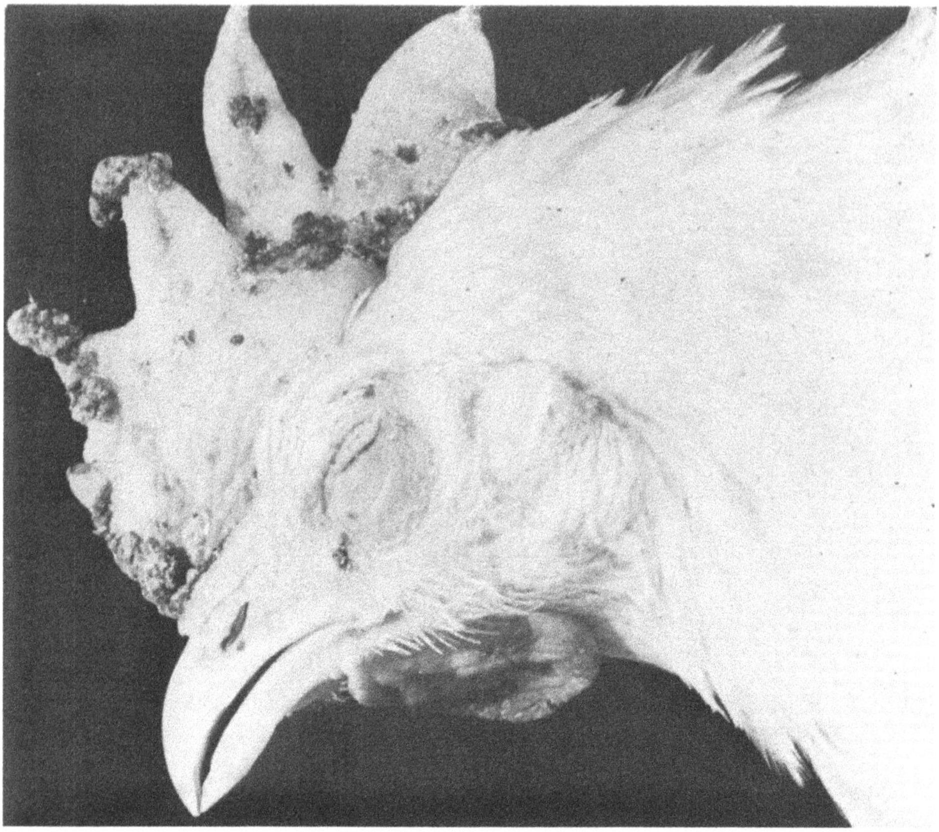

FIGURE 10. Epithelioproliferative lesions of fowl pox.

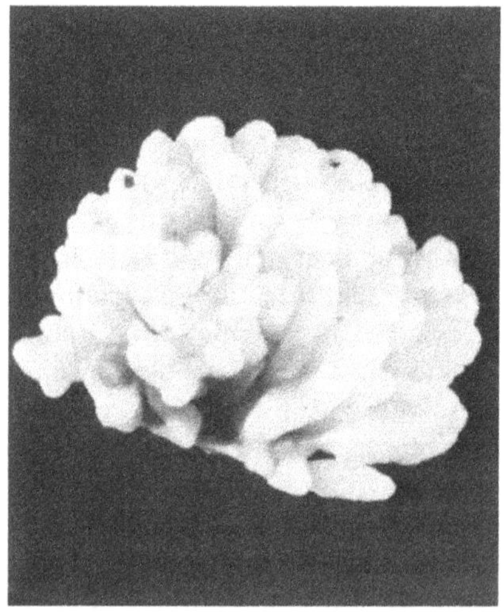

FIGURE 11. Equine cutaneous papilloma.

qualities of these transformed cells include a reduction in cytoplasmic to nuclear ratio, the development of large and pleomorphic nuclei with prominent nucleoli, the presence of many mitotic figures, and other bizarre cellular forms.

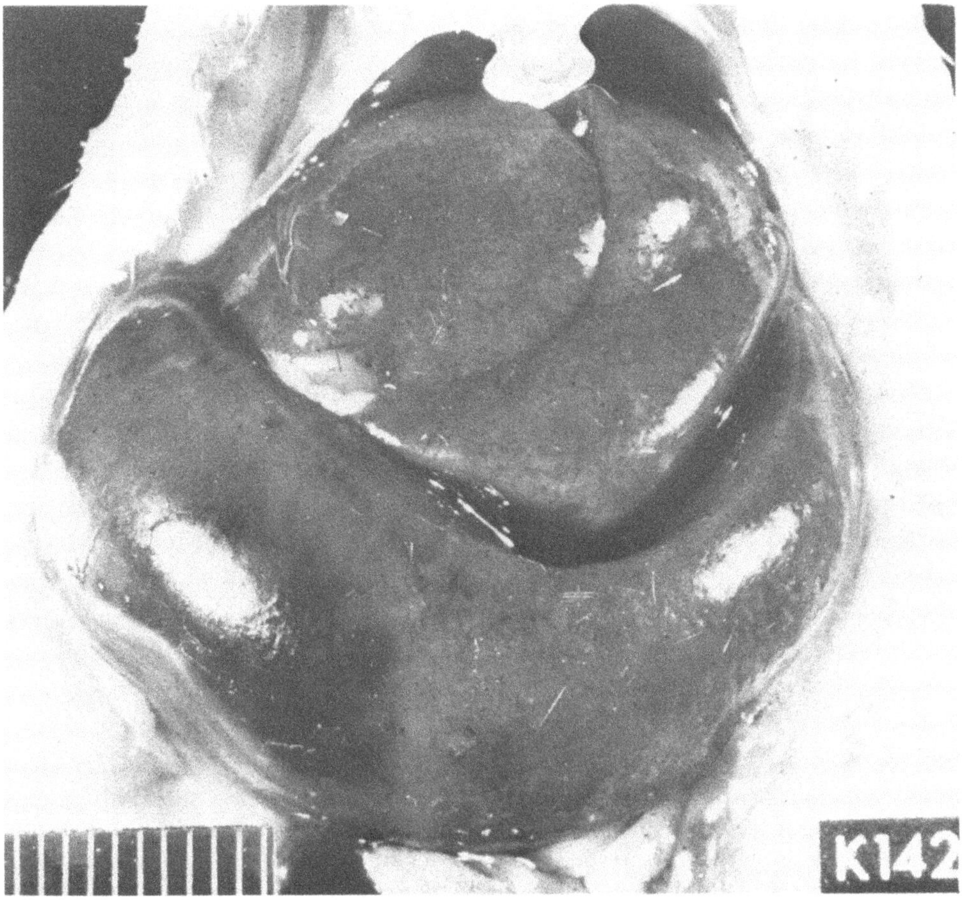

FIGURE 12. Hepatosplenomegaly in a mouse infected with Rauscher leukemia virus.

B. Local and Systemic Tissue Damage Attributable to Inflammatory Responses to Viral Infection

The virus induced cytopathic effects outlined above, if they occurred in isolation and/or without apparent host cellular responses, would be relatively easy to deal with diagnostically. However, both the specific and nonspecific inflammatory response to viral infection which invariably accompany these various types of cellular damage, complicate interpretation in the appearance of the gross and histopathologic findings. Of the major classes of pathogens affecting vertebrate species, it probably can be safely said that viral proteins *per se* do not exert a true chemotactic stimulus for phagocytic defense cells. While this may seem a theoretical and minor distinction, it does have some import in considering the range of inflammatory responses which viral infections experience. Thus, unlike bacteria, there are no endotoxin-associated effects and there are no manifestations of disease attributable to the chemotaxis of soluble products produced by the bacteria. Thus, in may respects the host tissue inflammatory response to viral infection is simplified and can be viewed largely as a consequence of virus-induced cellular degeneration and necrosis.

The local area will experience a nonspecific inflammatory response. Like many acute inflammations this has both vascular and cellular exudative components. At least in this early phase of host response to injury it is important to point out that this inflammatory response is mediated by released host cell constituents and not viral material.

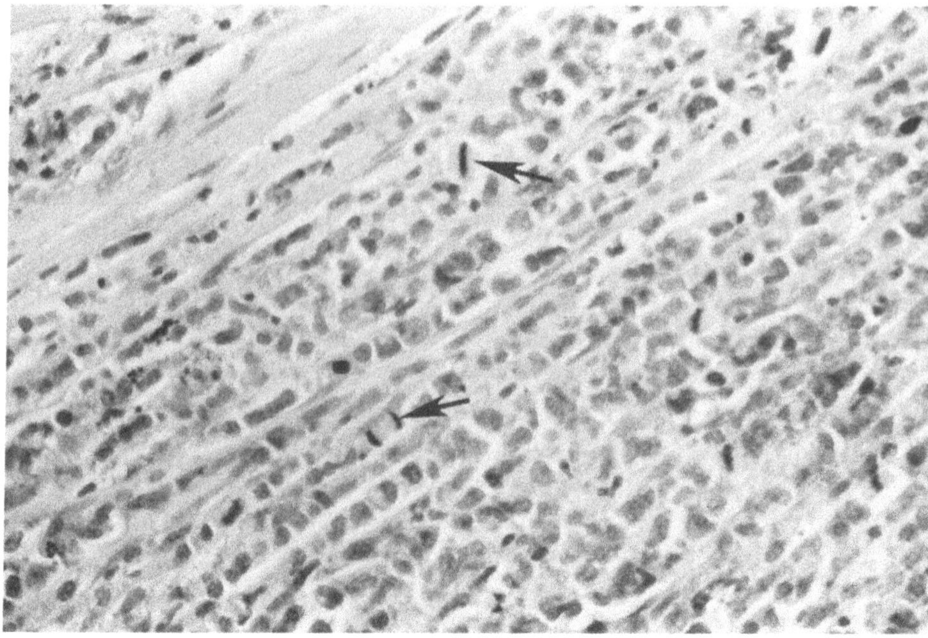

FIGURE 13. Bovine leukemia virus-induced lymphosarcoma. Note the mitotic figures (arrows).

Released lysozymes, proteases, and other substances may well affect local blood vessels producing vasoconstriction and/or increased capillary permeability, thereby providing a mechanism for leukocyte immigration into the tissues. Activation of the soluble mediators of the inflammatory response such as components of the blood coagulation system, complement components, and/or vascular kinins all contribute to fluid cellular extravasation (Figure 14).

In any event, this tissue response to cellular necrosis mediated chiefly by factors released from dead and dying host parenchymal cells will produce a transient immigration and appearance of inflammatory cells, chiefly polymorphonuclear (PMN) leukocytes. It is important to stress, however, that this neutrophilic response is transient and in most viral diseases a minor inflammatory cell component. A curious exception to this, however, are the prominent neutrophilic infiltrates in the brain of horses with eastern, western, or St. Louis encephalitis virus infection (Figure 15).

Since most viral proteins are immunogenic, this transient response is rapidly superceded by virus-specific immune inflammatory responses. Not only is an inflammatory response mediated by humoral and cellular factors possible as the disease progresses, but there exist numerous amplification pathways for enhancing the inflammatory response, thus rendering the final process somewhat nonspecific in appearance. The net effect of this developing and progressive antiviral response is the accumulation of leukocytes within these areas of viral production. However, unlike the transient inflammatory response induced by cellular necrosis, most of the leukocytes noted in this later phase of infection are of mononuclear cell origin (Figure 16). Lymphocytes of T-cell and B-cell lineage are often prominent at this stage of the disease. The major cell type active in this stage of inflammation is the mononuclear phagocyte or, in the brain, activated microglia (Figure 17). The macrophage component of the inflammatory response may well predominate producing a granulomatous reaction as occurs with visna virus infection in the brain of sheep (Figure 18). Both local and systemic antibody production will amplify this cell-associated inflammatory lesion. A

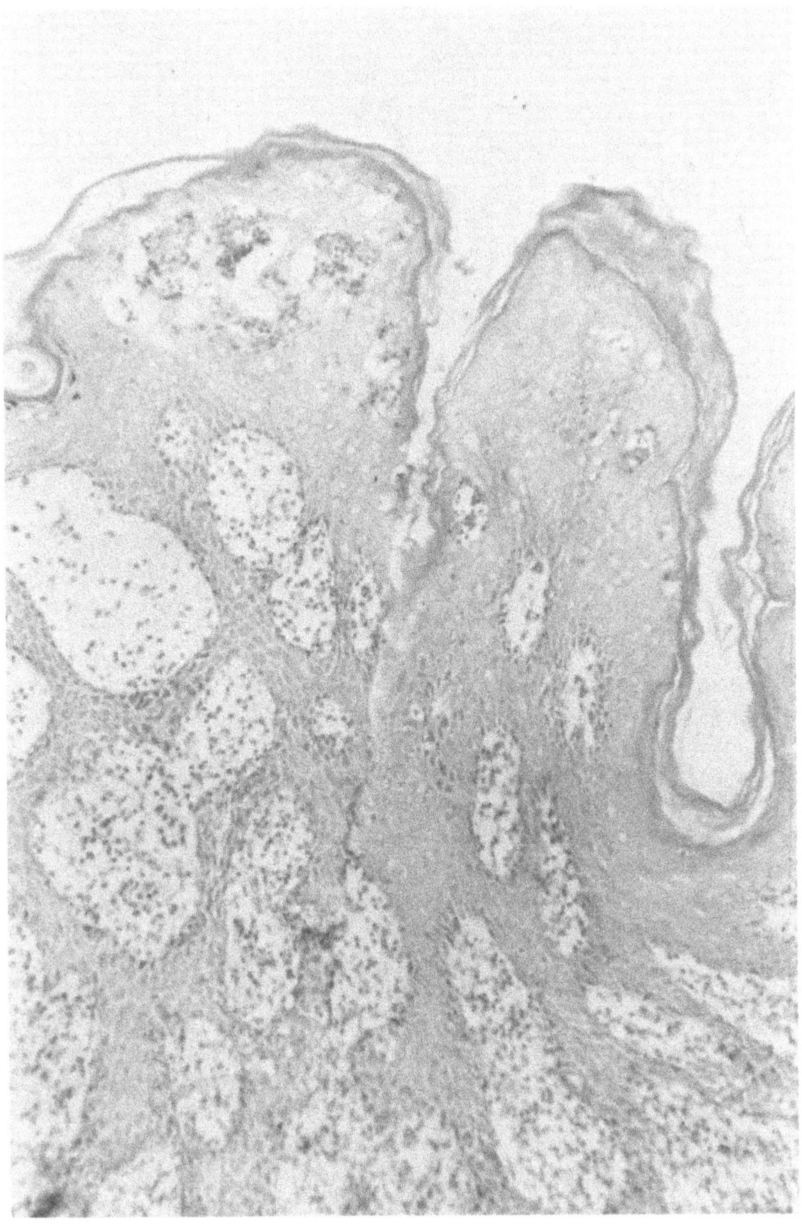

FIGURE 14. Subepithelial edema and related leukocytic transmigration in contagious ecthyma infection of sheep.

chief function of antibody is to neutralize extracellular infectious virus. Nonetheless, antibodies can function to direct the immune response to virus-infected cells through the mechanisms of an antibody-directed cellular cytotoxicity (ADCC) or by complement-dependent cytolysis. A third mechanism whereby antibody is capable of affecting the local inflammatory response is by modulating or removing viral proteins from the surfaces of infected cells without killing them.

A cytotoxic T-cell effector response is an integral part of the inflammatory process, again particularly in the acute selflimiting viral diseases. In general, cytotoxic T-cell

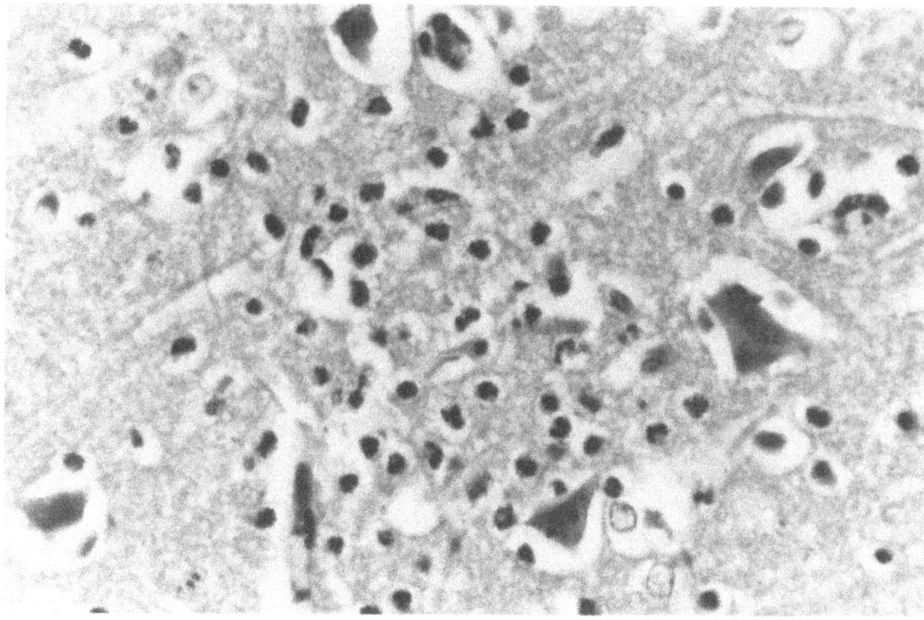

FIGURE 15. Neutrophilic infiltrates into the cerebral cortex of a horse infected with eastern equine encephalomyelitis virus.

responses tend to be short-lived and, of course, virus specific in nature. T-cells mediate their effect by direct or intimate contact with virus infected cells and subsequent release of factors which have the effect of lysing and destroying the affected cells, but also adjacent uninfected cells. This, of course, produces more cellular debris for incitement of other components of the inflammatory response and thus further, in general, is responsible for the nonspecific recruitment of inflammatory cells into a lesion of the disease progresses.

It is important to recognize that, in addition to this local epithelial or parenchymal cell effect, that viral infection, at least in the acute phases, is frequently accompanied by a vasculitis. This vasculitis may have the effect of inducing additional nonspecific tissue damage via hemorrhage and ischemia (Figure 18), but of more importance, the vasculitis may very well provide a mechanism for egress of inflammatory factors into the areas of virus-induced cellular damage. Vasculitis is a frequently overlooked component of a tissue response to viral infection and may be localized or generalized in nature. In most instances the vasculitis can be attributed to some direct viral interactions with the endothelial cells (Figure 19). However, it is known that vasculitis can occur indirectly by induction of disseminated intravascular coagulation-like phenomena, as occurs with Newcastle disease virus in chickens, and also Dengue virus infection in man. Also, vasculitis can occur indirectly in the form of circulating immune complexes in which viral antigen and antibody complex to complement are present in the circulation and produce local or systemic effects on the blood vascular system by deposition and subsequent generation of an acute inflammatory response.

Organs or tissues can be indirectly affected by viral disease if circulating immune complexes are produced. In this instance, the virus has gained access to systemic circulation and incited an acute intravascular inflammatory response. The net result is development of immune complex disease and related phenomena with subsequent deposition in tissues such as the renal glomerulus (Figure 20) far from the site or origin of production of this viral material. In the usual instance, immune complexes produce

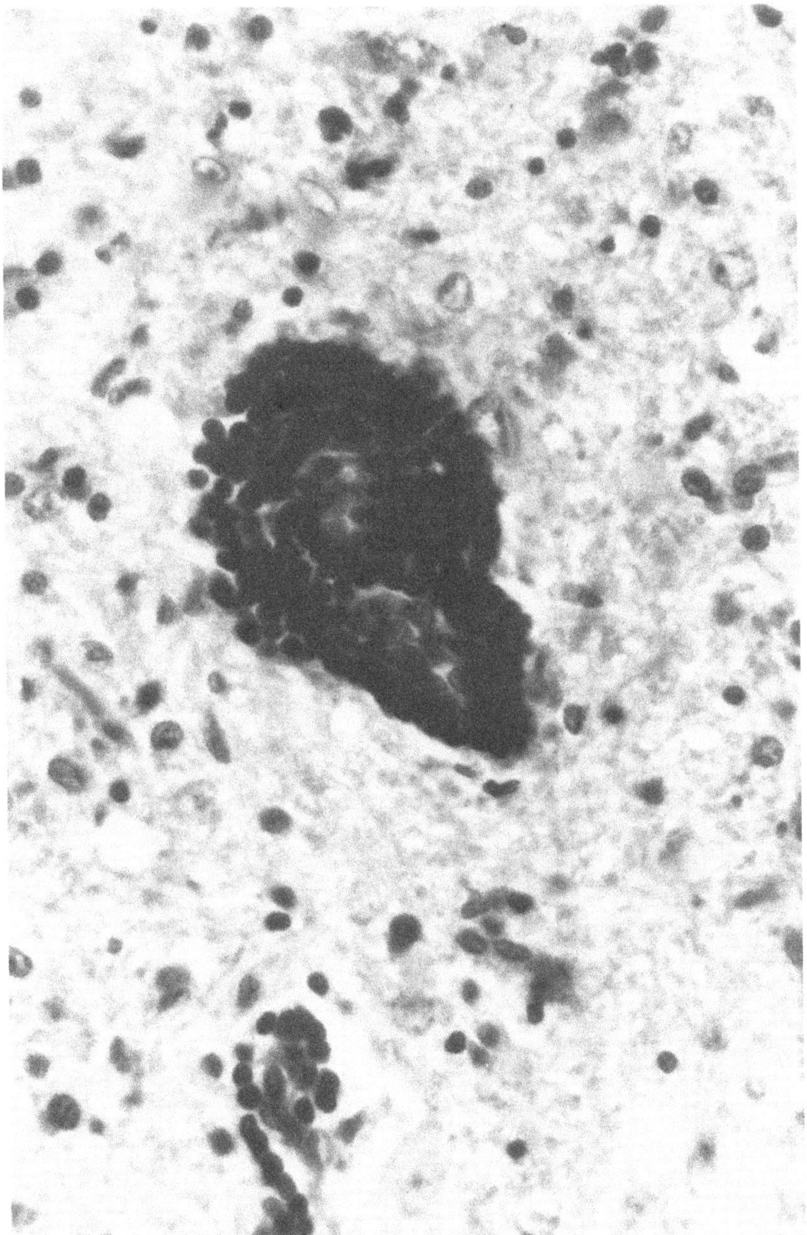

FIGURE 16. Prominent lymphophasmacytic perivascular cuffing in the brain of a dog with canine distemper virus-associated old dog encephalitis.

vasculitis, fibrinoid necrosis of blood vessels, segmental damage to capillary beds, and the appearance and presence of extravascular electron dense deposits consisting of both virus and associated immunoglobulin and complement components. These complexes are phlogistic in vivo and result in the attraction of inflammatory cells to these sites of deposition. Thus, in the usual instance, immune complex-associated disease results in segmental vasculitis, glomerulonephritis (Figure 21), and exceptionally, lesions attributable to viral infection in other capillary beds such as the choroid plexus of the brain.

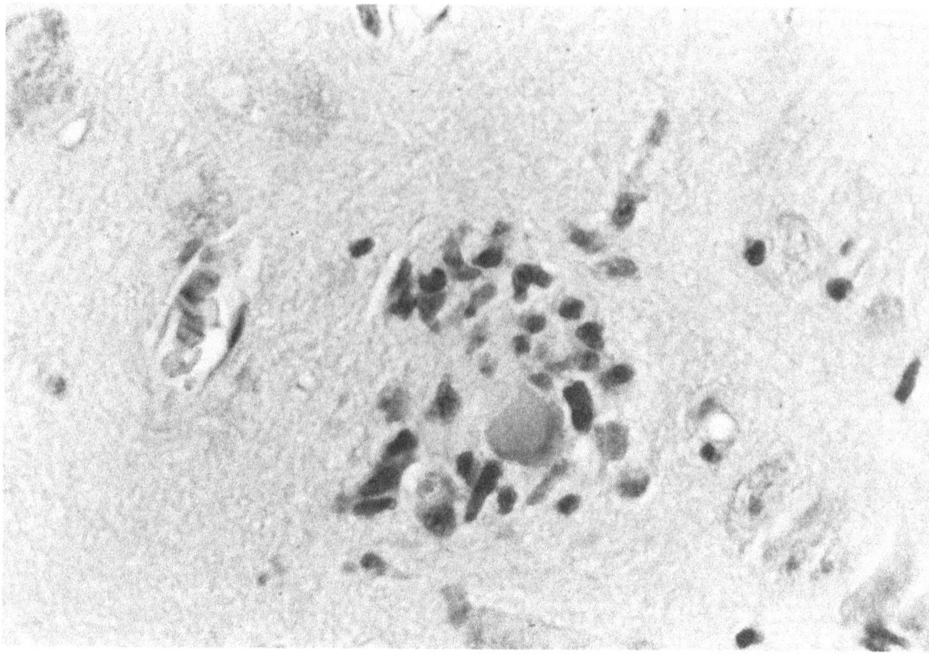

FIGURE 17. Focal microgliosis around a neuron infected with canine distemper virus.

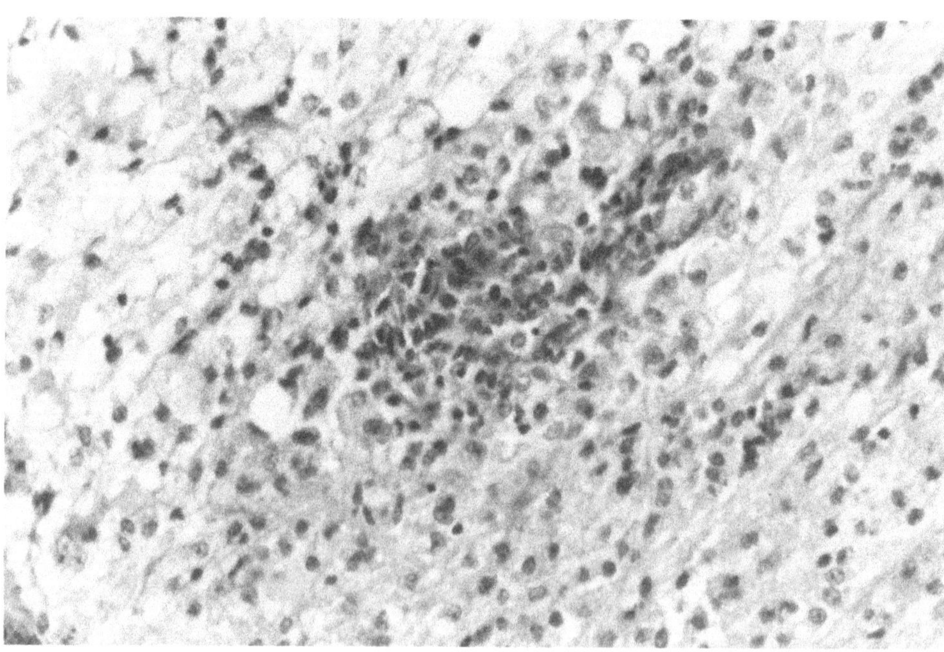

FIGURE 18. Focal granulomatous encephalitis induced by ovine progressive pneumonia (Visna) virus infection in a sheep.

In at least one viral disease of animals, notably feline infectious peritonitis (FIP) the immune complexes associated with this disease produce a lesion that is not like the typical immune complex phenomenon. The more typical FIP lesion contains a signifi-

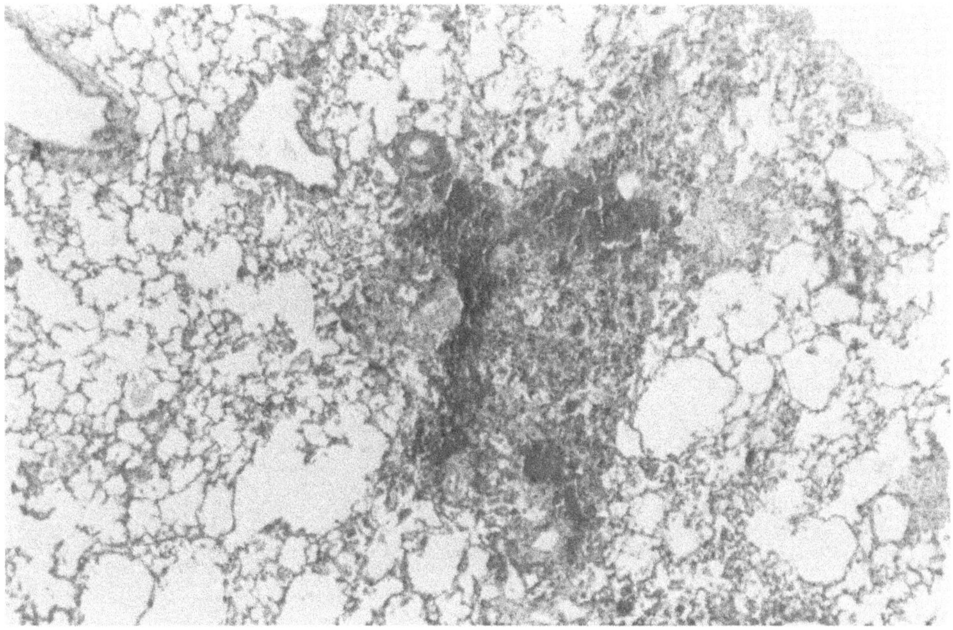

FIGURE 19. Acute segmental vasculitis with resultant hemorrhage in the lung of a dog infected with canine herpesvirus. (Courtesy of Daniel Morton, D.V.M., Department of Veterinary Pathobiology, The Ohio State University.)

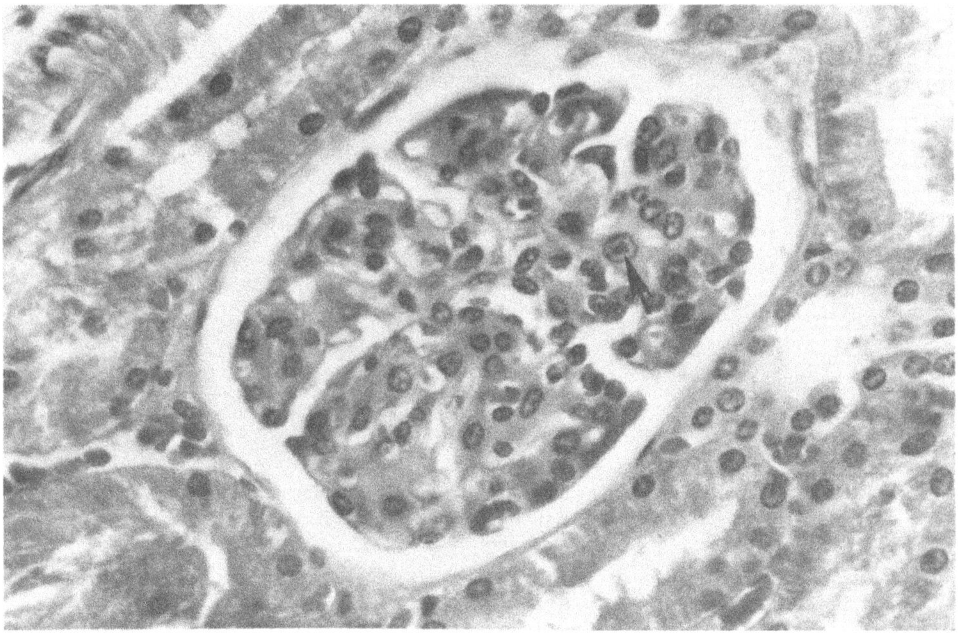

FIGURE 20. Endothelial cell infection in the renal glomerulus of a dog infected with infectious canine hepatitis virus (arrow).

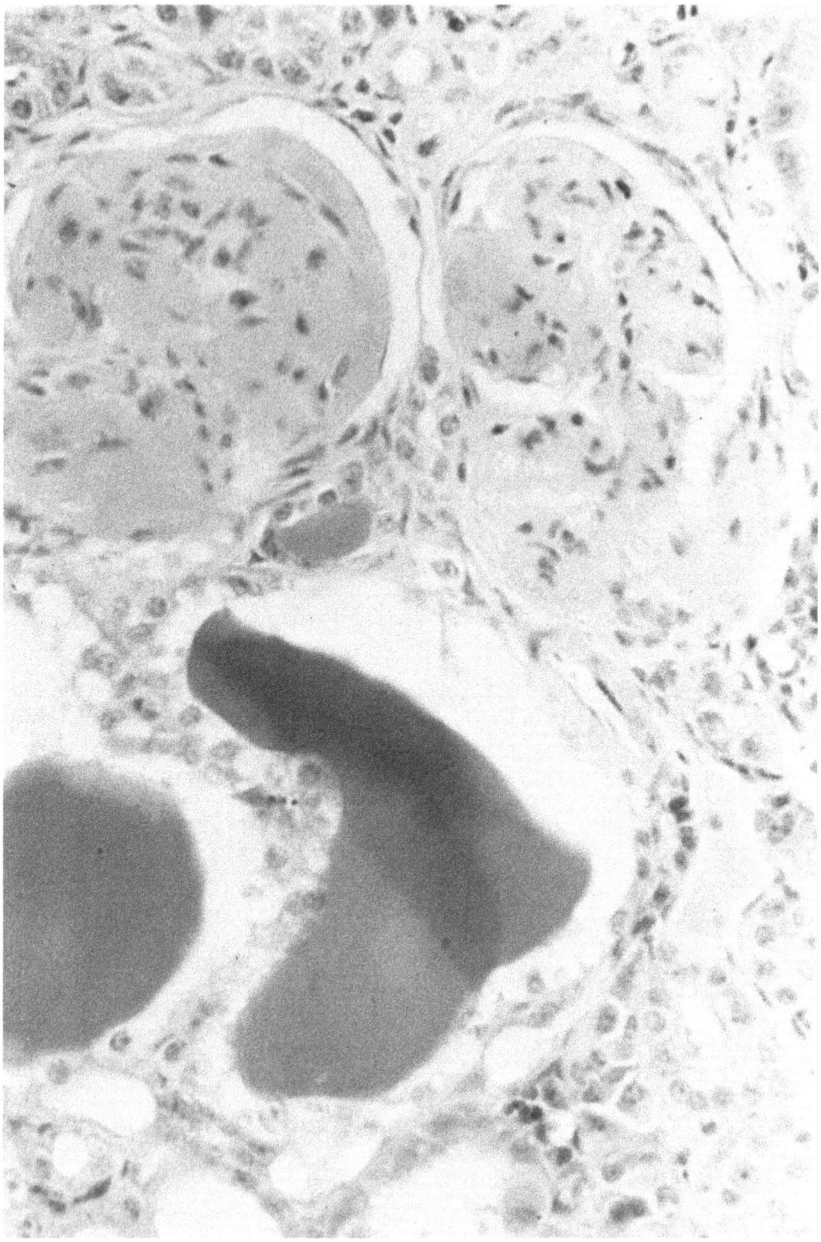

FIGURE 21. Immune complex glomerulonephritis with resultant damage and proteinuria induced by murine leukemia virus infection in the New Zealand Black (NZB) Mouse.

cant component of mononuclear inflammatory cell infiltration and is best described as granulomatous in nature. It is likely though not proven that the lesions observed can be attributed to the combined effects of deposition and interaction of immune complexes with host tissues and also from the fact that the FIP virus has been shown to replicate directly within macrophages. Conditions favoring development of immune complex disease occur when the virus has access to systemic circulation and is not effectively neutralized or cleared by normal host immune response mechanisms.

There are a number of viral diseases of animals in which an immune complex component has been demonstrated. FIP virus infection has been mentioned. Other candidate viruses include feline leukemia virus, equine infectious anemia virus, bovine virus diarrhea, Aleutian disease virus, infection in mink, infectious canine hepatitis infection in dogs, and the viruses of African swine fever and hog cholera, respectively.

IV. CONCLUSIONS

The pathologic effects of viral infection upon the host are complex and varied. Evaluation of tissues either grossly or at the histopathologic level is an essential part of the diagnostic process used to link viral infections to patterns of tissue damage. As the knowledge of the range and subtlety of virus' effect on the host has expanded, simplistic interpretation of lesions observed is no longer possible without concurrent virologic, biochemical, or immunological correlative data.

Virus-associated lesions, whether localized or systemic consist of several diagnostic elements. Manifestations of virus cytopathic effect (e.g. swelling and lysis, syncytium formation, and cellular proliferation) encompass the primary mechanisms of disease production. If these structureal changes are accompanied by visible virus structural products such as inclusion bodies detected by conventional histologic stains or viral antigens detected by the newer immunohistologic methods, the lesions observed can be quite convincingly linked to the direct effects of viral infection.

Since, almost without exception, the changes described above occur in the context of the host inflammatory response, the virus-specific damage may be frequently overshadowed by acute exudative (humoral and cellular) or chronic proliferative (granulomatous) inflammatory responses. Of course, local vascular responses ranging from biochemically mediated increased capillary permeability to fibrinoid necrosis of vessels with resultant hemorrhage further complicate the evolving lesion. Finally, during the healing phases, repair by substitution (e.g. fibrosis and scarring) or by proliferation of uninfected replacement cells may dominate the sites of viral infection.

In addition to these localized effects, many viruses produce a systemic phase of infection. Within broad limits, essentially the same sequence of events will occur in all tissues. The systemic effects of a viral infection can have independent long term and even metabolic consequences to the host. Neurological deficits (hydrocephalus, cerebellar hypoplasia, porencephaly, retinal dysplasia) are common. A common effect of sytemic viral disease is transient or permanent damage to lymphoreticular tissues. Thus, lesions of thymic or bursal atrophy and resultant depletion of lymphoid tissues will result in an immunocompromised host which is at great risk for development of secondary, opportunistic infections.

Certain virus diseases produce damage associated with circulating virus-antibody immune complexes. Examples of these include: Aleutian disease virus in mink, equine infectious anemia, feline infectious peritonitis and feline leukemia, infectious canine hepatitis, and virus-associated corneal edema. Still other viruses may produce secondary metabolic disease by virtue of their destructive effects upon critical host tissues. Thus, diabetes mellitus seen in both mice following infection with certain strains of encephalomyocarditis (EMC) virus infection and in cattle with foot and mouth disease infection is attributable to pancreatic damage. Recently, distemper virus infection in mice was shown to produce an obesity syndrome in mice surviving the acute infection. Finally lymphocytic choriomeningitis (LCM) virus infection in mice has been shown to interfere with growth hormone production and/or release thereby producing dwarfed hyperglycemic individuals. Thus, given all of these possible effects and outcomes, it is not surprising that the pathologic effects of viruses in host tissues are the net result of a complex biological process.

REFERENCES

1. Cheville, N. F., Cytopathology in viral diseases, *Monogr. Virol.*, 10, 1, 1975.
2. Cheville, N. F., *Cell Pathology*, 2nd ed., Iowa State University Press, Ames, 1983.
3. Fraenkel-Conrat, H. and Wagner, R. R., *Virus-Host Interactions: Immunity to Viruses*, Plenum Press, New York, 1979, 15.
4. Johnson, R. T., *Viral Infections of the Nervous System*, Raven Press, New York, 1982.
5. Malherbe, H. H. and Strickland-Cholmley, M., *Viral Cytopathology*, CRC Press, Boca Raton, Fla., 1980.
6. Notkins, A. L., *Viral Immunology and Immunopathology*, Academic Press, New York, 1975.
7. Poste, G. and Nicolson, G. L., *Virus Infection and the Cell Surface*, Vol. 2, Cell Surface Reviews, North-Holland, Amsterdam, 1977.
8. Rouse, B. T. and Babuik, L. A., Mechanisms of viral immunopathology, *Adv. Vet. Sci. Med.*, 23, 103, 1979.

Chapter 2

MEDICAL VIROLOGY

James R. Blakeslee, Jr.

TABLE OF CONTENTS

I. INTRODUCTION

The purpose of this chapter is to provide a broad overview of the concepts of disease production by animal viruses. Readers interested in more specific details are referred to the books and articles listed in the References. I have chosen not to discuss classification, physical properties, and replication schemes for each virus family because this text is concerned with the medical aspects of virus infection, the interaction(s) of the virus with living cells, and the outcome of this interaction leading to the disease state. It is in this context that the host-virus interactions will be discussed.

Essentially, a virus-cell interaction will lead to one of four different states:

1. Lysis of the infected cell
2. Steady state in which virus is synthesized with little cell death
3. Latent infection — cell is infected, no virus replication; genome can be activated
4. Transformation of the normal cell to a neoplastic cell

It is necessary to define a virus in order to understand the disease process that follows infection of susceptible cells. One of the original definitions of a virus was provided by Lwoff in which it was defined as a particle containing either DNA or RNA and carries all the genetic information needed for inducing the cell to replicate new virus particles. Later, it was shown that certain viruses contained either single or double stranded DNA or RNA. Thus, the virus is an obligate intracellular parasite that utilizes the energy and macromolecules of its host cell to synthesize new viral proteins and nucleic acid and directs the assembly of these components into new viral structures followed by release into the environment. The cell may die, be relatively unaffected, or be transformed as a result of the virus-specified products.

Many viruses are transmitted to susceptible hosts through the respiratory or alimentary tract. Others can be introduced directly to tissues by biting, and mechanical or biological transmission via insects. In the case of mechanical transmission, no virus replication occurs in the insect host, whereas with biological transmission, virus replication also occurs in the insect and the susceptible host.

Viruses that infect by way of the alimentary tract are the nonenveloped viruses that are able to withstand acidic conditions as low as pH 2.0 in the stomach, whereas with enveloped and nonenveloped viruses the mode of infection the respiratory tract or mucosa.

Infections may be localized where virus replication occurs in a general area or organ where the infection first occurred with no spread of infectious agents or conversely, virus replication may occur at site of entry, then disseminate via the lymphatics or the bloodstream to other organs (primary viremia). In certain diseases, a secondary replication may occur in other organs after the primary viremia (secondary viremia).

II. THE INFECTIOUS PROCESS

The infectious process is initiated by attachment of viruses through specific receptors on the cell controlled by the host cell genome. The initial attachment is electrostatic in nature and reversible with the negatively charged virions attaching to positive region(s) of the cell membrane. The second step of attachment usually requires energy and divalent cations, is temperature-dependent, and is irreversible. Viruses lacking envelopes or having envelopes dissimilar from cell membranes migrate into the cell by *pinocytosis*, whereas viruses with envelopes similar to cell membranes fuse with the cell membranes to form a continuous membrane in which the nucleocapsid enters the cell by a procedure called *viropexis*.

Cell membranes respond to attachment by invagination, forming a vacuole. Cellular lysosomes attach to the vacuole releasing enzymes which expose the nucleic acid for transcription or translation. If the infecting virus is an RNA-containing virus, the RNA can act as a messenger RNA (RNA+). Other RNA-containing viruses contain, or have attached, certain enzymes for initiation of viral RNA synthesis (RNA−). As previously stated, viruses also utilize cellular macromolecular synthetic processes such as energy generating systems, protein synthesis, protein phosphorylation, transcription, lipid and membrane synthesis. The viral genome provides information required for synthesis of specific enzymes for viral nucleic acid replication and structural proteins, the remainder of the enzymes are provided by the infected cell. Some viruses require specific enzymes for initiating the replication cycle not found in the host cell, but are carried in the virion. These enzymes are

1. RNA-dependent RNA polymerase (RDRP)
2. DNA-dependent RNA polymerase (DDRP)
3. RNA-dependent DNA polymerase (RDDP)

The RNA viruses lacking messenger RNA function produce copies of RNA that provide messenger function with RDRP. An example of a virus family containing RDRP are members of the Paramyxoviridae. DDRP is found in the Poxviridae and RDDP is found in the Retroviridae family.

III. LYTIC VIRUS EFFECTS

Cell death or cytocidal effect of viruses are usually initiated by the synthesis of "early" virus-coded proteins. These "early" proteins shut down cellular RNA and protein synthesis with DNA being secondarily affected. Early changes in the cell are pyknosis, margination of the chromatin, and changes in the nucleolar structure. Feline calicivirus, cowpox, pseudorabies, and infectious bovine rhinotracheitis are examples of viruses which induce these early events. Membrane changes usually follow. Plasma membrane functions decline most likely because turnover is affected by metabolic dysfunction and replacement of cellular proteins with viral proteins. The internal ionic environment is altered and nutrient transport and waste removal is diminished. Lysosomal enzymes begin to leak out and aid in destruction of the cell. Near the end of the replication cycle, large aggregates of virions may distort the cell, adding to changes in membrane permeability. These necrotic events aid in release of newly synthesized mature virus particles. An example of a lytic virus infection is transmissible gastroenteritis (TGE) of pigs; the agent is a member of the Coronaviridae family and the agent is porcine-transmissible gastroenteritis virus (TGEV). Infection occurs via the oral route and the initial site of infection and replication is the columnar epithelial cells of the duodenum and jejunum, with concomitant loss of villi. Following the destruction and loss of these cells, undifferentiated crypt epithelial cells undergo hypertrophy and hyperplasia. The loss of the villi leads to diarrhea with dehydration, electrolyte imbalance, and death. In pigs less than 10 days of age, it is a fatal disease, and older pigs are also susceptible to infection; however, the disease is less severe. The lungs and kidneys may be secondarily infected.

IV. PERSISTENT AND LATENT INFECTIONS

Latent viruses are found predominantly in the Herpetoviridae and Retroviridae families. They are characterized by a primary infection followed by disappearance of virus. Days, months, or years after the initial infection, recrudescence occurs with demonstrable infectious virus. An example of latent infection followed by reactivation has been demonstrated with calves naturally or experimentally infected with infectious bovine rhinotracheitis virus (IBR), a herpesvirus. Experimental reactivation occurred after corticosteroid or adrenocortical hormone injections. Stress, parturition, and estrus have also been associated with latent virus reactivation. Recrudesence occurs in the presence of detectable antibody levels. Experimental evidence suggests the virus is sequestered in the trigeminal ganglia and, under certain of the above mentioned conditions, reactivation with replication occurs. However, the state of the viral genome during the latent periods remains unclear as to whether it's integrated or not.

In some instances, cell damage is not the result of these cytotoxic changes due to these virus events, but due to the host's immunological response to virus infection as found in many persistent infections; an example is the disease Equine Infectious Anemia (EIA), a result of infection with the nontransforming EIA virus, a member of the Retroviridae family. Infection of a horse with EIA can be manifested as an acute, subacute, or chronic infection. The disease is characterized by recurrent fever, anemia, and a persistent infection for the life of the horse. Infected horses may die or survive and become an infectious carrier. The disease is manifested by diverse responses such as hemolytic anemia, hypergammaglobulinemia, hepatitis, widespread lymphoproliferative lesions, and glomerulonephritis.

Complement-fixing antibodies (CF), precipitating antibodies, and virus-neutralizing (VN) antibodies are detectable in the serum. In addition to CF antibodies, complement-fixation-inhibiting (CFI) antibodies of the IgG(T) class are found which vary inversely with the CF antibodies and are distinct from VN antibodies.

Several hypotheses have been proposed to explain the cyclic nature of the disease. Experimental evidence of rapid "antigenic drift" was provided by Kono et al. in 1971. During episodes of disease, virus-neutralizing antibodies specific for antigens on the virion surface, were specific for the virus shed during the most recent episode, but not the previous episodes. Thus, viral persistence with hypergammaglobulinemia and circulating immune complexes would be manifested. In addition to antigenic drift, it has been suggested that IgG and IgG(T) compete for antigenic sites, thereby preventing neutralization of infectious virus, and/or EIAV provirus is integrated into cellular DNA, thus escaping host immune surveillance.

The hemolytic anemia results from attachment of EIAV to specific receptors on the erythrocytes and in the presence of complement, lyses the cells. Other lesions seen in EIA are associated with the circulating immune complexes, i.e., glomerulonephritis and vasculitis. Thus, this disease is not a result of a lytic infection, but a disease resulting from viral persistence, the host's immune response to the agent with the lesions and the clinical symptoms are a consequence of this interaction.

V. NEOPLASTIC TRANSFORMATION

The fourth major result of viral infection is transformation of a normal cell to a neoplastic cell. All double-stranded DNA virus families contain members with oncogenic potential, whereas only the single-stranded RNA viruses in the Retroviridae family (with the exception of EIAV) are transforming viruses. The transforming gene(s) known as "oncogenes" are in the viral genome. However, some cancer-causing viruses lack oncogenes and induce tumors by a different mechanism and require a longer period (months) to induce tumors in contrast to viruses with oncogenes that are able to induce tumors in days or weeks. Regardless of the mechanism, the viral genome is integrated into host cell DNA for the life of the cell. The oncogenic RNA viruses synthesize a double-stranded DNA copy from the RNA genome by the previousy mentioned enzyme, RDDP, commonly called "reverse transcriptase". This copy is integrated into the infected cell and the viral genome replicates during normal cell metabolism and is expressed with cellular genes.

Studies showed these genes were also present in normal mammalian cells and were active in producing phosphorylating enzymes. Thus, the origin of these viral oncogenes appears to be cellular in origin. The function of the genes in normal cells was shown to play some part in regulating the normal growth of cells. Further studies showed when normal cellular oncogenes were introduced into other cells under certain conditions, the recipient cells were transformed as if they had been infected with a tumor virus. RNA viruses with no oncogenes appear to induce cellular oncogenes by integrating into cellular DNA near the cellular oncogene affecting the normal control of that gene which in turn leads to an amplified expression of the cellular gene leading to tumor formation. To date, 17 viral oncogenes related to cellular oncogenes have been identified in avian and mammalian species.

Feline leukemia virus (FeLV) is an example of a viral-induced disease by a retrovirus lacking oncogenes, as described above, whereas feline sarcoma virus (FeSV) contains viral oncogenes. For example, kittens infected with FeLV may require up to 18 months incubation before frank leukemias are seen, whereas FeSV produces tumors as early as 3 weeks postinoculation. This difference in time required for disease demonstrates the two theories involved in disease production by oncogene (+) and oncogene (−) tumor viruses. Disease responses in kittens infected with FeLV, in addition to malignancies of the hematopoietic system, often have nonneoplastic diseases such as anemia, glomerulonephritis, fetal reabsorptions and abortions, and bacterial diseases associated with immunosuppression. The immunosuppression is mediated by direct infection of lymphocytes and immunosuppressive viral glycoproteins. Lymphomas tend to develop in deep lymphoid tissues of the gut, spleen and mediastinum, and the thymus. The infected lymphocytes are of T-cell origin. Circulating neutrophils in infected cats contain viral antigens that are detectable by immunofluorescence and provide the basis of a test to detect viremic cats.

Feline sarcoma virus (FeSV) isolated from spontaneous fibrosarcomas, injected into young kittens, can cause death in a few days, whereas in young cats 6 to 8 weeks of age, it induces progressive tumors at the inoculation site within 3 to 10 weeks. Cats may die without formation of tumors or regression may occur. Regressor cats usually have antibodies to a virus-associated antigen found on membrances of infected cells.

VI. SUMMARY

This description of virus-host interactions with several illustrative diseases is a brief introduction to the text. Diseases resulting from virus infection are both unique and complex, based on the nature of the virus and the host's response to the invading virus. Clinically, viral infections may result in inapparent infections, acute, subacute or chronic infections, or malignant disease with death or recovery of the infected host.

The following chapters and subsequent volumes will provide the reader with specific examples of these complex interactions and the pathobiology associated with a broad spectrum of viral-induced diseases.

REFERENCES

1. Becker, Y., *Molecular Virology*, Martinus Nijhoff, Publ., The Hague, 1983.
2. Gillespie, J. and Timoney, J., *Hagen and Bruner's Infectious Diseases of Domestic Animals*, 7th ed., Comstock Publ. (a division of Cornell University Press), Ithaca, N.Y., 1981.
3. Joklik, W., Willett, H., and Amos, D., *Zinsser Microbiology*, 17th ed., Appleton-Century-Crofts, New York, 1980.
4. Klein, G., *Viral Oncology*, Raven Press, New York, 1980.

5. Mims, C. A., *The Pathogenesis of Infectious Disease*, 2nd ed., Adademic Press, New York, 1982.
6. Olsen, R. G., *Feline Leukemia*, CRC Press, Boca Raton, Fla., 1981.
7. Olsen, R. and Krakowka, S., *Immunology and Immunopathology of Domestic Animals*, Charles C Thomas, Springfield, Ill., 1979.
8. Bishop, J. M. Oncogenes, *Sci. Am.*, 246(3), 81, 1982.
9. Hardy, W., Old., L., Hess, M., Essex, M., and Cotter, Horizontal transmission of feline leukemia virus, *Nature (London)*, 244, 266, 1973.
10. Kahrs, R., Infectious bovine rhinotracheitis virus: a review and update, *J. Am. Vet. Med. Assoc.*, 171(10), 1055, 1977.
11. Kono, Y., Kobayashi, K., and Fukunaga, Y., Serological comparisons among various strains of equine infectious anemia virus, *Arch. Gesamte Virusforsh.*, 34, 202, 1971.
12. Lwoff, A., The concept of virus, *J. Gen. Microbiol.*, 17, 239, 1957.
13. Nakajima, J., Suguira, T., and Ushima, C., Immunodiffusion test for the diagnosis of equine infectious anemia, in, *Proc. of the 4th Int. Conf. on Equine Infectious Diseases*, Bryans, J. and Gerber, H., Veterinary Publ., Princeton, N.J., 1978.

Chapter 3

IMMUNE FACTORS IN VIRAL DISEASE

Richard G. Olsen, Darrell Tuomari, and Mary Stiff

TABLE OF CONTENTS

I. INTRODUCTION

Historically, the principles of immunity were constructed from the basic observations of Jenner and later Pasteur. That animals rarely develop a recurring specific disease is central to our understanding of immunity. Efforts in attempting to identify specific immune factors that bestow protection from viral disease are still unclear. For example, the presence or absence of humoral antibody may not correlate with resistance to viral disease. Today, the immune system is considered to be comprised of a network of intercommunicating lymphoid and nonlymphoid cells. The sum total of these cell functions results in the immune response. Therefore, prevention of viral disease may be considered to be compromised by all these cell functions.

The objective of this chapter is to discuss the role of this cellular network and to briefly review the role of the separate parts in the protection from disease.

II. HISTOCOMPATIBILITY SYSTEMS

Histocompatibility systems consist of genes coding for the set of cell surface molecules which not only characterize "self" but serve as functional molecules during cellular interactions of an immune response.[1] Recent advances in mouse genetics have enabled the study of the biological significance of these "self" antigens in mouse strains that differ at a single genetic locus within the MHC (congenic mice and their intra-H-2 recombinants).[2] The principal function of these gene products with the exception of genes within the complex coding for complement and complement-related compounds is in facilitating the presentation of antigen to the immune system.[1] Two major gene classes can be readily distinguished on the basis of structure of their products. Genes in the K and D regions (analogous to the human HLA-A,B,C) code for 45,000-dalton glycoproteins which are found in the plasma membrane always associated with β-2 microglobulin. These glycoproteins are involved in the antigen presentation during cell-mediated cytotoxicity which is an important immune mechanism in graft-type reactions and in clearing and rejection of virus-infected cells and tumor cells. These are referred to as Class I genes and gene products.[3] Genes in the murine 1 region (analogous to the HLA-D/DR) code for two types of peptides: (34,000 daltons and 28,000 daltons) that function in antigen presentation in the humoral immune response of antibody production. These are referred to as the Class II genes and their gene products.[3] The Class III genes produce complement or complement-related components.[4]

Before studying the influence of genetics on virus-host susceptibility, research with inbred animals demonstrated that the ability to mount an antibody response was under genetic control. McDevitt and co-workers working with congenic mice found the immune response to defined polypeptide antigens was under the control of the MHC.[5] This research has been extended to include studies of the MHC and virus susceptibility.

Gross leukemia virus (GV) which causes a spontaneous leukemia in the inbred AKR mouse was studied to determine if virus susceptibility was related to H-2 composition. Inbred mice carrying the $H\text{-}2^k$ haplotype were found highly susceptible to infection whereas mice carrying the $H\text{-}2^b$ haplotype were highly resistant.[6] To demonstrate that a gene within the MHC was responsible, congenic mice which were genetically identical except as the H-2 region were used.[1] These studies revealed that the genes controlling GV susceptibility were within the H-2, and that mice which possessed the $H\text{-}2^k$ haplotype were highly susceptible compared to resistent mice carrying the $H\text{-}2^b$ haplotype.

The course of infection of radiation leukemia virus (RadLV) has also been shown to be regulated by a gene within the MHC. Mice possessing the $H\text{-}2D^d$ haplotype were resistant to RadLV whereas $H\text{-}2D^q$ and $H\text{-}2D^a$ strains were susceptible. This gene has been correlated with resistant strains which demonstrate significantly fewer virus and viral lesions beyond 5 weeks of challenge compared to susceptible strains.[6] In the resistant mice, Meruelo and colleagues[7] demonstrated an increase in expression of both H-2^d and H-2^k thymocyte cell membrane proteins whereas suceptible mice did not show a similar increase in H-2^d expression. Since these Class I molecules are important in cellular interactions to destroy virally infected cells, the difference in H-2D protein expression may have altered the host immune response to viral invasion.

In still another mouse model with a nonleukemogenic virus, Grandy et al.[8] demonstrated an association between susceptibility to the herpesvirus murine cytomegalovirus (MCMV) and the H-2 haplotype. In congenic mice, possession of the H-2^k haplotype was associated with resistance compared to the H-2^b and H-2^d haplotypes. Studies with intra H-2 recombinants demonstrated that the pattern of resistance was due to two loci within the MHC; one mapping at the K end and the other mapping to the D end once again implicating a role of the Class I molecules in resistance to viral diseases.[8]

A. MHC and Viral Immunity

Immune mechanisms of disease to viral infection occur through both a specific humoral and cell-mediated immune response requiring the involvement of the MHC. Antibody production against free virus requires that T cells recognize simultaneously the viral antigenic determinants and "self" Class II MHC molecules on antigen processing cells (i.e., macrophages). This stimulation of T lymphocyte (helper) and interaction with B cells by way of MHC Class II molecules induces the production of an array of antiviral antibodies.[11] These antiviral antibodies then function to neutralize and eliminate free virus from the blood and lymph.[12]

Once host cells have been invaded by virus, the cellular immune response works to destroy the cells replicating virus. The function of the immune cytotoxic T lymphocyte (CTL) along with the appropriate accessory cells must be able to discriminate between infected and uninfected host cells.[13] Zinkernagel and Doherty[14] studied this discriminating behavior of the CTL in association with the MHC during lymphocytic choriomeningitis virus (LCM) infection in mice. They discovered that CTLs would destroy self (or MHC syngeneic) LCM-infected cells but not (or only poorly) allogeneic-infected cells which implied the essential requirement of recognizing "self" to stimulate destruction.[14] Further studies with congenic mice and their intra-H-2 recombinants have shown that restriction of CTL activity required syngeneity at the K and/or D regions of the MHC (Class I molecules). Syngeneity between CTL and virus-infected cells at the I region was neither necessary nor sufficient to allow cytolysis.[14]

Since histocompatible accessory cells are an absolute requirement for CTL function,[15,16] their role in MHC directed CTL activity against virus-infected cells was also investigated. In studies with herpes simplex virus (HSV-1)[17] and reovirus[18] in mice, it was found that these accessory cells must be expressing the MHC Class II molecules and producing interleukin 1 to activate the CTL recognition and destruction of virus-infected cells. The role of these cells in killing may be through their capabilities of processing antigen. Therefore, these studies demonstrate the requirement of functionally intact antigen processing cells which display Class II molecules along with Class I molecular interactions with virus to induce the cell mediated immune mechanisms of CTL.

B. Role of MHC in Viral Pathogenesis

Successful virus entry into a host is sometimes associated with the ability of a virus to bind to cell membrane receptors. The influenza viruses contain hemagglutinin proteins which bind to the neuraminic acid residues in the cell membrane.[19] The cell membrane components which contain these residues include the MHC molecules. Another virus, the Semliki Forest Virus (an alphavirus) has been found to gain entry into human and mouse cells very specifically by binding to the MHC Class I molecules.[20] Therefore, while the MHC molecules serve functionally to limit viral infections, these viruses are making use of the MHC as a receptor system to gain entry into the cell.

The budding viruses, such as vesicular stomatitis virus,[21] feline leukemia virus,[20] and Friend virus[23] have been found to incorporate host MHC determinants into their envelopes. During Friend virus infection, selective incorporation of the MHC Class I antigens occurred during budding of the virus.[23] This association at the cell membrane is thought to represent the "altered self" marker required for stimulation of CTL activity in destruction of the virally infected cells.

Besides association of the mature virus particle with their MHC, questions arise as to whether viral pathogenesis includes alteration of the expression and, therefore, function of the MHC during viral immunity. In a study by Mann et al., expression of human MHC antigens was examined on cells pre- and post-HTLV (human T-cell leukemia/lymphoma virus) infection. Their data demonstrated a consistent alteration of MHC antigen expression with a productive HTLV infection.[24] Due to this consistent finding of MHC alteration, hybridization experiments were done between the HTLV *env* (envelope) genes and the human MHC genes to investigate whether this change is virally coded or due to viral particle integration in the membrane expressing the MHC. Their experiments demonstrated a homology of the MHC-HLA-B locus to HTLV *env* genes.[25] The biological significance of these viral genes which mimic and possibly destroy MHC communications may allow a virus-infected cell to escape the CTL immune mechanism of elimination and, thereby, result in viral spread of tumorigenesis.

III. NONSPECIFIC FACTORS IN VIRAL IMMUNITY

A. Complement

The role of complement (C) in the virolysis,[28] lysis of virus-infected cells,[29] neutralization of virus,[30] and control and clearance of infectious virus in vivo[26] has been shown to occur independently of antibody and leukocyte function.

Complement-deficient mice, either complement depleted by treatment with cobra venom[29] or C5-deficient mice,[26,27] exhibit a higher mortality and an impaired virus clearance to Sendai virus infection than normal control animals. It has been suggested from these studies that C5 is important in the clearance of virus from the blood.[31,32] Terminal complement components (C5) appear to limit viral replication in infected tissues.[26] In the respiratory syncytial virus-infected cells, both nonsyncytial and syncytial cells activate complement by both classical and alternate pathways.[29] The presence of complement components in the respiratory tract[33] suggest that complement may mediate some control over respiratory syncytial virus infections in vivo.

Direct lysis of C-type retroviruses by human complement[28] has been suggested to be a natural defense mechanism against retroviral disease.[34] The high efficiency of primate complement-mediated virolysis contrasts with the low efficiency of complement-mediated virolysis in lower mammals (rats, mice, chickens).[35] It has been suggested that the inefficient virolysis of feline retrovirus by normal cat serum may be one of the factors influencing the high incidence of retrovirus-disease in cats.[34]

An autosomal recessive-associated C_3 deficiency has been identified in an inbred colony of Brittany Spaniels.[36] Serum opsonizing activity and chemotactic properties are reduced in these dogs.[37]

B. Antiviral Macrophage Functions

Macrophages are generally the first inflammatory cells to encounter infectious virus and virus-infected cells, and it is becoming apparent that the status of macrophage functions in a given individual animal or animal species is a major factor in determining resistance or susceptibility to a viral disease.[38,39] Resistance to viral disease is thought to be conferred by two macrophage mechanisms: release of extrinsic factor(s) which suppresses virus replication in other cells supports virus and intrinsic factor which is the inability of infectious virus to replicate in macrophages.[40] Extrinsic factor activity in mice to herpesvirus has been attributed to release and subsequent action of interferon.[41] The extrinsic effect has also been associated with activated macrophage depletion of arginine, an essential factor for replication of several animal viruses.[42-44] Recent evidence indicates that the extrinsic viral resistance factor may be produced to mouse hepatitis virus by macrophages from both resistant and susceptible mice. By contrast, the intrinsic factor is only evident from resistent mice. It is suggested that intrinsic antiviral activity may be the dominant mechanism for in vivo resistance to murine coronavirus.[40] The in vivo resistance to murine herpesvirus, however, correlates with the expression of extrinsic antiviral factor(s)[40] The implication of these studies is that macrophage resistance may be different between envelope and nonenvelope viruses.

It is becoming well accepted that virus infection of macrophages predisposes individuals to secondary infections.[45] Infectious bovine rhinotracheitis virus,[45] influenza A[46,47] and cytomegalovirus[48] are but a few examples of viruses that infect macrophages and lead to subsequent secondary bacterial infections. In some viral models, viral infection clearly results in macrophage dysfunctions. Infectious bovine rhinotracheitis virus infection reduces Fc-mediated receptor activity and phagocytosis of bovine alveolar macrophages within lesions postinfection. Moreover, the ability of these macrophages to function in the antibody-dependent cellular cytotoxicity assay is reduced.[8] Sendai virus likewise suppressed pulmonary antibacterial functions against *Staphylococcus aureus* but did not decrease the removal of inhaled bacteria from the lungs.[49] Influenza virus has been reported to suppress both phagocytic and bactericidal activity of macrophages. The mode of action of influenza interference varies with different populations of macrophages. Mouse peritoneal and alveolar macrophages differ in their response to influenza, i.e., alveolar macrophage is more susceptible.[50]

Numerous studies have shown that age-related susceptibility to viral disease correlates with macrophage functions.[38] Macrophages from adult cats do not support feline retrovirus replication whereas macrophages from kittens do.[51] This correlates with the same pattern (i.e., kittens being more susceptible) of susceptibility to feline retrovirus disease.[14]

IV. SPECIFIC FACTORS IN VIRAL IMMUNITY

A. Cytotoxic Thymus-Derived Lymphocytes

Thymus-derived lymphocytes play a key role in the immune destruction of microbial-infected cells,[52] allograft rejection, and tumor cytotoxicity.[53] Cytotoxic T-lymphocytes (CTL) exert their antiviral effect by direct effects but also by the secretion of soluble mediators (i.e., lymphokines) that induce the participation of monocytes, macrophages, and B-lymphocytes.[54] The evidence overwhelmingly demonstrates that virus-infected cells are lysed by direct contact by CTL and do not require secondary cells, complement, or immunoglobulin.[55,56]

The histocompatibility restriction of CTL lysis of virus-infected target cells is reviewed (vida supra). The sequence of events in CTL lysis of target cells involves: (1) recognition of viral-infected target cell by CTL, (2) CTL attack on target cell in which binding occurs by way of antigen receptor, and (3) change in target cell permeability resulting in cell lysis. The use of inhibitors indicates that CTL killing of target cells is energy dependent.

The viral epitopes recognized by CTL and how these differ from antibody recognition sites are unknown. With the use of monoclonal antibodies directed toward reovirus antigenic determinants, it has been shown that CTL response is directed toward a region of the hemagglutinin (HA) that is also responsible for tissue tropism and binding to neutralizing antibody.[57] CTL activity has been demonstrated in the lysis of target cells infected with either budding or nonbudding virus. Budding viruses display the virus-specific polypeptides (glycoproteins) located in the outer cell membrane during viral assembly and are the target epitopes recognized by CTL.[58] With nonbudding viruses, the CTL viral target antigens are poorly understood. The HA epitopes of reovirus serve as the target of CTL activity,[57] however, whether the reovirus HA protein is inserted into the membrane during viral synthesis or passively absorbs to the target cells is unknown.

Impairment of CTL functions during viral pathogenesis is associated with the severest form of disease. Appel and colleagues[59] demonstrated that virulent strains of canine distemper virus (CDV) induced a severe encephalitis. This was associated with little or no CTL activity to CDV-infected cells. During CDV infection, dogs that develop a delayed or reduced CTL response developed persistent central nervous system (CNS) infection.

The complete definition of the role of CTL in domestic animal species along with the regulation of and interaction with nonspecific defense mechanisms such as lymphokines, interferon, natural killer cells, and antibody formation is yet to be clarified. It is evident, however, that the cellular immune response, as well as the humoral response, is important in the recovery of some viral diseases.

B. Humoral Antibody

The role of virus neutralizing antibody in the immunoprotection of the animal host from infectious virus is well-documented.[60] The biophysical mechanisms by which antibody renders virus particles noninfectious has attracted much interest.[60] It is known that the interaction of four antibody molecules per poliovirion inhibits poliovirus infectivity.[61] This inhibition is mediated by antibody-induced changes in the conformation of the entire virion. This antibody-mediated physical change in the poliovirus particle rendered it noninfectious even after viral penetration of target cell. The poliovirus system serves to illustrate how serum antibodies may potentially inhibit or augment viral disease. Nonneutralizing antibody has long been suspected of playing a key role in Dengue virus disease,[62] feline infectious peritonitis,[63] and others.[64] The evolution of low-binding affinity antibody to type-specific epitopes may well constitute what has been relegated as nonneutralizing antibody. Though this concept may not be extended to all viral diseases, it illustrates the potential role of antibody in the viral enhancement.

C. Antibody-Dependent Cellular Cytotoxicity (ADCC)

The synergistic killing effect of antibody and leukocyte was noted when herpes simplex virus-infected cells specifically sensitized by way of passive uptake of specific anti-herpes simplex virus antibody were killed when subsequently treated with nonimmune lymphocytes (killer-cells [K-cells]).[65] The virus-infected cells exhibit virus epitopes that serve as receptor for antibody (either IgG or IgM). The target cells for ADCC, therefore, bear Fc aggregates that in turn serve as "docking" receptors for cells that possess Fc receptors (i.e., T-cells, B-cells, polymorphonuclear leukocytes, and macrophages). In fact, ADCC is known to be mediated by lymphocytes, neutrophils,[66] and monocyte-macrophages.[67,68]

The role of ADCC in the control of viral disease in various domestic animal species is poorly understood. In the bovine system, neutrophil appears to be an active ADCC effector cell whereas the peritoneal exudate and alveolar macrophage are most active in the rabbit.[69]

The exact roles of ADCC in the in vivo control of viral disease are not apparent.

However, studies with the respiratory syncytial virus in humans shed some light on its importance. In general, it has been noted that serum antibody plays no role in protection against infection with respiratory syncytial virus.[70] Specific ADCC responses in nasopharyngeal secretions were reported as early as 3 days postinfection. In addition, ADCC responses after reinfection were reported greater in both the acute and convalescent phases than after primary infection.[71]

D. Natural Killer Cells

Natural resistance to viral disease has been, at least in part, attributed to the presence of natural killer cells (NK cells). NK cells are mononuclear leukocytes of unknown lineage that, in some animal species, exhibit both macrophage[72] and T-lymphocyte markers[73] and exhibit natural cytotoxicity to target cells in the absence of previous immunogenic exposure.[74] The precise role NK cells play in resistance to viral disease in domestic animals is yet to be fully elucidated. Experimental studies with murine cytomegalovirus suggest that resistance in mice correlates with NK activity and not with the B- or T-lymphocyte response.[76] Interferon production has been shown to modulate NK cell induction.[75]

E. Local Immunity

1. Association of Resistance to Disease with Local Immunity

It is now recognized that a division of labor exists among peripheral lymphoid tissues from various anatomical sites. Particular attention has been focused on the mucosal-associated lymphoid tissues (MALT) which comprise the gut-associated lymphoid tissues (GALT) and bronchial-associated lymphoid tissues (BALT) as well as the mammary gland and urino-genital system.

The division of labor concept of MALT and lymphoid tissues associated with systemic immunity (i.e., spleen, peripheral lymph nodes) was given credibility by viral pathogenesis studies that attempted to associate immune factors with resistance to disease. Research on transmissible gastroenteritis (TGE) virus has illustrated exquisitely the role of MALT immunity to the virus as opposed to systemic immunity in this gut-born disease.[77] Transmissible gastroenteritis virus is a coronavirus disease of pigs in which the pig exhibits a striking age-related resistance and susceptibility. That is, young piglets are extremely susceptible to TGE disease exhibiting virtually 100% mortality, whereas the adult animal is resistant to TGE-induced pathologic lesions. Piglets that suckle immune dams are more resistant to TGE disease than piglets that suckle from nonimmune dams. Analysis of the colostrum from immune and nonimmune sows detected that only the former produced an abundance of IgA antibody to TGE.[77] Subsequent studies found that only sows that experienced a gut-born TGE infection produce TGE IgA activity in colostrum. This is in contrast to sows immunized by way of a subcutis route with TGE vaccine. These sows produced high serum neutralizing antibody to TGE virus, however, piglets that suckled these dams were not protected from disease. Studies suggest that a GALT immune response to TGE was a prerequisite for the subsequent appearance of IgA secreting B-lymphocytes in the mammary gland which culminates in the production of colostrum milk with specific secretory immunoglobulin with neutralizing properties to TGE virus.

In recent times, the canine parvovirus (CPV) disease re-emphasized the unique properties of the MALT system and its essential role in the control of disease. The CPV is a gut-born disease that may culminate in severe enteritis and death. In an attempt to associate immune factors with resistance to disease, it was found that only coproantibody to CPV of the IgA class correlated with resistance to CPV disease.[78] Serum antibodies and serum antibody titer did not.

It is evident that induction of local immunity is essential for the prevention of disease in select sites particularly those of mucosal surface. In nearly all cases, secretory immunoglobulin (IgA) is a prominent feature of mucosal immunity. However, as of late, immune cells in the lumen of the gut and in colostrum have been suggested to play an important role in the prevention from disease.

F. The Biologic Effect of Viral Disease on MALT

1. Induction of MALT Immunity

It is a well-accepted principle that fed immunogen leads to two immunologic events: (1) induction of a secretory antibody response in the gut and (2) induction of immunogen-specific tolerance (immunosuppression). The manifestations of the former include clonal expansion of IgA producing B-lymphocytes in mesenteric lymph nodes and subsequent populating of these cells in the gut lamina propria and Peyer's patches.[79]

The mechanism by which immune cells (i.e., IgA secreting B-lymphocytes) traffic to other MALT sites is unknown. However, experiments utilizing tagged GALT-associated cells definitely conclude that they traffic to BALT and MALT (lymphoid tissues of mammary gland, urinogential-associated lymphoid tissues and possibly other routes). Lactogenic hormones appear to play a key role in trafficking of GALT cells to mammary glands.[80] No doubt other physiological and humoral factors influence GALT cell trafficking to BALT and other MALT sites.

The high endothelial cell (HEC) of the post capillary venule is the site of lymphocyte re-entry into the lymphoid environment from systemic circulation. It is proposed that specific docking receptors (perhaps Ia) exist of certain HEC that may control the specific trafficking of GALT origin lymphocyte cells.[81]

The tolerance phenomena as a result of oral administration of immunogen is demonstrated by the specific absence of systemic antibody response. The suppression as sociated with oral feeding of immunogen implies that suppressor cells generated in the GALT exert a controlling influence on the systemic immune response. This suppressor cell mechanism is independent of control mechanisms that regulate secretory antibody response in MALT.[79]

At this time, only conjecture and speculation can be made about how viral infection may interfere with MALT functions that led to disease. The feline infectious peritonitis (see Volume II, Chapter 7) and canine parvovirus (see Chapter 10) are two viral candidates in which the pathogenesis may involve viral-induced dysfunctions of GALT. That feline infectious peritonitis virus interferes with suppressor cell functions can be suggested by several lines of evidence. First, the virus-induced pathology is not induced in cats previously immunosuppressed by corticosteroids.[125] Secondly, feline infectious peritonitis pathology is associated by previous immune exposure to the virus and a prominent feature of the disease in a poly- or monoclonal gammapathy (see Volume II, Chapter 7).

Canine parvovirus infection in gnotobiotic dogs produces no clinical lesions (e.g., acute gastroenteritis). However, CPV infection suppresses blastogenesis of lymphocytes of both systemic and Peyer's patch origin.[82] Moreover, CPV infection predisposes dogs to canine distemper.[83] Clinical data indicates a higher than usual incidence of CDV disease of the alimentary and respiratory type in dogs that have been exposed to naturally occurring CPV or attenuated CPV vaccines.

It is conceivable that disruption of lymphocyte trafficking to other MALT sites as well as GALT-derived suppressor function may be a prominent feature of many gut-associated viral diseases.

V. IMMUNOPATHOLOGY OF VIRAL DISEASE

Viral infections causing cell lysis can obviously lead to tissue destruction. The purpose of this section is to discuss a variety of mechanisms by which the host's immune

Table 1
TYPES OF HYPERSENSITIVITY

Type	Mediators	Example
Immediate		
Type I	IgE and mast cells	Bee sting allergy
Type II	Anticellular antibody and complement	Anti-RBC antibody
Type III	Immune complex deposition and complement	Serum sickness
Delayed		
Type IV	Immune cells and lymphokines	Tuberculin reaction

Table 2
SOME POSSIBLE MECHANISMS OF VIRAL-INDUCED AUTOIMMUNITY[94]

1. Release of sequestered antigens
2. Cell membrane alterations
   a. Alteration by viral enzymes
   b. Expression of development antigens
   c. Loss of tolerance-modulating antigens
3. Cross reactivity between virus and host antigens
4. Alteration of T-cell function: suppressor or helper

response results in greater tissue destruction or physiological alteration than that caused by viral cytolysis. In general, this is termed hypersensitivity. Table 1 lists four different hypersensitivity reactions and the classic examples. All four types will be discussed in relation to viral diseases.

Type 1 hypersensitivity is mediated by IgE (occasionally IgG) bound to mast cells or basophils. The primary antigen exposure results in the production of IgE which binds to mast cells. Secondary antigen exposure which binds to the IgE induces degranulation of the cells and release of vasoactive amines such as histamine. Type 1 hypersensitivity has been implicated in the development of asthma in human infants following a variety of respiratory viral infections.[84,85,86] More recent work involving epidemiological studies, determination of IgE levels, and other antibody titers indicates that type I hypersensitivity is not involved in viral-associated asthma.[87-90] Some reports dispute the previous reports on the association of some viruses and asthma.[91] Thus, the role of type I hypersensitivity during viral infection is controversial.

Type II hypersensitivity is mediated by antibody, usually IgG, directed against antigens on the cell surface. The viral-infected cell is destroyed following the activation of the complement cascade or cell-mediated lysis. Complement constitutes a group of proteins in which activation can be initiated by cell antibody leading to sequential activation of components culminating in cell lysis. Activated complement components can adhere to any cell membranes without antibody resulting in lysis of those cells. In addition, fragments of activated complement component can result in leukocyte chemotaxis, vasodilation, and degranulation of mast cells.[92,93] Often in the literature, type II hypersensitivity is referred to as viral-induced autoimmunity. A preferable phrase would be viral-induced immune-mediated cell lysis. Only in those cases where the immune response is at least in part directed against host antigens is it true autoimmunity. Table 2 is a review of some proposed mechanisms of autoimmunity in viral infections.[94] This phenomenon will be treated as a special case of type II hypersensitivity.

Table 3
FACTORS WHICH AFFECT THE PATHOGENIC POTENTIAL OF IMMUNE COMPLEXES[101]

1. Molar ratio of antigen and antibody
2. Chemical composition of antigen
3. Class/subclass of antibody
4. Affinity/avidity of the antibody
5. State of the phagocytic and complement system

The first three points in Table 2 are commonly proposed mechanism for autoimmunity discussed in many immunology texts.[92,93] Only point four will be discussed in detail using viral-induced diabetes mellitus as an example.

Diabetes mellitus in mice has been associated with a number of viral infections including reovirus types 1 and 3, encephalomyocarditis virus, and coxsackie B4.[95-97] In reovirus, type 3 infection results in autoantibodies against pancreatic islet cells. Reovirus type 1 infection results in autoantibodies against islet cells, insulin, anterior pituitary cells, and growth hormone.[96] The tissue tropism is determined by the viral hemagglutinin.[98] Infection with a recombinant virus of types 1 and 3 containing HA 1 resulted in similar autoantibody production as type 1 reovirus.[93] All four points in Table 2 may be involved, but point four is of particular interest. The high specificity conferred by HA 1 may allow it to infect very specific T-cell populations altering either suppressor or helper T-cell function resulting in autoantibody production. Thus, the virus could induce a very specific T-cell defect. Some investigators propose autoimmune disease due to loss of suppressor cell function is unlikely because generalized suppressor cell loss should result in a wide variety of autoantibody production, not a few specific autoantibodies.[99] However, the above discussion indicates that highly specific T-cell defects are possible, leading to the production of one or several autoantibodies.

An immune reaction directed against viral antigens on the cell membrane resulting in viral-induced, immune-mediated lysis may also be classified as type II hypersensitivity when certain critical cell populations are involved such as neurons or pancreatic islet cells. The physiological alteration resulting from viral-induced immune-mediated lysis of 1 million pancreatic islet cells is potentially far greater than that resulting from similar lysis of 1 million intestinal epithelial cells. Thus, the immune response resulting in destruction of pancreatic islet cells constitutes a hypersensitivity reaction. The severity of this reaction can be exacerbated by local complement activation leading to lysis of the target cell as well as number of neighboring cells.

Complement is also important in the third type of hypersensitivity which involves antigen combining with antibody to form an immune complex which subsequently activates complement. Often the immune complex (IC) forms in the blood and circulates leading to local tissue destruction when it becomes deposited in vascular basement membranes. Two common sites of deposition are the capillary beds of the renal glomerulus and the lung. Some investigators propose that the antigen deposits first with subsequent antibody binding and complement activation.[100,101] Whichever occurs, the end result is the same: local vascular tissue destruction at the site of deposition. Type III hypersensitivity is one component of immune complex-mediated disease. The following discussion will be on immune complex disease phenomenon in general.[101] A variety but not a complete list of important parameters will be discussed briefly; see Table 3. The importance of these phenomena will be related to several viral diseases.

First, the molar ratio of antigen and antibody can vary greatly depending on the rate of production of each. The antibody can combine with a multivalent antigen to form a lattice. The lattice formation is greatest at molar equivalence. This corresponds to the zone of precipitation in agarose immunodiffusion assays. Excess antigen or antibody results in smaller lattices. The size of the lattice is an important determinant of pathogenic potential. Large lattices are readily cleared from circulation by reticuloendothelial cells. Small lattices are inefficient in activating complement. Generally, moderate size lattices which can readily circulate and fix complement are formed at modest antigen excess and have the highest pathogenic potential.[101]

Second, the chemical composition of the antigen can have a multitude of effects. The size, shape, and valence of the antigen determine the number and position of antibody binding which may facilitate or deter the activation of complement. For example, the juxtaposition of antibodies on a globular protein will more readily activate complement than widely spaced antibodies on a linear protein.[101] In addition, the antigen itself may be immunostimulating like the bacterial antigens of *Mycobacterium bovis* (Bacillus-Calmette-Guerin [BCG]) and *Corynebacterium parvum*.[92,93,102] Also, antigens may be immunosuppressive such as feline leukemia virus (FeLV).[103] The immunosuppression associated with FeLV has a number of clinical manifestations including generalized increase in other infectious diseases, decreased lymphocyte blastogenesis, impairment of skin allograft rejections, and alter lymphocyte receptor capping. This suppression is mediated at least in part by a 15,000 dalton envelope protein (p15E). The mechanism of the immunosuppression is still uncertain, however, several hypotheses have been proposed: induction of suppressor cell, blockade of helper cell function, and direct suppression of lymphoid cells.[104] Thus, the role of p15E as an antigen may be very relevant in retrovirus-induced acquired immune deficiency syndrome.

Third, the class and subclass of the antibody affect the ability to activate complement. Immunoglobulin M (IgM) is very efficient in activating complement. However, only some subclasses of immuglobulin (IgG) can fix complement.[101] Immunoglobulins which do not fix complement have greatly reduced phlogistic potential.

Fourth, the affinity/avidity of the immunoglobulin is biologically important. Affinity is a measure of the strength of attraction between antigen and antibody during immune complex formation. Avidity is a measure of strength of the antigen-antibody bond in an immune complex. Low avidity immune complex will readily dissociate. High affinity, high avidity indicate the immune complex will readily form and are not readily dissociated.[92] Low affinity, low avidity, immune complex once formed, will readily dissociate, which tends to impair clearance of these materials by the reticuloendothelial system. In mice, it has been shown that strains which produce low avidity antibody following infection with lymphocytic choriomeningitis virus develop glomerulonephritis, whereas strains producing high avidity antibody do not develop nephritis.[105] Some success has been reported in removing immune complex by high affinity antibody infusions in the treatment of neoplastic and nonneoplastic diseases in man.[107,108]

Finally, the state of phagocytic or reticuloendothelial system and the complement system, in part, determine the level of circulating immune complex. Factors which impair the phagocytic system such as corticosteroids, silica, and cachexia reduce immune complex clearance.[109] Attempts to stimulate the reticuloendothelial system to decrease circulating immune complex levels have led to variable results.[107] In addition, hypocomplementemia due to excessive complement consumption, decreased production (liver disease), or experimentally induced with cobra venom factors decreases complement receptor-mediated clearance of immune complex.[110] Thus, impairment of the reticuloendothelial system and complement system tend to increase circulating immune complex levels.

In addition to the factors listed in Table 3, immune complex may interact with a variety of cells via Fc and/or complement receptors. These include leukocytes, platelets, endothelial cells, and primate erythrocytes.[101,111-117] The interaction of immune complex with the cells of the immune system may have profound effects on the immune response to many antigens. The immune complex may enhance or suppress the humoral response in an antigen specific or nonspecific manner. Also, immune complex tends to suppress cell-mediated immune reactions.[191,113-115] Immune complexes may also interact with Fc and complement receptors on tumor cells and viral-infected cells. Some herpesvirus-infected fibroblasts acquire an Fc-like receptor. Binding of antibody to these receptors tends to inhibit viral replication.[101,117,118] The possible role of immune complex in persistent viral infections is currently being studied by many investigators. From the above discussion, it is apparent that immune complex may modulate the immune response which may be in a local environment or systemically and, thus, also modulate viral replication. The biological importance of immune complexes in persistent feline leukemia virus infection was demonstrated by the use of extracorporeal immunosorption.[119] These cats were able to develop clearance of the viremia, regression of leukemia and tumors, and in some cats, effect an apparent cure.[119] Thus, immune complex may interact with a wide variety of cells and induce diverse biological responses.

The preceding discussion has been on immune complex disease in general. It is obvious that immune complex may have many biological effects and mediate many pathological reactions during viral infections. However, generally type III hypersensitivity deals only with the tissue destruction mediated by immune complex deposition and complement activation. An example of type III hypersensitivity during a viral infection is Aleutian disease of mink.

An important component of Aleutian disease of mink is a necrotizing arteritis. The histopathogenic progression of the arteritis first starts as a slight intimal thickening and small areas of granularity in the media. The lesion progresses to medial necrosis with neutrophil infiltrate then to more severe medial necrosis with necrosis of muscle layer and mononuclear cell infiltrate. The late stage lesion is characterized by fibrinoid necrosis which may involve only one area or the entire circumference of the vessels. Immunofluorescent studies indicate viral immune complexes and complement first in the intimal and later in the medial lesions.[120] A similar arteritis is seen in equine viral arteritis, but there are some important differences.

The histopathologic progression of the necrotizing arteritis of experimental equine viral arteritis starts after equine viral arteritis virus infection. The endothelial cell and the endothelium become degenerative and necrotic. There is an increased vascular permeability, leukocyte infiltration into the intima, and disruption of internal elastic lamina. The inflammatory cell infiltrate increases extending into the media and muscular layer. The lesion progresses to fibrinoid necrosis which may be severe. At this point the pattern of disease is suggestive of type II hypersensitivity. However, immunofluorescent studies fail to indicate any involvement of antibody.[120] More recent studies of this arteritis report a different histopathologic progression during natural infections indicating degeneration and fibrinoid necrosis of tunica media and inflammatory cell infiltrate without damage to the endothelium.[122] The initial report proposes that the pathogenesis is mediated by chemotactic factor released by the endothelial cells or platelets and complement.[120] The chemotactic factors released may resemble the lymphokines produced during type IV hypersensitivity which will be discussed next.

Type IV, delayed type hypersensitivity reactions, are considered cell-mediated reactions.[92,93] These factors, released predominantely from lymphocytes activated by antigen, result in the accumulation of inflammatory cells which are mostly mononuclear cells. Dense cell infiltrate coupled with lysosomal enzyme release can cause extensive tissue injury. In addition, cytotoxic T-cells and NK cells may lyse many host cells. Cell-mediated immunity (CMI) is important in elimination of intracellular parasites including viruses. However, as discussed for type II hypersensitivity, type IV reactions in a critical cell population such as nervous system as an example can have devastating biological consequences. Using the nervous system as an example, cell-mediated destruction of tissue has been implicated in many viral diseases including lymphocytic choriomeningitis virus, visna, measles-induced subacute sclerosing panencephalitis in man, and canine distemper virus-induced "old dog" encephalitis.[132-134]

In review, the four types of hypersensitivity reactions are not mutually exclusive, more than one may be operating in addition to the immune complex phenomenon described. The discussion on equine viral arteritis and the following chapters will indicate that the immunopathologic mechanism of many viral diseases are still unknown. The reader should be able to recognize the possible involvement of the hypersensitivity reactions and immune complex phenomenon which we have described in the discussion of the chapters that follow.

REFERENCES

1. Doherty, P. C. and Zinkernagel, R. M., A biological role for the major histocompatibility antigens, *Lancet*, 1, 1406, 1970.
2. Clark, W. R., Histocompatibility systems, in *The Experimental Foundations of Modern Immunology*, 2nd ed., John Wiley & Sons, New York, 1983, 274.
3. Clark, W. R., Histocompatibility systems, in *The Experimental Foundations of Modern Immunology*, 2nd ed., John Wiley & Sons, New York, 1983, 279.
4. Clark, W. R., Histocompatibility systems, in *The Experimental Foundations of Modern Immunology*, 2nd ed., John Wiley & Sons, New York, 1983, 290.
5. McDevitt, H. O., Deak, D. B., Shreffler, D. C., Klein, J., Stimpfling, J. H., and Snell, G. D., Genetic control of immune response mapping of the Ir. 1 focus, *J. Exp. Med.*, 135, 1259, 1972.
6. Clark, W. R., Genetic regulation of immune responsiveness, in *The Experimental Foundations of Modern Immunology*, 2nd ed., John Wiley & Sons, New York, 1983, 429.
7. Meruelo, D., A role for elevated H-2 antigen expression in resistance to neoplasia caused by radiation-induced leukemia virus. Enhancement of effective tumor surveillance by killer lymphocyte, *J. Exp. Med.*, 149, 989, 1979.
8. Grundy (Chalmer), J. E., MacKenzie, J. S., and Stanley, N. F., Influence of H-2 and non-H-2 genes on resistance to murine cytomegalovirus infection, *Infect. Immun.*, 32, 277, 1981.
9. Todd, R. F., Reinherz, E. L., and Schlossman, S. F., Human macrophage-lymphocyte interaction in proliferation to soluble antigen, *Cell Immunol.*, 59, 114, 1980.
10. Rosenthal, A. S. and Shevach, E. M., Function of macrophages in antigen recognition by guinea pig T lymphocytes. I. Requirement for histocompatible macrophages and lymphocytes, *J. Exp. Med.*, 138, 1194, 1973.
11. Wylie, D. E., Sherman, L. A., and Klinman, N. R., Participation of the major histocompatibility complex in antibody recognition of viral antigens expressed on infected cells, *J. Exp. Med.*, 155, 403, 1982.
12. Joklik, W. K., Viruses and viral proteins as antigens, in *Principles of Animal Virology*, Appleton-Century-Crofts, Prentice-Hall, New York, 1980, 66.
13. Zinkernagel, R. M. and Doherty, P. C., MHC-restricted cytotoxic T cells. Studies on the biological role of polymorphic major transplantation antigens determining T cell restriction-specificity function and responsiveness, *Adv. Immunol.*, 27, 51, 1979.
14. Zinkernagel, R. M. and Doherty, P. C., H-2 compatibility requirement for T-cell-mediated lysis of targets infected with lymphocytic choriomeningitis virus. Different cytotoxic T cell specificities are associated with structures coded in H-2K or H-2D, *J. Exp. Med.*, 141, 1427, 1975.
15. Finberg, R. and Benacerraf, B., Induction, control and consequences of virus specific cytotoxic T cells, *Immunol. Rev.*, 58, 157, 1981.
16. Zinkernagel, R. M., Major transplantation antigens in host response to infection, *Hosp. Pract.*, 13, 83, 1978.

17. Schmid, D. S., Larsen, H. S., and Rouse, B. T., Role of Ia antigen expression and secretory function of accessory cells in the induction of cytotoxic T lymphocyte responses against Herpes Simplex Virus, *Infect. Immun.*, 37, 1138, 1982.
18. Letvin, N. L., Kauffman, R. S., and Finberg, R., T lymphocyte immunity to reovirus. Cellular requirements for generation and role in clearance of primary infections, *J. Immunol.*, 127, 2334, 1981.
19. Lauer, W. G., Bachmager, H., and Weil, R., The influenza virus hemagglutinin, in *Topics in Infectious Diseases*, Vol. 3, Springer-Verlag, Basel, 1977.
20. Helenius, A., Morein, B., Fries, E., Simons, K., Robinson, P., Schirrmacher, V., Terhost, C., and Strominger, J. L., Human (HLA-A and HLA-B) and murine (H-2K and H-2D) histocompatible antigens and cell-surface receptors for Semliki Forest Virus, *Proc. Natl. Acad. Sci. U.S.A.*, 5, 3846, 1978.
21. Hecht, T. T. and Summers, D. F., Interactions of vesicular stomatitis virus with murine cell surface antigens, *J. Virol.*, 19, 833, 1976.
22. Azocar J. and Essex, M., Incorporation of HLA antigens into the envelope of RNA tumor viruses shown in human cells, *Cancer Res.*, 39, 3388, 1979.
23. Bubbers, J. E. and Lilly, F., Selective incorporation of H-2 antigenic determinants into Friend virus particles, *Nature (London)*, 266, 458, 1977.
24. Mann, D. L., Popovic, M., Sarin, P., Murray, C., Reitz, M., Strong, D., Haynes, B., Gallo, R., and Blattner, W., Cell lines producing human T-cell lymphoma virus show altered HLA expression, *Nature (London)*, 305, 58, 1983.
25. Clarke, M., Gelmann, E., and Reitz, M., Jr., Homology of human T-cell leukaemia virus envelope gene with class I HLA gene, *Nature (London)*, 305, 60, 1983.
26. Hirsch, R. L., Griffin, D. E., and Winkelstein, J. A., Role of complement in viral infections: participation of terminal complement components (C5 to C9) in recovery of mice from Sindbis virus infection, *Inf. Immun.*, 30, 899, 1980.
27. Hicks, J. T., Ernis, F. A., Kims, E., and Verbovitz, M., The importance of an intact complement pathway in recovery from a primary viral infection: influenza in decomplemental and in C5-deficient mice, *J. Immunol.*, 121, 1276, 1978.
28. Welsh, R. M., Jr., Jensen, I. C., Cooper, H. R., and Oldstone, M. B. A., Inactivation and lysis of oncornavirus by human serum, *Virology*, 24, 432, 1976.
29. Smith, T. F., McIntosh, K., Fishant, M., and Henson, P. M., Activation of complement by cells infected with respiratory syncytial virus, *Infect. Immun.*, 33, 43, 1981.
30. Mills, B. J. and Cooper, N. R., Antibody-independent neutralization of vesicular stomatitis virus by human complement, *J. Immunol.*, 121, 1549, 1978.
31. Hirsch, R. L., Griffin, D. E., and Winkelstein, J. H., The effect of complement depletion on the course of Sindbis virus infection in mice, *J. Immunol.*, 121, 1276, 1978.
32. Hirsch, R. L., Griffin, D. E., and Winkelstein, J. A., The role of complement in viral infections. II. The clearance of Sindbis virus from the bloodstream and central nervous system of mice depleted of complement, *J. Inf. Dis.*, 141, 212, 1980.
33. Hunninghake, G. W., Gadek, J. E., Kawanami, O., Ferrans, V. J., and Crystal, R. G., Inflammatory and immune processes in the human lung in health and disease: evaluation by bronchoalveolar lavage, *Am. J. Pathol.*, 97, 149, 1979.
34. Kobilinsky, L., Hardy, W. D., Jr., Ellis, R., Witkins, S. S., and Day, N. R., *In vitro* activation of feline complement by feline leukemia virus, *Infect. Immun.*, 29, 165, 1980.
35. Sherwin, S. A., Benveniste, R. E., and Todaro, G. J., Complement-mediated lysis of type-C virus: effect of primate and human sera on various retroviruses, *Inf. J. Cancer*, 21, 6, 1978.
36. Winkelstein, J. A., Cork, L. C., Griffin, D. E., Adams, R. J., and Price, D. L., Genetically determined deficiency of the third component of complement in the dog, *Science*, 212, 1169, 1981.
37. Winkelstein, J. A., Johnson, J. P., Swift, A. J., Ferry, F., Yolken, R., and Cork, L. C., Genetically determined deficiency of the third component of complement in the dog: *in vitro* studies of the complement system and complement-mediated serum activities, *J. Immunol.*, 129, 2598, 1982.
38. Mogenson, S. C., Role of macrophages in natural resistance to virus infection, *Microbiol. Rev.*, 43, 1, 1979.
39. Mims, C. A., Aspects of the pathogenesis of virus diseases, *Bacteriol. Rev.*, 28, 30, 1964.
40. Stohlmen, S. A., Woodward, J. G., and Felinger, J. A., Macrophage antiviral activity: extrinsic versus intrinsic activity, *Infect. Immun.*, 36, 672, 1982.
41. Martinex, D., Lynch, R. J., Maeher, J. G., and Field, A. K., Macrophage dependence of polyriboinosinic acid-polyribocytidylic acid-induced resistance to herpes simplex virus infection in mice, *Infect. Immun.*, 28, 147, 1980.
42. Archard, L. C. and Williams, J. D., The effect of arginine deprivation on the replication of vaccinia virus, *J. Gen. Virol.*, 12, 249, 1971.

43. Rouse, H. C., Bonifas, V. H., and Schlesinger, R. W., Dependence of adenovirus replication on arginine and inhibition of plaque formation by pleuropneumonia-like organisms, *Virology*, 20, 357, 1963.
44. Winters, A. L. and Consigli, R. A., Effects of arginine deprivation on polyoma virus infection of mouse embryo cultures, *J. Gen. Virol.*, 10, 53, 1971.
45. Forman, A. J. and Babuik, L. A., Effect of infectious bovine rhinotracheitis virus infection on bovine alveolar macrophage function, *Infect. Immun.*, 35, 1041, 1982.
46. Fischer, J. J. and Walker, D. H., In vasine pulmonary aspergillosis associated with influenza, *J. Am. Med. Assoc.*, 241, 1493, 1979.
47. Young, L. S., Laforce, M., Head, J. J., Fealey, B. S., and Bennet, B. V., A simultaneous outbreak of meningococcal and influenza infections, *N. Engl. J. Med.*, 287, 5, 1972.
48. Shanley, J. D. and Pesanti, E. L., Replication of murine cytomegalovirus in lung macrophages: effect on phagocytosis of bacteria, *Infect. Immun.*, 29, 1152, 1980.
49. Jakab, G. J. and Green, G. M., Defects in intracellular killing of *Staphylococcus aureus* within alveolar macrophages in Sendai virus-infected murine lungs, *J. Clin. Invest.*, 57, 1533, 1976.
50. Rodgers, B. and Mims, C. A., Interaction of influenza virus with mouse macrophages, *Inf. Immun.*, 31, 751, 1981.
51. Hoover, E. A., Rojko, J. L., Wilson, P. L., and Olsen, R. G., Macrophages and susceptibility of cats to feline leukemia virus infection, in *Feline Leukemia Virus*, Hardy, W. D., Jr., Essex, M., and McClelland, A. J., Eds, Elsevier/North Holland, Amsterdam, 1980, 195.
52. Blanden, R. V., T-cell response to viral and bacterial infection, *Transplant. Rev.*, 19, 56, 1974.
53. Cerottini, J. C. and Brunner, K. T., Cell mediated cytotoxicity, allograft rejection and tumor immunity, *Adv. Immunol.*, 18, 67, 1974.
54. Blanden, R. V., Hapel, A. J., Doherty, P. C. and Zinkernagel, R. M., Lymphocyte-macrophage interactions and macrophage activation in the expression of antimicrobial immunity *in vivo*, in *Immunobiology of the Macrophage*, Nelson, D. S., Ed., Academic Press, New York, 1976, 367.
55. Blanden, R. V., Pang, T. E., and Dunlap, M. B. C., T-cell recognition of virus-infected cells, in *Virus Infection and the Cell Surface*, Poste, G. and Nicolson, G., Ed., North Holland, Amsterdam, 1977, 249.
56. Sell, S., *Immunology, Immunopathology and Immunity*, 3rd ed., Harper & Row, New York, 1980, 147.
57. Finberg, R., Spriggs, D. R. and Fields, B. N., Immune response to reovirus: CTL recognize the major neutralizing domain of the viral hemagglutinin, *J. Immunol.*, 129, 2235, 1982.
58. Nai-Ki, M., Yang-He, Z., Ada, G. L., and Tannock, G. A., Humoral and cellular responses of mice to infection with a cold-adapted influenza A virus variant, *Infect. Immun.*, 38, 218, 1982.
59. Appel, M. Z. G., Shek, W. R., and Summers, B. A., Lymphocyte-mediated immune cytotoxicity in dogs infected with virulent canine distemper virus, *Infect. Immun.*, 37, 592, 1982.
60. Onions, D. E., The immune response to virus infections, *Vet. Immunol. Immunopath.*, 4, 237, 1983.
61. Mandel, B., Characterization of type 1 poliovirus by electrophoretic analysis, *Virology*, 44, 554, 1971.
62. Halstead, S. B. and O'Rourke, E. J., Antibody enhanced dengue virus infection in primate leukocytes, *Nature (London)*, 265, 739, 1977.
63. Pederson, N. C., Feline infectious peritonitis, *Comparative Pathobiology of Viral Diseases*, Vol. 2, Olsen, R. G., Krakowka, G. S., and Blakeslee, J. R., Eds., CRC Press, Boca Raton, Fla., 1985, Chap. 7.
64. Halstead, S. B., Immune enhancement of viral infection, in *Prog. Allergy*, 31, 301, 1982.
65. Shore, S. L., Nahonias, A. J., Starr, S. E., Wood, P. A., and McFarlin, D. C., Detection of cell-dependent cytotoxic antibody to cells infected with herpes simplex virus, *Nature (London)*, 251, 350, 1974.
66. Fujimiya, T., Rouse, B. T., and Babuik, L. A., Human neutrophil-mediated destruction of antibody sensitized herpes simplex virus type 1 infected cells, *Can. J. Microbiol.*, 24, 182, 1977.
67. Kohl, S., Starr, S. E., Okeske, J. M., Shore, S. L., Ashman, R. B., and Nahmias, A. J. Human monocyte-macrophage mediated antibody dependent cytotoxicity to herpes simplex virus-infected cells, *J. Immunol.*, 118, 729, 1977.
68. Russell, A. S. and Miller, C., A possible role for polymorphonuclear leukocytes in defense against recrudescent herpes simplex infection in man, *Immunology*, 34, 371, 1978.
69. Smith, J. and Sheppard, A. M., Activity of rabbit monocytes, macrophages, and neutrophils in antibody-dependent cellular cytotoxicity of herpes simplex virus-infected corneal cells, *Infect. Immun.*, 36, 685, 1982.
70. Beem, M., Repeated infection with respiratory syncytial virus, *J. Immunol.*, 98, 1115, 1967.
71. Karl, T. N., Welliner, R. C., and Ogra, P. L., Development of antibody-dependent cell-mediated cytotoxicity in the respiratory tract after natural infection with respiratory syncytial virus, *Inf. Immun.*, 37, 492, 1982.
72. Ault, K. A. and Spranger, T. A., Cross-reaction of a rat anti-mouse phagocyte-specific monoclonal antibody (anti-mac-1) with human monocytes and natural killer cells, *J. Immunol.*, 126, 359, 1981.

73. Fast, L. D., Hansen, J. A., and Newman, W., Evidence for T-cell nature and heterogeneity within natural killer (NK) and antibody-dependent cellular cytotoxicity (ADCC) effectors: a comparison with cytolytic T-lymphocytes (CTL), *J. Immunol.*, 127, 1448, 1981.
74. Anderson, J. J., Innate cytotoxicity of CBA mice spleen cells to Sendai virus-infected L cells., *Infect. Immun.*, 20, 608, 1978.
75. Santoli, D. G., Trinchieri, G., and Koprowski, H., Cell-mediated cytotoxicity against virus-infected target cells in humans. II. Interferon induction and activation of natural killer cells, *J. Immunol.*, 121, 532, 1978.
76. Selgrade, M. K., Daniels, M. J., Hu, P. C., Miller, F. J., and Graham, J. A., Effects of immunosuppression with cyclophosphamide on acute murine cytomegalovirus infection and virus-augmented natural killer cell activity, *Infect. Immun.*, 38, 1046, 1982.
77. Bohl, E. H., Gupta, R. K. P., McCloskey, L. W., and Saif, L. W., Immunology of transmissible gastroenteritis, *J. Am. Vet. Med. Assoc.*, 160, 543, 1972.
78. Rice, J. B., Winters, K. A., Krakowka, K., and Olsen, R. G., Comparison of systemic and local immunity in dogs with canine parvovirus gastroenteritis, *Infect. Immun.*, 38, 1003, 1982.
79. Mattingly, J., Cellular circuiting involved in orally induced systemic tolerance and local antibody production, *Ann. NY Acad. Sci.*, 409, 204, 1983.
80. Jackson, D. E., Lally, E. T., Nakamura, M. C., and Montgomery, P. C., Migration of IgA-bearing lymphocytes into salivary glands, *Cell. Immunol.*, 63, 203, 1981.
81. Wiman, K. B., Curman, U., Forsum, L., Klareskog, L., Malmnas-Tjernlund, U., Rask, L., Tragardh, L., and Paterson, P., Occurrence of Ia antigens in tissues of nonlymphoid origin, *Nature (London)*, 276, 711, 1978.
82. Olsen, C. G., Stiff, Mi. I., and Olsen, R. G., Comparison of the blastogenic response of peripheral blood lymphocytes from canine-positive and -negative dogs, *Vet. Immunol. Immunopathol.*, 6, 285, 1984.
83. Krakowka, S., Olsen, R. G., Axthelm, M. K., Rice, J., and Winters, K., Canine parvovirus infection potentiates canine distemper encephalitis attributable to modified live-virus vaccine, *J. Am. Vet. Med. Assoc.*, 180, 137, 1982.
84. Gardner, P. S., McQuillin, J., and Court, S. D. M., Speculation on pathogenesis in death from respiratory syncytial virus infection, *Br. Med. J.*, 2, 237, 1970.
85. Minor, T. E., Dick, E. C., Baker, J. W., Oullette, J. J., Cohen, M., and Reed, C. E., Rhinovirus and influenza type A infections as precipitants of asthma, *Am. Rev. Respir. Dis.*, 113, 149, 1976.
86. Becroft, D. M. O., Bronchiolitis obliterans, bronchiectasis, and other sequelae of adenovirus type 21 infections in young children, *J. Clin. Pathol.*, 24, 72, 1971.
87. Similar, S., Linna, O., Lanning, P., Heikkinen, E., and Ala-Houhala, M., Chronic lung damage caused by adenovirus type 7: a ten-year follow-up study, *Chest*, 80, 127, 1981.
88. McIntosh, K. and Fishaut, J. M., Immunopathologic mechanisms in lower respiratory tract disease of infants due to respiratory syncytial virus, *Prog. Med. Virol.*, 26, 94, 1980.
89. Pullan, C. R. and Hey, E. N., Wheezing, asthma, and pulmonary dysfunction 10 years after infection with respiratory syncytial virus in infancy, *Br. Med. J.*, 284, 1665, 1982.
90. Dick, E. C. and Reed, C. E., Effects of experimental rhinovirus 16 infection on airways and leukocyte function in normal subjects, *J. Aller. Clin. Immunol.*, 61, 80, 1978.
91. Twiggs, J. T., Larson, L. A., O'Connel, E. J., and Elstrup, D. M., Respiratory syncytial virus infection: ten-year follow-up, *Clin. Pediatr. (Philadelphia)*, 20, 187, 1981.
92. Sell, S., *Immunology, Immunopathology, and Immunity*, 3rd ed., Harper & Row, New York, 1980.
93. Bellanti, J. A., *Immunology II*, W. B. Saunders, Philadelphia 1978.
94. Hirsch, M. S. and Profitt, M. R., Autoimmunity in viral infections, in *Viral Immunology and Immunopathology*, Notkins, A. L., Ed., Academic Press, New York, 1975, 419.
95. Onodera, T., Roniolo, A., Ray. U. R., Jenson, A. B., Knazek, R. A., and Notkins, A. L., Virus-induced diabetes mellitus. XX. Polyendocrinopathy and autoimmunity, *J. Exp. Med.*, 153, 1457, 1981.
96. Vialettes, B., Baume, D., Charpin, C., DeMaeyer-Guignard, J., and Vague, Ph., Assessment of viral and immune factors in EMC virus-induced diabetes: effect of cyclosporin A and interferon, *J. Clin. Lab. Immunol.*, 10, 35, 1983.
97. Yoon, J. W., Onodera, T., and Notkins, A. L., Virus-induced diabetes mellitus. XV. Beta cell damage and insulin-dependent hyperglycemia in mice infected with Coxsackie B4, *J. Exp. Med.*, 148, 1068, 1978.
98. Weiner, H. L., Powers, M. L., and Fields, B. N., Absolute linkage of virulence and central nervous system tropism of reovirus for viral hemagglutinin, *J. Infect. Dis.*, 141, 609, 1980.
99. Knight, J. G., Autoimmune disease: defects in immune specificity rather than a loss of suppressor cells, *Immunol. Today*, 3(12), 326, 1982.
100. Border, W. A., Ward, H. J., Kamil, E. S., and Cohen, A. H., Induction of membranous nephropathy in rabbits by administration of an exogenous cationic antigen: demonstration of a pathogenic role for electrical charge, *J. Clin Invest.*, 69, 451, 1982.

101. Theofilopoulos, A. N. and Dixon, F. J., The biology and detection of immune complexes, *Adv. Immunol.*, 28, 89, 1979.
102. Milas, L. and Scott, M. T., Antitumor activity of *Corynebacterium parvulum*, *Adv. Cancer Res.*, 26, 257, 1977.
103. Mathes, L. E., Olsen, R. G., Hebebrand, L. C., Hoover, E. A., Schaller, J. P., Adams, P. W., and Nichols, W. S., Immunosuppressive properties of a virion polypeptide, a 15,000-Dalton protein from feline leukemia virus, *Cancer Res.*, 39, 950, 1979.
104. Olsen, R. G., Mathes, L. E., and Nichols, W. S., FeLV-related immunosuppression, in *Feline Luekemia*, Olsen, R. G., Ed., CRC Press, Boca Raton, Fla., 1981, 149.
105. Alpers, J. H., Steward, M. W., and Soothill, J. F., Differences in immune elimination in inbred mice. The role of low affinity antibody, *Clin. Exp. Immunol.*, 12, 121, 1972.
106. Pincus, T., Haberkein, R., and Christian, C. L., Experimental chronic glomerulonephritis, *J. Exp. Med.*, 127, 819, 1968.
107. Hersey, P. and Isbister, J. P., Developments in immune complex therapy and its application to cancer, in *Immune Complexes and Plasma Exchange in Cancer Patients*, Serrou, B. and Rosenfeld, C., Eds., Elsevier/North Holland, Amsterdam 1981, 135.
108. Delire, M. and Massen, P. L., The detection of circulating immune complexes in children with recurrent infections and their treatment with human immunoglobulins, *Clin. Exp. Immunol.*, 29, 385, 1977.
109. Hoover, E. A., Rojko, J. L., and Olsen, R. G., Factors influencing host resistance to feline leukemia virus, in *Feline Leukemia*, Olsen, R. G., Ed., CRC Press, Boca Raton, Fla., 1981, 69.
110. Kobilinsky, L., Hardy, W. D., Jr., and Day, N. K., Hypocomplementemia associated with naturally occurring lymphosarcoma in pet cats, *J. Immunol.*, 122, 2139, 1979.
111. Gauci, L., Caraux, J., and Serrou, B., Immune complexes in the context of the immune response in cancer patients, in *Immune Complexes and Plasma Exchange in Cancer Patients*, Serrou, B. and Rosenfeld, C., Eds., Elsevier/North Holland, Amsterdam, 1981, 37.
112. Schmitt, M., Mussel, H. H., and Dierich, M. P., Qualitative and quantitative assessment of C_3-receptor activities on lympoid and phagocytic cells, *J. Immunol.*, 126, 2041, 1981.
113. Miyama-Inaba, M., Suzuki, T., Paku, Y-H., and Masuda, T., Feedback regulation of immune responses by immune complexes: possible involvement of a suppressive lymphokine by FcR-bearing B cell, *J. Immunol.*, 128, 882, 1982.
114. Gorden, J. and Murgita, R. A., Suppression and augmentation of the primary *in vitro* immune response by different classes of antibodies, *Cell Immunol.*, 15, 392, 1975.
115. Kolsh, E. J., Oberbarnscheidt, J., Bruner, K., and Heuer, J., The Fc receptor: its role in the transmission of different signals, *Immunol. Rev.*, 49, 61, 1980.
116. Iida, K., Mornaghi, R., and Nussenzweig, V., Complement receptor (CR 1) deficiency in erythrocytes from patients with systemic lupus erythematosus, *J. Exp. Med.*, 155, 1427, 1982.
117. Westmoreland, D., St. Jear, S., and Rapp, F., The development by cytomegalovirus-infected cells of binding affinity for normal human immunoglobulin, *J. Immunol.*, 116, 1566 1976.
118. Costa, J. Rabson, A. S., Yee, C., and Tralka, T. S., Immunoglobulin binding to herpes virus-induced Fc receptor inhibits virus growth, *Nature (London)*, 269, 251, 1977.
119. Jones, F. R., Yoshida, L. H., Ladiges, W. C., Zeidner, N. S., Kenny, M. A., and McClelland, A. J., Treatment of feline lymphosarcoma by extracorporeal immunosorption, in *Feline Leukemia Virus*, Hard, W. D., Jr., Essex, M., and McClelland, A. J., Eds., Elsevier/North Holland, Amsterdam, 1980, 235.
120. Henson, J. B. and Crawford, T. B., The pathogenesis of virus-induced arterial disease — Aleutian disease and equine viral arteritis, *Adv. Cardiol.*, 13, 183, 1974.
121. Golnik, W., Michalska, Z., and Michalak, T., Natural equine viral arteritis in foals, *Schweiz. Arch. Tierheilk.*, 123, 523, 1981.
122. Nathanson, N., Monjan, A. A., Panitch, H. S., Johnson, E. D., Petursson, G., and Cole, G. A., Virus-induced cell-mediated immunopathological disease, in *Viral Immunology and Immunopathology*, Notkins, A. L., Ed., Academic Press, New York, 1975, 357.
123. Lampert, P. W., Autoimmune and virus-induced demyelinating disease: a review, *Am. J. Pathol.*, 91, 176, 1978.
124. Dal canto, M. C. and Rabinowitz, S. C., Experimental models of virus-induced demyelination of the central nervous system, *Ann. Neurol.*, 11, 109, 1982.
125. Davis, E., personal communication.

Chapter 4

CHEMICAL ALTERATION OF HOST SUSCEPTIBILITY TO VIRAL INFECTION

Melinda J. Tarr

TABLE OF CONTENTS

I. INTRODUCTION

Exposure to environmental drugs, hormones, or any other nonmicrobial agent can have a profound influence upon a host's susceptibility to infectious or neoplastic disease. During the past few years, there have been numerous reports of increased or decreased susceptibility of experimental animal hosts to viral infections associated with exposure to various compounds. Epidemiologically, there are some instances of disease outbreaks or increased incidence of disease associated with exposure of man or animals to environmental contaminants. Increased susceptibility to viral infection of individuals treated with certain drugs or hormones is also well documented.

There are several ways that exposure to a chemical can alter a host's susceptibility to viral infections. The most important mechanism is via a direct effect on the immune system, particularly the cell-mediated immune (CMI) system, which is the most important defense against viral infections[1] as viruses are obligate intracellular parasites. However, suppression of the humoral antibody response may allow a higher or prolonged viremic state or allow spread of a viral infection from a primary to a secondary site which would normally be prevented by viral neutralizing antibody.[2] Interferon synthesis and activity is a third important nonspecific viral defense mechanism which can be altered by chemical agents. Finally, chemicals can alter susceptibility to enteric and respiratory viral infections by interfering with either specific or nonspecific mucosal immune systems such as pulmonary macrophage function or IgA synthesis and activity.

It must be remembered that chemicals can enhance as well as reduce a host's resistance to viral infection by stimulating all or part of the immune system. Immunostimulation is not always beneficial to the host, as autoimmune diseases may develop due to interference with immunoregulation or excessive stimulation of the helper or effector arms of the immune system.

Chemicals may also directly alter a viral target cell, thus making the cell more or less susceptible to viral infection or replication. The best known and best publicized example of this type would be chemical viral cocarcinogensis in which chemical alteration of deoxyribonucleic acid permits insertion and/or expression of viral oncogenes. Volumes have been written about this topic so it will not be reviewed in this chapter. Other examples which will be discussed include damage to respiratory epithelium by toxic or irritating inhalants resulting in enhancement of susceptibility to viral infections.

Thirdly, chemicals may interact directly with viral particles resulting in enhanced chemical toxicity and/or enhanced viral replication or virulence.

Finally, endocrine imbalances, resulting either from exposure to toxic chemicals or hormone analogs or from endogenous disturbances, could alter susceptibility to viral infections. This is usually secondary to hormonal effects on the immune system.

The intent of this chapter is to emphasize the ubiquity and importance of environmental or pathophysiological factors which may influence a host's susceptibility to viral infection by citing documented examples or extrapolating from other studies.

II. EFFECTS OF ENVIRONMENTAL POLLUTANTS ON SUSCEPTIBILITY TO VIRAL INFECTIONS

A. Heavy Metals

A recent review by Koller[3] cites several examples of altered susceptibility to viral infections resulting from exposure to compounds containing heavy metals. Mice which were exposed to compounds containing lead,[4,5] mercury,[4,6] nickel,[4] arsenic,[7] or cobalt[8] were more susceptible to infection with encephalomyocarditis virus (EMCV) at doses of metal compounds which otherwise caused no overt signs of toxicity. Zinc had no effect on the same model[4] while cadmium caused increased susceptibility to EMCV infection in one report,[4] and decreased susceptibility in a different report.[5] This discrepancy for cadmium is also noted in investigations of the effects of cadmium on the immune response which report either enhancement,[9,10] suppression,[11,12] or a biphasic effect (enhancement early in exposure, then suppression).[13,14] Most likely, these differences are due to differences in dose, time and route of exposure, or parameters evaluated. Other compounds show similar effects.

In other host-virus models, Thind et al. reported increased susceptibility to Langet virus in mice exposed to lead.[15] Sheffy and Schultz found that dogs fed a diet deficient in selenium and vitamin E mounted a lower antibody response to canine distemper virus vaccine;[16] and Koller found that rabbits exposed to lead, cadmium or mercury developed lower neutralizing antibody titers to pseudorabies virus inoculation.[17]

The mechanism by which susceptibility to viral infection is altered by heavy metals undoubtedly relates to their effects on the immune system. The specific immunotoxic effects which are responsible for this alteration of susceptibility are not delineated in most studies. Lead suppresses most immune mechanisms so it increases susceptibility by depressing cell-mediated, humoral, and phagocytic immune functions.[3] Cobalt, sulfate, and high doses of arsenicals[18] inhibited interferon synthesis in cell culture experiments, so this may be at least one mechanism by which these chemicals exert their enhancement of viral susceptibility. The reports of effects of cadmium on immune functions are conflicting. At least two investigations noted enhanced macrophage clearance[19] and phagocytosis[20] associated with cadmium exposure which could explain the results of Exon et al.[5] which, as mentioned, noted a decreased susceptibility to viral infection.

In conclusion, exposure to subtoxic levels of most heavy metals results in increased susceptibility to viral infections. Possible exposure to these compounds should be considered when interpreting results of epidemiological or experimental studies involving viral diseases.

B. Pesticides

The use of pesticides is becoming more and more prevalent in both argriculture and industry. Many of these compounds are highly toxic and degrade very slowly or not at all in the environment. There are few studies which have specifically evaluated viral susceptibility in association with pesticide exposure. An early in vitro study by Gabliks[21] indicated that HeLa cells grown in the presence of subcytotoxic concentrations of dimethoate, Dipterex, trichlorfan, chlordane, disulfoton, or dinocap were subsequently more susceptible to infection with polio virus. In another study, Thigpen et al.[22] reported that oral treatment with tetrachlorodibenzo-*p*-dioxin (TCDD) did not affect the susceptibility of mice to *Herpes suis* virus infection. Barnet et al.[23] treated pregnant mice with diazinon and carbofuran, and found that the suckling offspring suffered higher mortality due to "respiratory infections" than controls; however, the cause of the infection was not mentioned. Ducks exposed to the insecticide dieldrin in their diet suffered higher and more prolonged mortality when challenged with duck hepatitis virus.[24] Finally, two studies by Crocker and colleagues revealed that mice exposed to a combination of the insecticides DDT and fenitrothion,[25] or the solvents and emulsifiers used in the actual spray mixture of these insecticides,[26] suffered more severe disease and higher mortality when inoculated with normally sublethal doses of encephalomyocarditis virus. The mice developed fatty livers and severe neurological signs, a syndrome similar to Reye's syndrome. The authors suggested that an increased incidence of Reye's syndrome seen in the vicinity of heavy forest spraying operations could be due to insecticide-viral interactions.[26] This model also provides an example of either direct chemical-viral interaction or chemical damage to target tissue as being responsible for the enhanced susceptibility to viral infections.[26]

There are numerous other reports of increased susceptibility to bacterial infections and/or depressed immunity resulting from exposure to pesticides, many of which were summarized recently by Street.[27] It is safe to assume that many of these chemicals would also alter susceptibility to viral diseases, particularly those which preferentially suppress the CMI system such as 2,3,7,8-tetrachlorodibenzo-*p*-dioxin (TCDD).[28]

C. Industrial Chemicals and Food Additives — Systemic Exposure

Recently, the extensive use of polychlorinated and polybrominated biphenyls (PCBs, PBBs) has caused much concern among environmentalists due to their extreme toxicity, and to their persistence and abundance in the environment.[29] They have been shown to enhance susceptibility to viral infections and cause immune dysfunctions both experimentally and under natural conditions. In one study by Friend and Trainer,[30] mallard ducklings which were fed subtoxic doses of PCBs suffered earlier and higher mortality when they were challenged with duck hepatitis virus.

Another experiment by Koller and Thigpen[31] indicated that rabbits fed various formulations of PCBs all mounted significantly lower serum-neutralizing antibody titers compared to corn oil-fed controls when inoculated with formalized pseudorabies virus. The decreased titers did not correlate with differences in hepatotoxicity of the three compounds, and none of the formulations caused any pathologic changes in the spleen, thymus gland, or lymph nodes.

An accident in Michigan in which PBBs were inadvertently used as a feed supplement for dairy cattle resulted in exposure of many Michigan residents to PBB in dairy products. Farm families suffered the heaviest exposure, and many of these individuals suffered a higher incidence of diseases[32] as well as immunologic dysfunctions.[33]

Methylnitrosourea (MNU), a carcinogenic nitrosamine which, under certain conditions, may form as a result of sodium nitrite ingestion,[34] abrogates the natural resistance of adult cats to feline leukemia virus infection and disease.[35] This is most likely related to its immunosuppressive (as opposed to carcinogenic) properties in the cat[36] as it is lethal in this species at carcinogenic doses. Virtually all carcinogens are immunosuppressive either during the preneoplastic period or at subcarcinogenic exposure levels, so increased susceptibility to viral infections would be expected due to either immunosuppression or chemical alteration of target cell DNA or metabolism.

Many other environmental contaminants are reported to have immunosuppressive or immunomodulating effects at otherwise nontoxic levels, such as hydrazine compounds,[37] pentachlorophenol,[38] or hexachlorobenzene.[39] Others have immunotoxic or subtle organ-specific toxic effects at federally approved levels including PCB[40] and formaldehyde.[41] Exposure to any of these compounds could, thus, conceivably result in decreased resistance to viral infections.

D. Inhaled Toxicants

Air pollutants are considered in a separate category because of the importance of their localized effect on the respiratory system, although many of them can be absorbed into the body through the respiratory epithelium, and exert significant systemic toxic effects. The effects of airborne toxicants will vary depending on the nature (gas or particulate), particle size, configuration, and composition.[42] They may alter susceptibility to viral infection by affecting either the respiratory epithelial cells directly (mucociliary function, epithelial cell necrosis) or the respiratory immune system (alveolar macrophages, bronchus-associated lymphoid tissue).[41]

Both increased and decreased susceptibility to respiratory viral infections upon exposure to air pollutants have been reported. Mice exposed to 20 ppm of sulfur dioxide for 7 days developed more pneumonia when infected with influenza virus than unexposed mice.[43] Also, short-term exposure to nitrogen dioxide decreased the resistance of alveolar macrophages to viral infection.[44] A third experiment by Spurgash et al. revealed that mice infected with influenza virus and then exposed to tobacco smoke suffered higher mortality than control mice.[45] Epidemiologically, there is a positive correlation between high ambient air levels of sulfur dioxide and increased incidence of respiratory symptoms suggestive of viral respiratory infections,[45] and between high air pollution indexes and short-term respiratory infections.[46,47]

On the other hand, studies by Goldring et al.[49] indicated that multiple exposure of Syrian hamsters to high levels of sulfur dioxide did not affect the course or severity of influenza infection. Squirrel monkeys which were exposed to 1 ppm nitrogen dioxide (NO_2) for several months mounted higher and earlier serum neutralizing antibody titers in response to influenza viral challenge.[50] However, the NO_2-treated, influenza-infected monkeys developed more severe pulmonary lesions than either group alone. Studies with formaldehyde showed that mice exposed to 15 ppm formaldehyde for 21 days were actually more resistant to bacterial challenge (*Listeria monocytogenes*) than control animals.[51]

Respiratory and systemic immune functions have also been evaluated following exposure to various air pollutants. Gregson and Prentice[52] looked at the amount of bronchus-associated lymphoid tissue (BALT), pulmonary alveolar macrophage (PAM) lysosomal enzyme activity, and bronchial immunoglobulin levels in rats chronically exposed to tobacco smoke. They found that BALT increased to above controls by week 5, then decreased by week 14, then increased again and remained higher than controls from months 12 through 20 (the end of the experiment). The PAM lysosomal enzyme activity was higher in exposed animals for the first 14 weeks of exposure (only time of assessment) and bronchial immunoglobulin first increased, then decreased during the first 14 weeks. Other experiments show, in general, a depression of systemic immunity in animals exposed to tobacco smoke, including antibody response (plaque-forming cells)[53] and pulmonary lymphocyte response to phytohemagglutinin.[54]

Chronic nitrogen oxide inhalation resulted in an initial enhancement, then suppression of several immunologic functions in mice, including serum antibody response to T-dependent antigens and graft vs. host response. Other functions were suppressed (lymphocyte proliferation in response to phytohemagglutinin and resistance to tumor challenge).[55]

In summary, in a review of the literature on the effects of air pollutants on host immunity and resistance, Holt and Keast[56] concluded that continuous exposure to air contaminants elicits a triphasic response in host resistance: an initial acute depression, followed by stimulation (increased resistance), and finally, a severe long-lasting depression.

III. OTHER CHEMICALS AFFECTING HOST-VIRAL INTERACTION

A. Hormones

Fluctuations in hormone balance or administration of exogenous hormones may definitely have an effect on host-viral interaction. Just a few examples will be cited here.

Most literature dealing with hormones and host susceptibility involves corticosteroids. It has long been known that pharmacological doses of corticosteroids render host animals more susceptible to any type of pathogenic agent. For instance, an early study by Schwartzman[57] showed that mice and hamsters given cortisone were more susceptible to poliomyelitis viral infection. More recently, Cummins and Rosenquist[58] reported that calves injected with hydrocortisone and infected with infectious bovine rhinotracheitis (IBR) virus developed higher and more persistent viremias than the control calves. Exogenous steroid administration may also cause reactivation of latent IBR viral infections in cattle.[59]

Stress, which is characteristically associated with endogenous rises in corticosteroid levels, also influences host susceptibility to viral infection. The "cold sore", or recurrent herpes simplex infection in man, has long been associated with emotional stress.[60] A higher incidence of Marek's disease has also been shown in chickens stressed by being introduced into new pecking orders.[61] There are numerous similar examples.

The mechanism of corticosteroid-mediated reduction of susceptibility undoubtedly relates at least in part to their anti-inflammatory and immunosuppressive effects, but may also be due to a decrease in interferon synthesis which has been demonstrated in some studies.[62]

Many other hormones, including gonadal steroids, thyroid hormones, growth hormone, insulin, prolactin, and prostaglandins have been shown to alter immune function and/or alter host-viral interaction (reviewed by Alqvist[63] and Hellman and Weislow[64]).

B. Therapeutic Drugs

Although there are few documented examples (except for steroid hormones), many therapeutic agents used for nonviral diseases can undoubtedly influence host susceptibility to viral infections. Most drugs which are used to treat cancer, organ transplant rejection, or autoimmune diseases are moderately to highly immunosuppressive. Cyclophosphamide, a commonly used drug in this category, has been shown to increase viral persistence and mortality due to Tahyna virus infection when given to mice at only slightly immunosuppressive doses.[65] Other studies indicated that psychotropic drugs are immunosuppressive based on their suppression of lymphocyte proliferative responses to mitogens.[66] Amphotericin B, an antibiotic used to treat fungal infections, has been reported to either augment[67] or suppress[68] immune functions in mice.

Immunotherapy, or use of nonspecific immunopotentiating compounds, is now common in treatment of many chronic diseases associated with immunosuppression including viral diseases. One of these compounds, levamisole, was used successfully in the treatment of herpetic keratitis.[69] Two other immunostimulatory compounds, muramyl dipeptide and mycobacterial fractions, rendered mice more resistant to aerogenic influenza virus infection.[70]

The drug inosiplex has recently been extensively investigated for use as an immunostimulatory compound as well as an antiviral compound. In a comprehensive study by Ohnishi et al.,[71] inosiplex was demonstrated to have direct antiviral properties along with a variety of immunoenhancing effects including increased gamma interferon production. It has been used successfully in the treatment of herpes simplex.[72]

IV. CONCLUSIONS

This brief review has hopefully pointed out that there are many exogenous and endogenous chemicals which can influence an animal's susceptibility to viral infection, other than properties of the virus or host alone. Many of the cited references involve experiments performed under artificial or high-exposure conditions, so one might argue that the results are unrealistic and would not occur under natural conditions. However, altered viral susceptibility does occur with subtoxic and, in some cases, less than legally allowed doses of many of the compounds tested. Most of the reported experiments were also short term; since many environmental pollutants are poorly biodegradable or metabolized, they can accumulate both in the environment and in the host over prolonged periods of time and build up to significant levels. Some of them may also exert their effects long after the time of the exposure. Most importantly, a host in a natural environment is exposed to many chemicals, including numerous pollutants as well as its own hormones. This results in a potentially additive or even synergistic effect on alteration of host susceptibility to any type of infectious organism. It is clear that the rapidly expanding interest and knowledge in environmental toxicology and immunotoxicology will play a central role in our understanding of the pathobiology of viral diseases.

ACKNOWLEDGMENTS

The preparation of this chapter was supported in part by the Air Force Office of Scientific Research, contract no. F49620-83-C-0114. Special thanks goes to Ms. Teri Roberts for prompt editing and typing services.

REFERENCES

1. Allison, A. C., Immunity against virus, in *The Scientific Basis of Medicine; Annual Review*, Athlone Press, London, 1972.
2. Svehag, S. and Mandel, B., The formation and properties of poliovirus neutralizing antibody, *J. Exp. Med.*, 119, 1, 1964.
3. Koller, L. D., Immunotoxicology of heavy metals, review/commentary, *Int. J. Immunopharmacol*, 2, 269, 1980.
4. Gainer, J. H., Effects of heavy metals and of deficiency of zinc on mortality rates in mice infected with encephalomyocarditis virus, *Am. J. Vet. Res.*, 38, 869, 1977.
5. Exon, J. H., Koller, L. K., and Kerkvliet, N. I., Lead-cadmium interaction: effects on viral-induced mortality and tissue residues in mice, *Arch. Environ. Health*, 34, 469, 1979.
6. Koller, L. D., Methylmercury: effect on oncogenic and non-oncogenic viruses in mice, *Am. J. Vet. Res.*, 36, 1501, 1975.
7. Gainer, J. H. and Pry, T. W., Effects of arsenicals on viral infections of mice, *Am. J. Vet. Res.*, 33, 2299, 1972.
8. Gainer, J. H., Increased mortality in encephalomyocarditis virus-infected mice consuming cobalt sulfate: tissue concentrations of cobalt, *Am. J. Vet. Res.*, 33, 2067, 1972.
9. Koller, L. D., Roan, J. G., and Kerkvliet, N. I., Mitogen stimulation of lymphocytes in CBA mice exposed to lead and cadmium, *Environ. Res.*, 19, 177, 1979.
10. Fujimaki, H., Murakami, M., and Kubota, K., *In vitro* evaluation of cadmium-induced augmentation of the antibody response, *Toxicol. Appl. Pharmacol.*, 62, 288, 1982.
11. Gaworski, C. L. and Sharma, R. P., The effects of heavy metals on ^{3}H thymidine uptake in lymphocytes, *Toxicol. Appl. Pharmacol.*, 46, 305, 1978.
12. Graham, J. A., Miller, F. J., Daniels, M. J., Payne, E. A., and Gardner, D. E., Influence of cadmium, nickel, and chromium on primary immunity in mice, *Environ. Res.*, 16, 77, 1978.
13. Koller, L. D., Roan, J. G., and Exon, J. H., Humoral antibody response in mice after exposure to lead or cadmium, *Proc. Soc. Exp. Biol. Med.*, 151, 339, 1976.
14. Jones, R. H., Williams, R. L., and Jones, A. M., Effects of heavy metals on the immune response. Preliminary findings for cadmium in rats, *Proc. Soc. Exp. Biol. Med.*, 137, 1231, 1971.
15. Thind, I. S., Thind, G. S., and Louria, D. B., Potentiation of viral infections due to trace metal intoxication, Int. Symp. on Clin. Chem. and Chem. Toxicol. of Metals, Monte Carlo, 1977.
16. Sheffy, B. E. and Schultz, R. D., Influence of vitamin E and selenium on immune response mechanisms, *Fed. Proc. Fed. Am. Soc. Exp. Biol.*, 38, 2139, 1979.
17. Koller, L. D., Immunosuppression produced by lead, cadmium, and mercury, *Am. J. Vet. Res.*, 34, 1457, 1973.
18. Gainer, J. H., Effects of arsenicals on interferon formation and action, *Am. J. Vet. Res.*, 33, 2579, 1972.
19. Cook, J. A., Marconi, E. A., and DiLuzio, N. R., Lead, cadmium endotoxin interaction: effects on mortality and hepatic function, *Toxicol. Appl. Pharmacol.*, 28, 292, 1974.
20. Koller, L. D. and Roan, J. G., Effects of lead and cadmium on mouse peritoneal macrophages, *J. Reticuloendothel. Soc.*, 21, 7, 1977.
21. Gabliks, J., Responses of cell culture to insecticides. III. Altered susceptibility to poliovirus and diphtheria toxin, *Proc. Soc. Exp. Biol. Med.*, 120, 172, 1965.
22. Thigpen, J. E., Faith, R. E., McConnell, E. E., and Moore, J. A., Increased susceptibility to bacterial infection as a sequela of exposure to an environmental contaminant 2,3,7,8-tetrachlorodibenzo-p-dioxin (TCDD), *Infect. Immun.*, 12, 1319, 1975.

23. Barnett, J. B., Spyker-Cranmer, J. M., Avery, D. L., and Hoberman, A. M., Immunocompetence over the lifespan of mice exposed in utero to carbofuran or diazinon. I. Changes in serum immunoglobulin concentrations, *J. Environ. Pathol. Toxicol.*, 4, 53, 1980.
24. Friend, M. and Trainer, D. O., Experimental dieldrin-duck hepatitis virus interaction studies, *J. Wildl. Manage.*, 38, 896, 1974.
25. Crocker, J. F. S., Rózee, K. R., Ozere, R. L., Digout, S. C., and Hutzinger, O., Insecticide and viral interaction as a cause of fatty visceral changes and encephalopathy in the mouse, *Lancet*, 2, 22, 1974.
26. Crocker, J. F. S., Ozere, R. L., Safe, S. H., Digout, S. C., Rozee, K. R., and Hutzinger, O., Lethal interactions of ubiquitous insecticide carriers with virus, *Science*, 192, 1351, 1976.
27. Street, J. C., Pesticides and the immune system, in *Immunologic Considerations in Toxicology*, Sharma, R. P., Ed., CRC Press, Boca Raton, Fla., 1981.
28. Sharma, R. P., Effects of tetrachloradibenzo-p-dioxin (TCDD) on immunologic systems, in *Immunologic Considerations in Toxicology*, Sharma, R. P., Ed., CRC Press, Boca Raton, Fla., 1981.
29. Risebrough, R. W., Rieche, P., Peakall, D. B., Herman, S. G., and Kirven, M. N., Polychlorinated biphenyls in the global ecosystem, *Nature (London)*, 220, 1098, 1968.
30. Friend, M. and Trainer, D. O., Polychlorinated biphenyl: interaction with duck hepatitis virus, *Science*, 170, 1314, 1970.
31. Koller, L. D. and Thigpen, J. E., Biphenyl-exposed rabbits, *Am. J. Vet. Res.*, 34, 1605, 1973.
32. Anderson, H. A., Lilis, R., and Selikoff, I. J., Unanticipated prevalance of symptoms among dairy farmers in Michigan and Wisconsin, *Environ. Health Perspect.*, 23, 217, 1978.
33. Bekesi, J. G., Anderson, H. A., Roboz, J. P., Roboz, J., Fischbein, A., Selikoff, I. J., and Holland, J. F., Immunologic dysfunction among PBB-exposed Michigan dairy farmers, *Ann. N.Y. Acad. Sci.*, 320, 717, 1979.
34. Montesano, R. and Magee, P. N., Evidence of formation of N-methyl-N-nitrosourea in rats given N-methylurea and sodium nitrite, *Int. J. Cancer*, 7, 249, 1971.
35. Schaller, J. P., Mathes, L. E., Hoover, E. A., and Olsen, R. G., Enhancement of feline leukemia virus-induced leukemogenesis in cats exposed to methylnitrosourea, *Int. J. Cancer*, 24, 700, 1979.
36. Tarr, M. J., Olsen, R. G., Hoover, E. A., Kociba, G. J., and Schaller, J. P., The effects of methylnitrosourea on the immune system and hematopoietic system of adult specific-pathogen-free cats, *Chem. Biol. Interact.*, 28, 181, 1979.
37. Tarr, M. J., Olsen, R. G., and Jacobs, D. L., *In vivo* and *in vitro* effects of 1,1-dimethylhydrazine on selected immune functions, *Immunopharmacology*, 4, 139, 1982.
38. Exon, J. H. and Koller, L. D., Effects of chlorinated phenols on immunity in rats, *Int. J. Immunopharmacol.*, 5, 131, 1983.
39. Loose, L. D., Pittman, K. A., Benitz, K.-F., Silkworth, J. B., Muller, W., and Coulston, F., Environmental chemical-induced immune dysfunction, *Ecotoxicol. Environ. Safety*, 2, 173, 1978.
40. Thomas, P. T. and Hinsdill, R. D., Effect of polychlorinated biphenyls on the immune responses of rhesus monkeys and mice, *Toxicol. Appl. Pharmacol.*, 44, 41, 1978.
41. Loomis, T. A., Formaldehyde toxicity, *Arch. Pathol. Lab. Med.*, 103, 321, 1979.
42. Dannenberg, A. M., Jr., Influence of environmental factors on the respiratory tract: summary and perspectives, *J. Reticuloendothel. Soc.*, 22, 273, 1977.
43. Fairchild, G. A., Roan, J., and McCarroll, J., Atmospheric pollutants and the pathogenesis of viral respiratory infection, *Arch. Environ. Health*, 25, 174, 1972.
44. Valand, S. B., Acton, J. D., and Myrvik, Q. N., Nitrogen dioxide inhibition of viral-induced resistance in alveolar monocytes, *Arch. Environ. Health*, 20, 303, 1970.
45. Spurgash, A., Ehrlich, R., and Petzold, R., Effect on resistance to bacterial and viral infection, *Arch. Environ. Health*, 16, 385, 1968.
46. Thompson, D. J., Lebowitz, M., Cassell, E. J., Wolter, D., and McCarroll, J., Health and the urban environment. VII. Air pollution, weather, and the common cold, *Am. J. Public Health*, 60, 731, 1970.
47. Dohan, F. C., Everts, G. S., and Smith, R., Variations in air pollution and the incidence of respiratory disease, *J. Air Pollut. Control Assoc.*, 12, 418, 1962.
48. Ipsen, J., Deane, M., and Igenito, F. E., Relationships of acute respitatory disease to atmospheric pollution and meteorological conditions, *Arch. Environ. Health*, 18, 462, 1969.
49. Goldring, I. P., Cooper, P., Rathner, I. M. et al., Pulmonary effects of sulfur dioxide exposure in the Syrian hamster. I. Combined with viral respiratory disease, *Arch. Environ. Health*, 15, 167, 1967.
50. Fenters, J. D., Findlay, J. C., Port, C. D., Ehrlich, R., and Coffin, D. L., Chronic exposure to nitrogen dioxide. Immunologic, physiologic, and pathologic effects in virus-challenged squirrel monkeys, *Arch. Environ. Health*, 27, 85, 1973.
51. Dean, J. H., Lauer, L. D., House, R. V., Murray, M. J., Stillman, W. S., Irons, R. D., Steinhagen, W. H., Phelps, M. C., and Adams, D. O., Studies of immune function and host resistance in B6C3F1 mice exposed to formaldehyde, *Toxicol. Appl. Pharmacol.*, 72, 519, 1984.
52. Gregson, R. L. and Prentice, D. E., Aspects of immunotoxicity of chronic tobacco smoke exposure of the rat, *Toxicology*, 22, 23, 1981.

53. Thomas, W., Holt, P. G., and Keast, D., Cellular immunity in mice chronically exposed to fresh cigarette smoke, *Arch. Environ. Health,* 27, 372, 1973.
54. Thomas, W. R., Holt, P. G., and Keast, D., Effect of cigarette smoking on primary and secondary humoral responses of mice, *Nature (London),* 243, 240, 1973.
55. Holt, P. G., Finlay-Jones, L. M., Keast, D., and Papdimitrou, J. M., Immunological function in mice chronically exposed to nitrogen oxides (NO_x), *Environ. Res.,* 19, 1979.
56. Holt, P G. and Keast, D., Environmentally-induced changes in immunological function: acute and chronic effects of inhalation of tobacco smoke and other atmospheric contaminants in man and experimental animals, *Bacteriol. Rev.,* 41, 205, 1977.
57. Schwartzman, G., Enhancing effect of cortisone upon poliomyelitis infection (strain MEFI) in hamsters and mice, *Proc. Soc. Exp. Biol. Med.,* 75, 835, 1950.
58. Cummins, J. M. and Rosenquist, B. D., Leukocyte changes and interferon production in calves injected with hydrocortisone and infected with infectious bovine rhinotracheitis virus, *Am. J. Vet. Res.,* 40, 238, 1979.
59. Davies, D. H. and Duncan, J. R., The pathogenesis of recurrent infections with infectious bovine rhinotracheitis virus induced in calves by treatment with corticosteroids, *Cornell Vet.,* 64, 340, 1974.
60. Schneck, J. M., The psychological components in a case of herpes simplex, *Psychomat. Med.,* 9, 62, 1947.
61. Gross, W. B., Effect of social stress on occurrence of Marek's disease in chickens, *Am. J. Vet. Res.,* 33, 2275, 1972.
62. Rytel, M. W. and Kilbourne, E. F., The influence of cortisone on experimental viral infection, *J. Exp. Med.,* 123, 767, 1966.
63. Alqvist, J., Hormonal influences on immunologic and related phenomena, in *Psychoneuroimmunology,* Ader, R., Ed., Academic Press, New York, 1981, 355.
64. Hellman, A. and Weislow, O. S., Potential biohazards associated with depressed cellular and humoral immunity, in *Naturally Occurring Biological Immunosuppressive Factors and Their Relationship to Disease,* Neubauer, R. H., Ed., CRC Press, Boca Raton, Fla., 1979, 259.
65. Hubalek, Z. and Bardos, V., Effect of cyclophosphamide on the infection of mice with Tahyna virus, *Zbl. Bakt. Abt. Orig. A,* 251, 145, 1981.
66. Nahas, G., Leger, C., Desoize, B., and Banchereau, J., Inhibitory effects of psychotropic drugs on blastogenesis of cultured lymphocytes, in *Immunologic Considerations in Toxicology,* Vol. 2, Sharma, R. P., Ed., CRC Press, Boca Raton, Fla., 1981, 83.
67. Blanke, T. J., Little, J. R., Shirley, S. F., and Lynch, R. G., Augmentation of murine immune responses by amphotericin B, *Cell Immunol.,* 33, 180, 1977.
68. Ferrante, A. B., Rowan-Kelly, B., and Thong, Y. H., Suppression of immunological responses in mice by treatment with amphotericin B, *Clin. Exp. Immunol.,* 38, 70, 1979.
69. Voci, M. C., Pivetti-Pezzi, P., Quinti, I., and Aiuti, F., Levamisole therapy of herpetic keratitis; a clinical and immunological study, *Int. J. Immunopharmacol.,* 2, 165, 1980.
70. Masihi, K. N., Brehmer, W., Lange, W., and Ribi, E., Effects of mycobacterial fractions and muramyl dipeptide on the resistance of mice to aerogenic influenza virus infection, *Int. J. Immunopharmacol.,* 5, 403, 1983.
71. Ohnishi, H., Kosuzume, H., Inaba, H., Ohkura, M., Shimada, S., and Suzuki, Y., The immunomodulatory action of inosiplex in relation to its effects in experimental viral infections, *Int. J. Immunopharmacol.,* 5, 181, 1983.
72. Bunta, S., Report of clinical testing of isoprinosine tablets in herpes simplex infections, *Int. J. Immunopharmacol.,* 2, 208, 1980.

Chapter 5*

BOVINE HERPESVIRUS-1: INTERACTIONS BETWEEN ANIMAL AND VIRUS

Gary A. Splitter, Linda Eskra, Michele Miller-Edge, and Jackie L. Splitter

TABLE OF CONTENTS

* This work was supported by USDA-SEA Grants 82CRSR21051 and 901-15-145 and the College of Agricultural and Life Sciences.

I. INTRODUCTION

Bovine herpesvirus type 1 (BHV-1) or infectious bovine rhinotracheitis (IBR) virus is an important pathogen of cattle and can cause severe respiratory infections, vulvovaginitis, abortions, conjunctivitis, meningoencephalitis, and generalized systemic infections.[1] Although numerous reports have documented the pathology caused by BHV-1, few studies have elucidated the mechanisms of interplay between host and virus which result in infection, latency, or host protection. BHV-1 is capable of causing disease by infecting host cells with resulting cell cytolysis.

Cytolysis of infected cells occurs from irreversible damage to host cell metabolism by the virus. Alternatively, cell cytolysis can occur when viral antigens are expressed on the surface of infected cells. These antigens are recognized and infected cells destroyed by the animal's immune system. Many of the potential mechanisms for host recognition and response to BHV-1 are discussed in the accompanying sections of this article. A unique attribute of BHV-1 and herpesviruses in general is their ability to reside latently in cells which neither damage the host cell nor activate potential cellular or humoral immune defense mechanisms. Latency has been demonstrated in trigeminal ganglia and can be reactivated under appropriate conditions.[2-3]

The viral-host mechanisms involved in selecting pathways of host cell destruction vs. latency of virus in host cells is currently poorly understood. Multiple mechanisms have been proposed for viral pathogenesis as well as for host recovery from infection. This review will focus on the immune potential of the host to interact with BHV-1. Since other herpesviruses often infect their appropriate animal species in a manner analogous to BHV-1 in cattle, parallel or supplemental evidence from such viruses will be included. Examples of similar viruses are herpes simplex virus (HSV) types 1 or 2, and varicella-zoster virus which are associated with common skin diseases in their natural hosts and persist as latent infections of the central nervous system. From latent infections these viruses can cause recurrences of mild or severe disease involving the skin or central nervous system in natural hosts or closely related species.

II. VIRAL RECOGNITION BY LYMPHOCYTES

A. Lymphocyte Proliferation

In the efferent arm of antibody- and cell-mediated immunity, lymphocytes through their antigen-specific receptors are triggered to differentiate and secrete soluble mediators, such as interleukin-2 and T cell replacing factor(s) by T lymphocytes, or antibody to BHV-1 antigens by B lymphocytes. Other T lymphocytes may mature to cytotoxic lymphocytes with the capability of killing virally infected host cells. An in vitro correlate occurs when lymphocytes are cultured in the presence of an antigen to which they have been previously sensitized. DNA synthesis increases and the cell enlarges and "transforms" into a lymphoblast. This process can be quantitated by measuring the uptake of radiolabeled thymidine and is used extensively in the study of cell-mediated immunity.

As a first prerequisite in the study of immune cells and BHV-1, bovine lymphocytes need to recognize and proliferate in response to the virus. Bovine lymphocytes have been found by several workers to respond to BHV-1. Data in Figure 1 illustrate the in vitro recognition of BHV-1 by cows previously primed to BHV-1 in vivo. In each of seven animals tested, DNA synthesis as measured by tritiated thymidine uptake increased in lymphocytes responding to BHV-1.

Other investigators have found similar in vitro lymphocyte responses to BHV-1. Lymphocyte blastogenesis occurred in culture with peripheral blood lymphocytes from cattle infected or immunized with this virus, but failed to respond to other herpesvi-

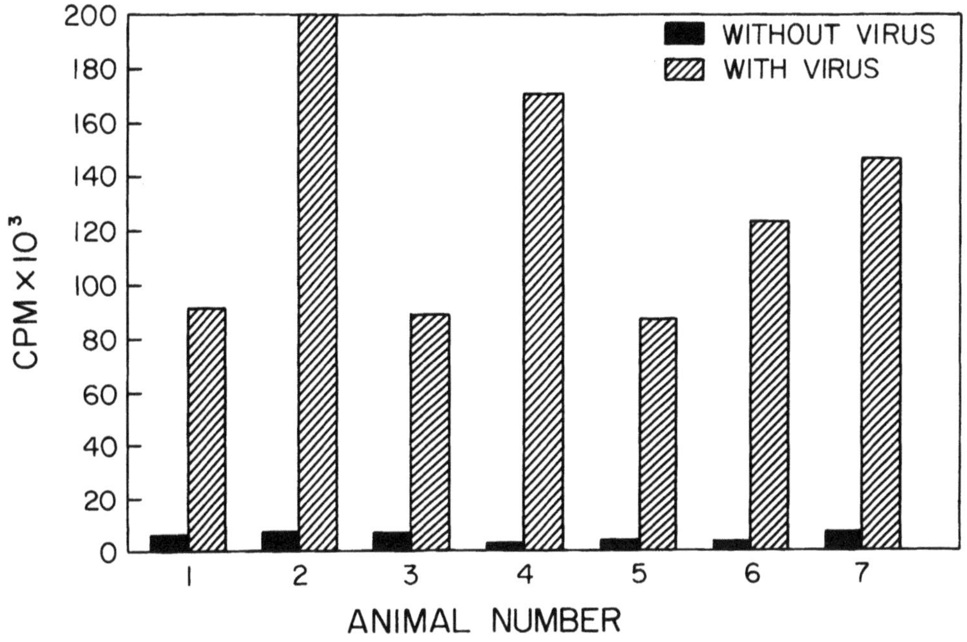

FIGURE 1. Response of Guernsey cattle to UV inactivated BHV-1 virus. Cattle were vaccinated with BHV-1 3 to 6 months earlier, peripheral blood lymphocytes were obtained and cultured with 1×10^6 PFU of UV inactivated BHV-1 virus/mℓ for 5 days. Six hr before harvest the cells were pulsed with 1 μCi ^{3}H-thymidine.

ruses, e.g., herpes simplex virus and equine rhinopneumonitis[4] indicating the antigen specificity of lymphocytes. In human studies, lymphocytes from donors with a history of previous HSV, varicella-zoster, and/or cytomegalovirus could respond to BHV-1 even though no donors had antibody to BHV-1.[5] This response is probably related to recognition of a common herpesvirus antigen. In summary, the above studies suggest that clones of bovine lymphocytes recognize specific viral antigens of BHV-1 and that common viral antigens may exist among herpesviruses as illustrated in human studies. Evidence for cattle recognizing common herpes viral antigens has not yet been demonstrated.

B. Macrophages and T Cells

A specific antigen receptor associated with self-major histocompatability (MHC) antigens is required for antigen recognition by T lymphocytes. Important MHC antigens in the mouse are Ia antigens coded for by genes in the H-2 complex. In the human Ia-like antigens are SB, DR, and DC coded for by genes in the HLA complex, and in cattle Ia-like antigens are tentatively coded for by genes in the BoLA-D complex. Blastogenic responses of lymphocytes in cattle to BHV-1 were prevented by reacting BHV-1 specific antigen with specific antibody before adding it to the cultures, but not by incorporating BHV-1 antibody in the culture medium after adding free antigen.[4] The requirement of viral antigen plus Ia-like (BoLA-D) antigen recognition by antigen specific T cell receptors probably accounts for the inability of antibody alone to block T cell proliferation. The necessity of antigen presenting cells to present the antigen in the context of self-major histocompatibility antigens, i.e., Ia is a central mechanism for triggering proliferation of naive specific lymphocytes. Recently it has been demonstrated that bovine macrophages are susceptible to BHV-1.[6] Macrophages serve as excellent antigen presenting cells in a variety of species. If macrophage functions were altered because of viral infections, then lymphocyte recognition of virus might be im-

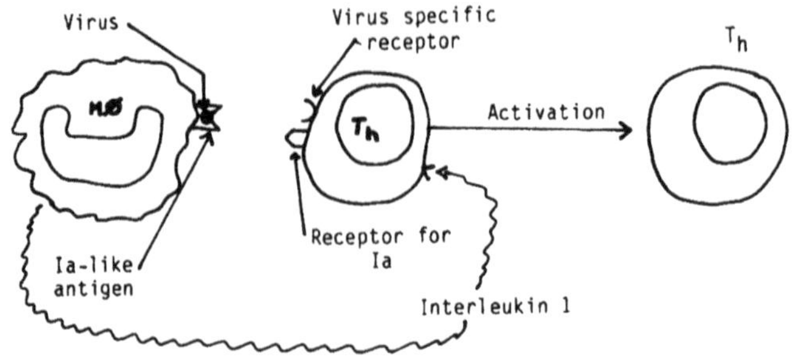

FIGURE 2. Interaction between an antigen presenting macrophage and antigen specific T cell. Antigen is presented by the macrophage in the context of an Ia-like antigen as a first signal to the T cell. A second soluble signal, interleukin 1, is important to maximally activate the T cell to proliferate.

paired. It has been reported that infection of macrophages with BHV-1 resulted in rapid reduction in Fc-mediated receptor activity, complement receptor (C3b) activity, and phagocytosis. These alterations suggest that antigen presentation, a process usually occurring after phagocytosis of antigen, might also be reduced.

In vivo documentation of macrophage importance has been difficult. Intranasal exposure of calves with BHV-1 indicated minimal evidence of viral antigen or virus in cells lavaged from lungs.[7] Also, macrophage Fc- and complement-receptor activities, phagocytic activity, and ability to mediate certain cell mediated protective mechanisms were unaltered.[7] In vivo, the importance of macrophages in HSV infection has been deomonstrated by pretreating mice with silica, trypan blue, or dextran sulfate to inhibit macrophage function. Mice treated in this manner were markedly susceptible to the lethality of systemic intravenous infection with HSV-2, but not to local virus growth after vaginal infection of mice with HSV-2.[8] Likewise, nude mice have increased resistance to HSV-2 hepatitis because of nonspecifically activated macrophages.[9]

To substantiate the role of macrophage involvement in lymphocyte proliferation to BHV-1, others[10] have shown that exogenous stimuli, such as levamisole, enhanced in vitro blastogenic response of bovine lymphocytes to nonspecific mitogens and BHV-1. These workers noted that levamisole-treated macrophages had increased numbers Fc receptors. Levamisole increased chemotaxis and phagocytosis by macrophages and has been claimed to produce clinical improvement in humans with severe herpes infections.[11] The immunopotentiator, levamisole, has been used in association with feedlot and laboratory immunization with BHV-1 and the drug appeared to have a beneficial effect on antibody responses of vaccinated cattle.[12] One mechanism of action of increased lymphocyte proliferation to BHV-1 by levamisole is probably increased production of interleukin-1 by macrophages. In other words, antigen sensitive T cells respond to BHV-1 presented by macrophages via specific receptor(s) which recognize viral antigen plus an Ia-like antigen on the macrophage. This would represent a first signal of activation to T cells. A second signal, interleukin-1, provided as a soluble factor to T cells maximally activates the cell, Figure 2. This maximal T cell activation could shift the balance of infectivity, accelerating the host's immune response as opposed to viral replication and cell destruction caused by BHV-1. The requirement for bovine macrophages to present virus to T lymphocytes has not been rigorously investigated.

Evidence for the importance of Ia-like antigen presenting cells to enhance T cell function has been shown in humans with herpes simplex virus.

In humans when HLA-DR positive Langerhans cells were present, HSV was presented in an immunogenic way to T cells resulting in their strong proliferation.[13] T cells from individuals with recurrent HSV-1 responded in vitro to HSV-1 antigen; interestingly, this occurred only in the presence of macrophages.[14] The intensity of response was closely related to the number of macrophages present. The optimal ratio of T cells to macrophages (adherent cells) was 10:1. In their later work, these workers found that cooperation between T cells and macrophages from a sensitized donor was restricted to self HLA-D/DR molecules.[15] In allogeneic mixtures of T cells and macrophages the lack of cooperation was not due to either suppression or generation of cytotoxic T cells.

Understanding the nature of soluble products produced by B and T lymphocytes may help elucidate cellular mechanisms involved in response and subsequent protection to BHV-1. Production of antibodies by B lymphocytes has been examined by a number of workers over the last 3 decades and will be addressed later in this chapter. The assessment of T lymphocyte involvement in BHV-1 infection has received less attention. This oversight has stemmed largely from the lack of markers to critically identify T cells in cattle and methodology to separate T cells from other cell populations. Further assays have not been adequately described to measure the products or functions from activated bovine T cells. Criteria for isolation and subsequent detection of T and B lymphocytes are important in assessing the roles of these cells.

Bovine T lymphocyte isolation has been accomplished by lysing all surface immunoglobulin bearing B cells in cultures using rabbit antibovine immunoglobulin, goat antirabbit immunoglobulin, and complement.[16] A panning technique has been used as an alternative method of T lymphocyte identification rather than E-rosettes because approximately 10% of rosetting T cells still exhibited fluorescence when treated with rhodamine-conjugated F(ab')2.[17] Peanut agglutinin (PNA) and rabbit anti-T cell antibody identified the same population of bovine T lymphocytes, and in dual fluorescence staining experiments, capping experiments, and differential inhibition of cellular binding of PNA by its carbohydrate ligand, unique and molecularly independent cell surface receptors were detected.

Using peanut agglutinin as a marker for bovine T cells, could a subpopulation of lymphocytes respond to BHV-1? Periheral blood mononuclear cells were treated with fluorescent labeled PNA, which binds D-galactose residues on bovine T lymphocytes. Labeled cells were separated into high fluorescent (positive) and low fluorescent (negative) populations by passage through a fluorescence activated cell sorter (FACS IV), Figure 3. Separated cell populations were then cultured with BHV-1 to assess their reactivity to the virus. Cells most responsive to BHV-1, were PNA^{+}T cells, Figure 4. Recent work in other species has indicated that antigen-sensitized T cells convert phenotypically from PNA^{-} to PNA^{+}[18] The soluble products of these activated cells and the mechanisms of how these cells or their products might provide protection against BHV-1 will require further study.

Several workers have found evidence that T cells respond to various strains of herpesvirus including BHV-1. Others[19] have shown that peripheral blood lymphocytes from cattle experimentally infected with BHV-1 undergo blast transformation after 3 to 5 days of in vitro contact with UV-inactivated virus. After 15 days of culture, 10% of the cells formed large rosettes with sheep red blood cells, which were thermostable. Responsive cells were detected in blood 7 days after intranasal infection but disappeared by 14 days. Similar transitory appearances of responsive lymphocytes were observed after each reinfection and virus specific lymphocytes were detected in the bronchial mucosa and spleen but not in regional lymph nodes or thymus.

Infectivity of bovine peripheral blood mononuclear cells vs. their responsiveness to BHV-1 is another matter. BHV-1 can cause abortion. It is believed that transmission of infection to the uterus is by way of monocytes since viral replication occurred in monocytes but not lymphocytes.[20] Although the virus was absorbed to lymphocytes,

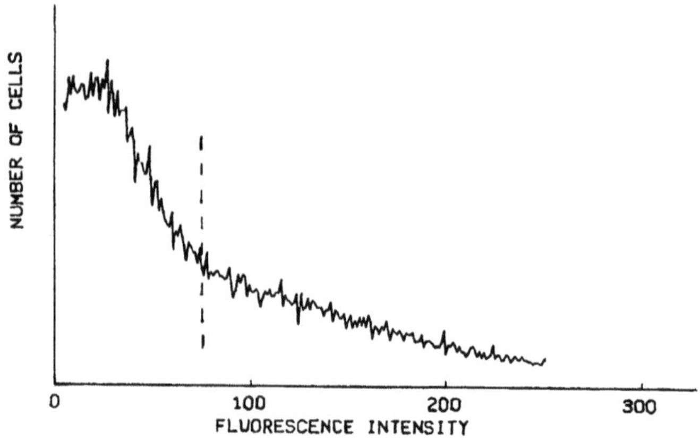

FIGURE 3. Peripheral blood mononuclear cells were labeled with FITC peanut agglutinin (25 μl). The brightest 60% of the PNA^+ cells were sorted into the PNA^+ population (---), the remainder into the PNA^- population.

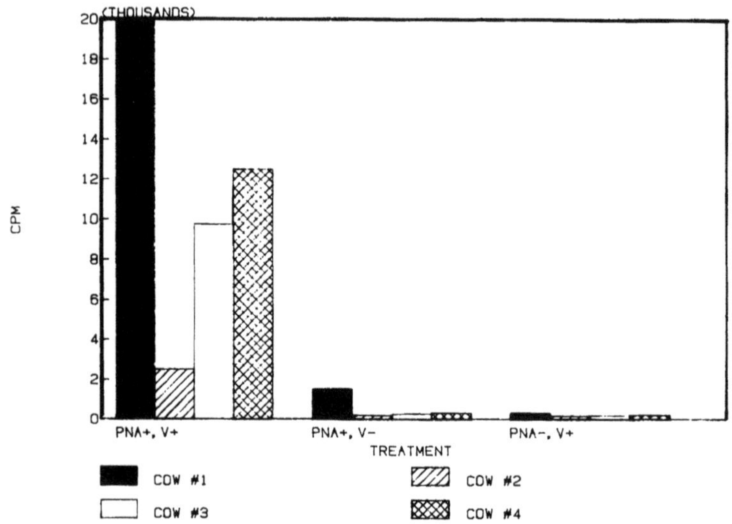

FIGURE 4. Response of lymphocytes separated by fluorescence activated cell sorting to UV inactivated BHV-1. Four cows were immunized to BHV-1 and cells separated as described in Figure 3. PNA^+ and PNA^- lymphocytes were then cultured with or without virus. Six hours before harvest the cells were pulsed with 1 μCi/well of ^{3}H-thymidine.

viral replication occurred only following phytohemagglutinin stimulation of lymphocytes. Similarly, others[21] observed that BHV-1 failed to replicate in normal or immune bovine peripheral blood mononuclear cells. Monocytes were present in the isolated cell preparations but there was no mention of infectivity.

Examining research on other herpesviruses, investigators have found that T cells from immunized rabbits were activated by HSV.[22] Others have shown that T cell competent mice could be immunized against HSV-1 or a HSV-1 envelope antigen and be successfully protected against HSV-1 and HSV-2.[23] T cell incompetent mice were not protected even though immunized. HSV grown on different lymphoid cell lines indicate T cells at all stages of growth were not permissive to HSV, but B cells and null cells supported the growth of HSV.[24] T cells blocked viral synthesis after viral adsorption. Also T cell lines appeared resistant to HSV infection in humans, but B cell and other cell lines expressed HSV cell surface antigens indicating a lack of resistance.[25] In contrast, others have been able to infect human T, B, and myeloid cell lines with HSV-1 for as long as 400 days.[26] The role of gamma interferon produced by the T lymphocytes may be to interfere with viral synthesis in these cells. None of these authors have explored this possibility.

Results with HSV indicated that sensitized rabbit spleen cells or peripheral blood lymphocytes respond and incorporate thymidine within 3 days although serum antibody was not detectable for another 4 days.[27,28] Response to the virus was found at the time of acute infection and declined during convalescence but remained at a detectable level. Lymphocyte transformation is, therefore, an indicator of sensitization to virus antigens, and lymphocyte multiplication will expand clones of cells which are sensitized to the stimulating antigen. This lymphocyte activation is a first step in immune-mediated antiviral response by the animal.

III. PHENOTYPIC ALTERATIONS AND SOLUBLE PRODUCTS OF BHV-1 ACTIVATED LYMPHOCYTES

Lymphocytes which proliferate in response to a virus may undergo certain functional changes. Examples of functional changes are the secretion of soluble products, such as IL-2 or T cell replacing factor(s), or killing of virally infected cells. As a first step in examing functional changes of responding cells, one might be able to detect phenotypic alterations on membranes of the cells which respond to virus. Since bovine T cells proliferate in response to BHV-1, perhaps these cells undergo such a phenotypic change. One such surface alteration might be an increase in Ia-like antigen. Ia antigens have been shown to play a role in foreign antigen presentation by macrophages. Ia antigens have been demonstrated on B cells as well. The expression of Ia antigens on T cells has been uncertain until recently; however, Ia is expressed on T cells which proliferate in response to mitogen or antigen.[29]

The expression of Ia-like antigens on bovine T cells responding to BHV-1 was examined in our own laboratory. T lymphocytes selected by panning on anti-Ig-coated plates were cultured for 5 days with or without BHV-1. Following culture, cells were then treated with monoclonal Ia-like antibody (provided by Dr. Paul Lalor), and complement. The remaining cells were evaluated for viability by dye exclusion and pulsed for 8 hr with ^{3}H-thymidine. As shown in Table 1, those cells cultured with BHV-1 proliferated to a greater extent. The addition of monoclonal antibody and complement killed the virally responding T lymphocytes. This indicates that such cells have Ia antigen on their surface and that a phenotypic change occurred in those cells responding to the virus. The possibility that macrophages may have been present in the cell population was not excluded. Even with this caveat, cells that proliferate to the virus expressed an Ia-like molecule.

The expression of Ia-like antigens in virally activated bovine T cells is similar to Ia-like antigen expression in other species. Ia-like antigens increase on human T cells when exposed to HSV-antigen pulsed macrophages, purified protein derivative from *Mycobacterium* spp., autologous B-lymphoid cell line, and alloantigen.[30] Although the exact function of Ia-like antigens on bovine T cells is unknown, others[31] have shown that an Ia-like positive T cell subset in humans has a pivotal role for suppressing B cell differentiation to immunoglobulin secreting plasma cells. Perhaps activated T helper cells which express Ia antigens function in the homeostatic balance of an immune response.

Phenotypic changes occur in T lymphocytes since 70% of T lymphocytes acquired Ia-like antigens after stimulation with PHA, compared to 1% before stimulation.[32]

Table 1
VIRALLY RESPONDING T CELLS EXPRESS AN IA-LIKE MOLECULE

Virus present	Complement	Antibody	Counts/min
+	−	−	72,386
+	−	+	73,400
+	+	−	33,294
+	+	+	7,559 (90)
−	−	−	30,518
−	−	+	28,434
−	+	−	15,437
−	+	+	6,463 (79)

Note: T lymphocytes were enriched prior to culture by Sephadex G-10 passage (macrophage depletion) followed by "panning" on antiimmunoglobulin plates (T cell enrichment). T lymphocytes were then cultured with or without UV treated IBR virus for (1×1^6 PFU/m*l*) 5 days, readjusted to 2×10^5 cells/well followed by treatment with monoclonal antibody for Ia antigen (provided by Dr. Paul Lalor) and complement. The remaining cells were evaluated for viability by dye exclusion and pulsed for 8 hr with ^{3}H-thymidine. Numbers in parentheses are the percent suppression compared to cells not treated with antibody or complement.

This occurs by induction or increased synthesis. These data suggest that T cells can have Ia-like antigens under centain conditions, just as B cells are positive for Ia-like antigens. These antigens may be important to T cell-activated functions.

Serum levels of Ia antigens can be altered by selected viral, bacterial, and protozoal infections of mice,[33] although these observations are controversial because of the unique carbohydrate nature of these Ia molecules. Serum levels of Ia antigens were enhanced or suppressed during the course of infection. Ia antigens in serum may represent a sensitive marker for certain types of immune responses to infectious agents. Suppression of Ia levels apparently correlates with macrophages dysfunction or T cell depletion, and an increase of Ia levels correlates with T cell proliferation.

An unusual example of a cell surface marker in herpesvirus infections is composed of virus plus immunoglobulin binding to the cell surface.[34] Workers have found that virions of HSV-1 have surface receptors which can bind to the Fc region of IgG.[34] Also, Fc-binding glycoprotein (gE) was shown to be present in extracts from purified virions. They demonstrated that Fc binding receptors from parent virions became incorporated into the surface membranes of infected cells. It has been suggested by others[35] that binding of aggregated IgC molecules bound to cell-surface Fc receptors protect HSV-infected cells against complement-dependent cell-mediated immune lysis.

As presented earlier, immune-mediated responses often result in the secretion of soluble factors by immunocompetent T cells. The production of several lymphokines in vitro has recently been described.[36-39] Interleukin 2 (IL-2), a product of antigen- or mitogen-activated lymphocytes, has been used in vitro for long-term growth of T lymphocytes, and this lymphokine has opened the door to examining the function of these cells[40-44] as well as assessing the degree of cell activation. The ability of IL-2 to amplify and maintain specific T cell responses in vitro suggests that it has an important role in vivo as well.[45-46]

The inherent capacity of an individual to produce and respond to IL-2 may provide a quantitative measure of specific immune responsiveness. Several investigators have

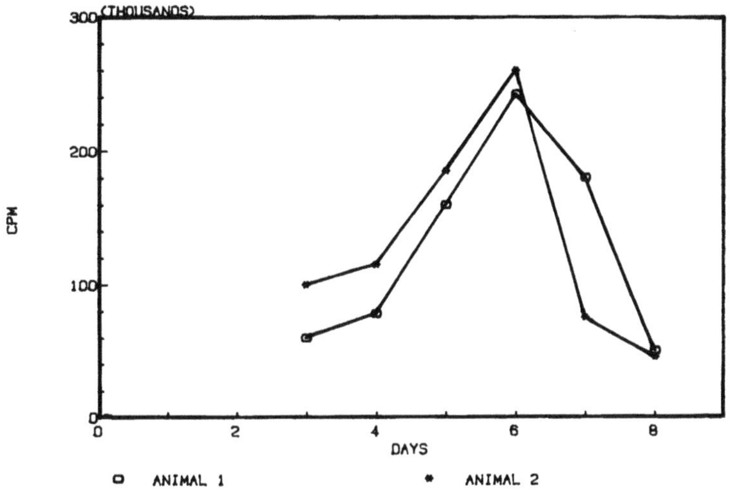

FIGURE 5. Kinetics of proliferation by virus-stimulated cells. Proliferation was measured in cultures of bovine mononuclear cells (1×10^6 cells/mℓ) from a number of animals with UV inactivated BHV-1 (1×10^6 PFU/mℓ). Cells were pulsed with ^{3}H-thymidine 6 hr prior to harvest on the indicated day. ^{3}H-thymidine incorporation is expressed as CPM. Similar responses were observed for seven different animals. Control values were less than 1,000 cpm.

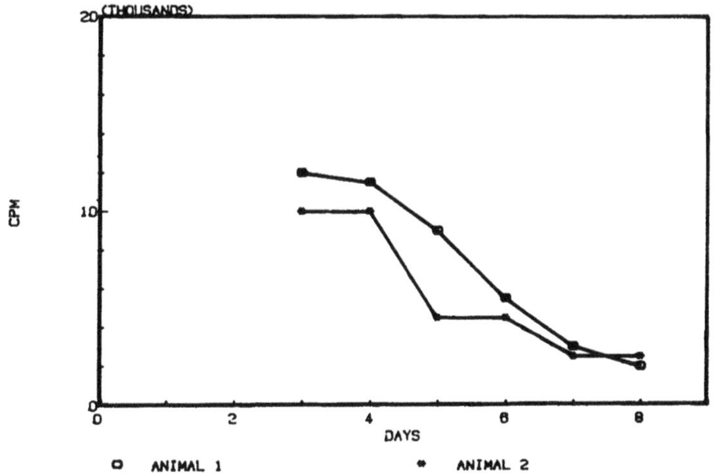

FIGURE 6. Kinetics of IL-2 production by virus stimulated bovine cells. Supernatants were harvested from the cultures of Figure 5. Samples were tested using bovine IL-2 dependent cells. ^{3}H-thymidine incorporation is expressed as CPM. Controls include medium and BHV-1 added to the IL-2 dependent cells and resulted in values less than 200 cpm. Similar responses were observed for seven different animals.

reported decreased proliferation and a related decrease in IL-2 synthesis in response to mitogen in aged or immunodeficient mice and humans.[47-52] Others have suggested that lymphokine production can be used to assess an individual's cellular immune capability.[53-56]

Bovine lymphocytes stimulated in vitro with BHV-1 were able to synthesize IL-2. Representative data in Figure 5 show that peak cell proliferation to BHV-1 occurred on day 5 and 6, and in Figure 6, IL-2 was maximal at 2 to 3 days of culture. If addi-

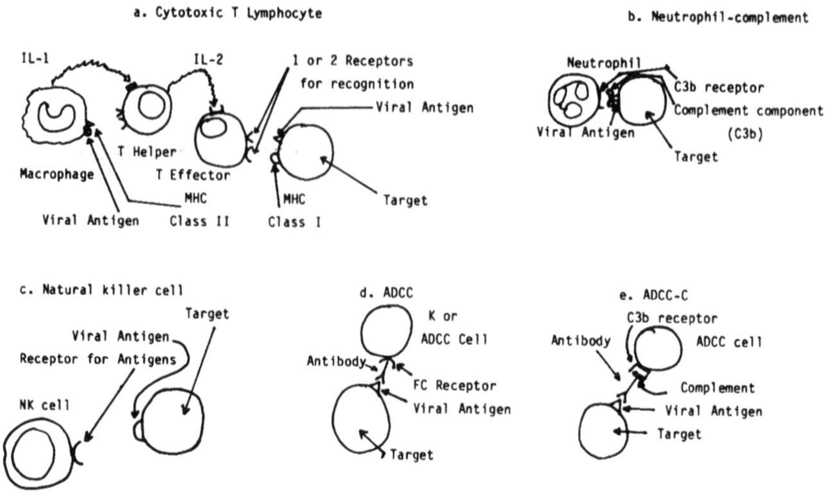

FIGURE 7. (a) Cooperative interaction of a T helper cell and T effector cell through interleukins (IL) results in lysis of virally infected cells. In this case the T effector cell recognizes the viral antigens in the context of an MHC class I-coded structure. T-cell recognition of target is depicted as resulting from an antigen receptor and an MHC receptor. (b) Lysis of a virally infected target by a neutrophil, in which the viral antigen has bound complement components and is recognized by the neutrophil through its C3b receptor. (c) NK cell destruction of a virally infected cell. A distinct surface antigen is recognized by the NK cell that is different from the viral antigen recognized by T cells. (d) Antibody-dependent cell-mediated.

tional exogenous IL-2 was added to responding lymphocytes on day 5, further proliferation was observed. Production of IL-2 in vitro indicates the level of activation of antigen-stimulated cells. The further expansion of previously sensitized lymphocytes by the addition of exogenous IL-2 suggest that the soluble product has a functional role in expanding antigen stimulated bovine clones. Additional work in our laboratory indicates that bovine IL-2 has a molecular weight between 18,000 and 30,000 on SDS-polyacrylamide gel electrophoresis. In viral infections IL-2 functions to trigger exponential proliferation of virus-specific T-cell clones and to induce cytotoxic T cells reactive to viral antigens. An interesting feature of IL-2 activity in the induction of cytotoxic T cells is the strict requirement for the presence of antigen.[41,57] This indicates that resting T cells do not express cellular receptors for IL-2. After interaction with antigen, T cells from each of the effector classes may respond by expressing a receptor for IL-2. Thereafter, the presence of sufficient concentrations of IL-2 will cause proliferation or clonal expansion of the activated T cells.

IV. EFFECTOR CELL MECHANISMS OF VIRAL DESTRUCTION

Virus presentation by macrophages to T helper lymphocytes could lead potentially to several immune-mediated responses. In fact, many cellular mechanisms have been reported in the destruction of BHV-1 and related herpesviruses. In this section, we will discuss briefly some of the seemingly ever expanding number of cellular mechanisms, discovered largely from in vitro models, which may function in virus infection. These mechanisms include antibody dependent cell-mediated cytotoxicity (ADCC), complement facilitated antibody dependent cell-mediated cytotoxicity (ADCC-C), natural killer (NK) cells, complement dependent neutrophil-mediated cytotoxicity (CDNC), and T cell-mediated cytotoxicity. The mechanisms of viral destruction by these cells are illustrated in Figure 7.

A. ADCC

Antibody-dependent cell-mediated cytotoxicity requires an effector cell acting in concert with antibody (IgG) to lyse the virally infected target cell. Effector cells in the ADCC system are usually of macrophage-monocyte or neutrophil lineages which possess receptor for the Fc portion of an immunoglobulin molecule. Specificity of killing by the effector cells is provided by the affinity of antibody for viral proteins present on target cells. ADCC is reportedly quite sensitive in vitro requiring low levels of antibody.[58] In fact, one or more orders of magnitude less antibody is required for ADCC than that required to mediate antibody complement lysis or viral neutralization. ADCC has been demonstrated in vitro against BHV-1 infected cells.[59]

Neutrophils have been found to be the most important agent in ADCC against BHV-1.[60] These cells required less antiserum to sensitize for cytotoxicity and destroyed BHV-1 infected target cells faster and more completely than macrophages. Although macrophages were somewhat effective, lymphocytes were ineffective in an ADCC-mediated response. This would suggest that the ADCC mechanism utilizing neutrophils may be one of the most important phenomenon in causing early recovery from BHV-1 infection.

In an earlier review on mechanisms of viral immunopathology it was noted that although ADCC has been shown in vitro it may not occur in vivo.[61] Many in vitro studies have been done, but to date few have been done in vivo. In vivo evidence for the role of human cells and antibody to protect against HSV via an ADCC mechanism has been presented in one intriguing study.[62] Adoptive transfer of specific antiviral IgG antibody and human leukocytes could protect infant mice from fatal HSV infection. Subneutralizing antibody concentrations as low as 10^{-8} combined with 5×10^6 leukocytes (or 1×10^6 monocyte-macrophages) were protective if given 1 day before viral challenge. Since newborn mice are exquisitely sensitive to HSV infections and lack the ability to mediate ADCC to HSV-infected cells, newborn mice were used to adoptively transfer human cells and to avoid MHC-dependent cellular cytotoxicity.[63] Protection afforded by antibody and monocyte-macrophage cells was striking because 100% of the treated mice survived 6 days vs. none of the control mice. Sixty percent of the mice survived 21 days, the length of the study. In vivo importance of ADCC in HSV infection has been implied by the optimal protection of cyclophosphamide-treated adult mice by nonimmune syngeneic spleen cells and antibody.[64] Protection offered by ADCC has not been demonstrated in vivo with BHV-1, but by analogy ADCC in cattle probably plays an important role in limiting the scope of infection.

Neutrophils were shown to be the effector cells in ADCC against erythrocytes and BHV-1 infected target cells in vitro.[65] Neutrophils also seemed to limit the spread of BHV-1 in tissue culture. In order for bovine neutrophils to be effective in ADCC close contact between the effector and target cells, presence of specific antiviral antibody, and expression of viral antigen on target cells membranes were required.[66] Antibody must recognize the viral antigen on cell membranes as a prerequisite for cytotoxicity. However, the mechanism of killing target cells which express viral antigens is not well described.

In earlier ADCC experiments done with bovine cells, peripheral blood leukocytes and mammary leukocytes killed BHV-1-infected bovine kidney target cells, with the mammary leukocytes having greater activity.[59] However, only mononuclear phagocytes were able to kill virus-infected target cells. Since BHV-1 affects the respiratory tract, it would have been interesting to evaluate the functional killing capacity of alveolar macrophages.

The time during infection when BHV-1 is susceptible to ADCC was examined,[67] and the results suggest that target cells were lysed by effector cells about the same time as intracellular transmission of virus occurred. Because the presence of effector cells and antibody in infected cultures markedly diminished viral transmission, another important role of ADCC may be to reduce infection.

Other workers have investigated ADCC with HSV.[68] Using three different target cell types (BHK-21, rabbit corneal cells, and stromal keratocytes) infected with HSV-1, it was found that all three cell types expressed surface antigen and were susceptible to lysis by antibody produced during the course of disease. These findings suggest a direct relationship between in vitro demonstration of complement dependent cytolysis in infected stromal cells and the disease process in experimental animals.

Antibodies directed to each of three viral glycoprotein antigens (Ag-11, Ag-8, and Ag-6) were able to mediate ADCC when tissue culture cells late in the infectious HSV cycle (18 hr post infection) were used.[69] ADCC provided a means to determine when individual viral antigens were expressed on the surface of infected cells. All three viral antigens first made their appearance from 3 to 4 hr post infection and quantities of these antigens increased with time. Complement was not required because heat-inactivated immune serum was found to augment damage of HSV-infected cells in vitro by human mononuclear cells.[70] This group of workers characterized the adhesion step. ADCC directed against HSV-infected or uninfected tissue culture cells was only modestly dependent on effector cell energy generation and that microfilaments but not microtubules were necessary. Minimal ambient temperature and energy requirements, independence of extracellular divalent cations, lack of sensitivity to colchicine, and relative resistance to supraphysiological temperature serve to distinguish the adhesion step from the lytic step in ADCC.[71]

B. ADCC-C

Complement-assisted antibody-dependent cell-mediated cytotoxicity (ADCC-C) utilizes complement components possibly C3b, which may bind to antibody (probably in the C_3H domain) and to the effector cell via a receptor for C3b on the effector cell. Killing of the virally infected cells by ADCC-C is more effective than ADCC alone[72] although there is some disagreement by workers in this area.[70] Interestingly, IgM antibody functions in ADCC-C but not in ADCC, thereby suggesting that this in vitro phenomenon may play a role early in in vivo defense.

When complement is added to in vitro ADCC assays (ADCC-C), there is an increase in the rate and extent of destruction of antibody sensitized BHV-1-infected cells.[73] Also, lymphocytes were shown to mediate ADCC as well, but only with the addition of complement. Perhaps this system more closely resembles in vivo circumstances because of the complexity of having complement, antibody, and various cell types present in the assay. Neutrophils which mediate ADCC can considerably enhance lysis upon the addition of low levels of complement that alone failed to lyse antibody-sensitized target cells.[72] This enhancement was not apparent under suboptimum conditions such as at low effector-to-target cell ratios, low levels of sensitizing antiserum, and short-duration assays. The action of complement appeared to form a more tenacious bond between effector and target cells. This mechanism may provide an early defense before the immune system is at optimal levels for protection, similar to the function of ADCC.

C. CDNC

Complement dependent neutrophil-mediated cytotoxicity includes neutrophils, complement, and virally infected target cells. The mechanism of complement dependent neutrophil-mediated cytotoxicity (CDNC) has been outlined in in vitro work.[74,75] In these studies, BHV-1 infected cells were destroyed by bovine neutrophils in the presence of complement. This mechanism required viable neutrophils and active complement. Bovine kidney cells infected with BHV-1 were destroyed more readily than vesicular stomatitis virus- or rotavirus-infected target cells. No explanation was offered by the authors regarding the variation in killing different virally infected targets, but the CDNC mechanism may be limited to certain viral infections. Alternatively, conditions for vesicular stomatitis virus and rotavirus infection of kidney cells may not have been optimal. In these experiments neutrophils were more effective than macrophages. Lymphocytes were ineffective as were neutrophils alone, but complement without neutrophils was slightly cytotoxic. If this mechanism occurs in vivo, the authors postulated that neutrophils may mediate recovery from infection before protective levels of antibody and immune cells are generated.

D. NK Cells

NK cell activity in BHV-1 has not been explored; however, NK function on other herpesviruses has received some investigation. NK cells have been identified in spleen, lymph nodes, bone marrow, and the blood of normal animals using a variety of cellular cytotoxicity assays. NK cells lack surface immunoglobulin and C3 receptors but appear to possess immunoglobulin Fc receptors in several animal species. NK cells fail to show phagocytic activity and lack surface phenotypes of T and B cells. NK cells are functionally distinct from K cells since the NK activity is not dependent on the presence of anitbody and is not inhibited by presence of antigen-antibody complexes. Levels of NK cytotoxic activity are different in different strains of mice suggesting that in outbred animals NK cell function may differ from individual to individual. NK activity can be augmented by exposure to bacterial adjuvants, tumor cells, virally infected cells, and interferon.[76-79] Further, activation does not induce immunologic memory.

Evidence that herpesviruses induce interferon in cattle as well as in other species suggests that NK activity may be augmented in BHV-1 infections; however, to our knowledge, no investigations regarding NK function have been reported. From studies with other herpesviruses there are only a few reports suggesting a relevant role for natural killer cells arising from viral infections.[78,80-83] Preferential lysis of virus-infected target cells by NK cells has been shown.[78,83] It has been proposed[84] that MHC nonrestricted NK cells arise early after viral infections and lyse these infected target cells and contribute to control of the virus infection. Hence, these findings tempt one to assign an important function to NK cells as a first-line defense system against virus infections. Interferon which has been documented to be necessary for induction of NK cells[85,89] apparently protects selectively healthy cells against attack by increasing the function of NK cells. Also, it has been suggested that interferon[90] or other mediators[91] might modulate permissiveness of host cells to virus replication in vivo.

In biological experiments of nature where the immune system is ontogenetically impaired, for example, including individuals with certain primary immunodeficiency disorders and newborn infants, patients often suffer herpesvirus infections of markedly increased severity. Since these patients usually have a depressed cell-mediated immune capacity, circumstantially it is thought that cell-mediated immunity plays a major role in host defense against herpesvirus infections. Humoral immune responses, on the other hand, are often normal or slightly exaggerated in these patients. However, specific mechanisms of cell-mediated immunity against herpesvirus infections are complex. For example, even though T cells play a central role in cell-mediated immunity, severe herpesvirus infections have only rarely been associated with specific T cell deficiences.[92] NK activity in patients with severe herpesvirus infections was below the normal mean by two standard deviations in one study.[93] Patients that recover from virus infections have higher natural killer activity. Low NK activity in severe HSV-1 infected individuals may reflect their susceptibility to virus infection.

E. Cytotoxic T Lymphocytes

T lymphocytes have been accepted as the important protective agent in direct cell cytotoxicity of virally infected cells. In vitro evidence of cytotoxic T lymphocytes

(CTL) has been found often in HSV, but little evidence of CTL killing has been found for BHV-1-infected cells. Perhaps this paucity of evidence in cattle is due to previous difficulty in T and B cell isolation and identification as well as the lack of understanding for the triggering mechanisms of this protection. Although it is an important mechanism in destruction of virus and disease control, it has been stated that in vivo many antigen-specific T cells do not tend to localize at the site of the antigen so the killing of all target cells in vivo is probably not likely.[61] This comment should be tempered with the knowledge that adoptive cell transfer experiments of mature cell types usually do not lead to optimal cell homing. In general, it has been shown that T cells are highly effective in lysing infected target cells.

Bovine lymphocytes were separated using a nylon wool column assay.[94] The effluent cells, thought to be mostly bovine T cells, were more reactive against BHV-1. The next extension of this work was to collect peripheral blood lymphocytes from BHV-1 immunized cattle and measure cytotoxicity against BHV-1 infected cells.[95] Cytotoxicity was expressed and found to be directly mediated by T cells and not attributable to ADCC. However, unlike other species, genetic restriction between CTL effectors and virally-infected target cells was not observed. Additional published work regarding CTL killing of BHV-1 infected target cells has not been reported.

Since herpesviruses spread from infected cells to susceptible cells by intercellular bridges, the virus need not enter the extracellular environment. Therefore, humoral antibody may have little part in the recovery process. Cellular defense mechanisms such as CTL activity may be the most important for defense against BHV-1 although bovine T cell participation in this process has not been thoroughly examined. Another lymphocyte-mediated process (in all probability T cell mediated) which has been demonstrated against BHV-1 is inhibition of viral plaque formation.[21] Immune lymphocytes could inhibit viral cytopathology of bovine kidney cell monolayers, whereas nonimmune lymphocytes were not protective. The inhibitory activity of immune lymphocytes declined 3 weeks after infection. Further, viral specificity was demonstrated by experiments showing that lymphocytes from animals immune to BHV-1 did not have inhibitory activity against herpes simplex. These workers hypothesized that the mechanism involved suppression of viral replication perhaps by soluble products including interferon rather than destruction. Although the role of interferon was not demonstrated, interferon is produced by activated T lymphocytes. It seems quite plausible that interferon was released by these lymphocytes, and with time following in vivo sensitization, there were less antigen sensitive lymphocytes capable of producing interferon.

In contrast to these findings, others[96] have shown suppression of HSV-1 plaque formation could be prevented by cocultivation of infected fibroblast monolayers with peritoneal macrophages, but not with splenic lymphocytes, from adult mice. Macrophages from weanling mice failed to surpress the development of plaques, suggesting that increased resistance to HSV-1 with age was a result of increased macrophage competence.

The importance of cell-mediated immunity in protection against herpes simplex virus has been reported.[97,98] Several groups of researchers have been able to demonstrate thymus-dependent T-killer responses (Lyt 2, 3-positive cells) against HSV-2-infected targets.[99-101] However, the assay systems reported require varying degrees of manipulation. For example, in one report, HSV-specific H-2-restricted T-killer cells were obtained only following in vitro culturing of in vivo sensitized lymph node cells.[99] Other workers reported virus-specific T cells after complex immunization protocol.[100] Different investigators had to inject high doses (10^6 to 10^8 PFU per mouse) of apathogenic HSV to gain only low levels of specific T killer cells, unless the animals were pretreated with cyclophosphamide.[101]

Later studies present a more convincing role for CTL involvement in herpes viral infections. Cytotoxic T lymphocytes have been recovered from lymph nodes of mice infected with HSV.[102] Also, spleen cells primed with HSV were able to differentiate into CTL after culture with infectious HSV.[103] In this report there was H-2 restriction between CTL and target cells. Cultures depleted of adherent cells did not respond unless peritoneal cells, L cell fibroblasts, or amplifying factor were added. Two signals were required for induction of the CTL response to HSV-1. One was provided by the virus and the other by stimulated helper T cells. These results suggest that H-2 restriction was required by helper cells and that a macrophage was needed for presentation of virus. T lymphocytes were shown to mediate the primary response of CTL in mice to HSV-1[101] and that cyclophosphamide enhanced the level of primary cytotoxicity, as well as augmenting secondary CTL responses. The mechanism of cyclophosphamide action has been shown by other workers to result from selective depletion of suppressor T cells.

That is, cytotoxic lymphocytes primed to HSV-1 have been found to kill both HSV-1 and HSV-2.[104] This phenomenon can be explained by two subpopulations of CTL which exist in anti-HSV immune spleen cells.[105] One subpopulation recognized type-specific determinants and the other recognized cross-reactive antigen determinants, present on the surface of HSV-infected cells, that are shared by HSV-1 and HSV-2.

Multiple clones are generated to the many viral proteins expressed by herpesviruses. Viral antigens need not be synthesized in host cells for CTL lysis of virally coated cells. The evidence suggests that early after HSV infection putative virus-specific antigens involved in CTL recognition are derived directly from viral antigens that attach to the cell surface. Later in infection, the relevant viral antigens are those which are newly synthesized.[106] If killing of virally infected cells by CTL occurs through recognition of viral antigen plus a self MHC antigen, then attachment of infectious virus occurs adjacent to MHC antigens or reorganization of cell surface proteins and virus occurs. Further, this same group has provided evidence that HLA-restricted and virus-specific T cell mediated cytotoxicity can be demonstrated in the human HSV system.[107]

Some workers have found T cell products to be important in host response to HSV. Recently, murine studies with HSV indicate that accessory cells are required for presentation of virus to specific helper cells.[108] Helper cells then respond by producing IL-2. One mechanism of virus evasion of the host's immune system postulated by these workers was that HSV-1 may trigger the response of CTL precursors but not of helper cells. However, there does not appear to be overwhelming evidence from other workers to support this concept which would imply that IL-2 is not generated from helper cells. Supporting evidence from others[109] has indicated that production of the T cell lymphokine, macrophage migration-inhibition factor (MIF), was completely suppressed during recrudescence, but during convalescence MIF returned to levels similar to seropositive controls. This would suggest that T cell help required to activate macrophages can be reduced or suppressed in active herpesvirus infections. In BHV-1 studies where bovine lymphokine-containing preparation enhanced ADCC,[110] not only were cells killed earlier in the presence of lymphokine, but the amount of infectious virus released was reduced. Lymphokine containing preparations were found to cause the activation of macrophages.

V. ANTIBODIES AND INTERFERON

A. Antibodies

Previously, we have discussed the importance of cell mediation in combating infectious agents. Let us now turn our attention to the role antibodies play in the immune response to BHV-1. Both cell-mediated and humoral immunity have been identified concurrently during recovery from herpesvirus infection.[111] Current immunological theory and evidence would support this finding.

Little work has been done to date on the importance of antibodies specific for BHV-1. Humoral immune mechanisms have been examined to determine if host cells could be destroyed by anti-BHV-1 antibody before or after transmission of virus to susceptible adjacent cells.[112] Viral antigens were detected on cell membranes 6 hr postinoculation and transmission of virus to adjacent cells began 10 hr post-inoculation. Extracellular virus was not detected until 12 to 13 hr post-inoculation. Continued addition of antibody and complement reduced virus dissemination 50-fold. In this study the humoral immune system played a significant role in reducing infection and limiting the spread of BHV-1.

Others have demonstrated in vivo that antibody to BHV-1 did not protect vaccinated cattle from respiratory and ocular disease thought to be caused by BHV-1.[113] Cattle from seven Holstein herds with a history of respiratory disease were studied. All animals were vaccinated and swabbed intranasally and bled monthly for approximately 1 year. BHV-1 was found in healthy as well as ill animals. In two herds BHV-1 was the cause of recurrent respiratory and ocular disease. BHV-1 was present, but not established as the cause of disease in the other five herds, and BHV-1 was persistent in six of seven herds. Vaccination did not raise antibody levels in animals with existing BHV-1 antibodies, but if no antibody was present vaccination produced a rise in antibody titers in adults, but not in calves. Lastly, it was observed that BHV-1 antibody titers were higher in vaccinated herds with recurrent respiratory disease. It appears that even though antibodies can be demonstrated to reduce infectious virus in vitro, and antibodies are evident in vivo, protection from infection does not occur and BHV-1 persists in animals.

The extent of cattle exposure to BHV-1[113] appears to be as widespread as human exposure to HSV.[114] In 84% of persons tested, antibodies to HSV were found. IgG and IgA antibodies were present in 52.4% and IgG antibody only in 47.7% of individuals examined. Neutralization of HSV infection in Vero cells by gingival fluid was demonstrated as well. No correlation was found between the recurrences of HSV infection and the type of antibody produced.

Use of antigenic viral components as immunogens for host protection continues to be an area of interest with herpesviruses. The ability of certain antigens (envelope antigen, EV, and chloroform-methanol antigen, CM) from BHV-1 to produce serum-virus neutralizing antibodies and function in protection was investigated.[115] The two antigens, EV and CM, when inoculated with adjuvant subsequently induced significant increases in serum-virus neutralizing antibody titers. Nonvaccinated controls failed to produce antibodies. Vaccinated and nonvaccinated animals were bred and challenged with a virulent strain of BHV-1. Although there was less abortigenic activity caused by BHV-1 in cattle immunized with EV and CM, the difference was not significant compared with nonvaccinated controls.

An interesting study illustrating the relationship of common viral antigens of BHV-1 and HSV has been done in humans.[5] Peripheral blood mononuclear (PBM) cells from humans with a history of HSV, varicella-zoster virus, and/or cytomegalovirus were cultured with antigens from these viruses and BHV-1. In all cases there was a response by 6 days. Sero-negative donors had a different cellular immune response than did sero-positive donors. Previous infection with herpesvirus did not influence the response of PBM cells cultured with human HSV antigens, but some donors' (none with antibodies to BHV-1) PBM cells had a proliferative response to BHV-1. This is a host cell-mediated function, and the authors speculated that there may be a common HSV-BHV-1 antigen.

Studies have examined antibodies directed against herpes virus type-common and type-specific antigens. Using crossed immunoelectrophoresis, two HSV type-common, membrane-bound glycoproteins, Ag-11 (gA + gB) and Ag-8 (gD) were identified which are localized on the surface of the viral envelope and in the membranes of infected cells.[116] These two type-common antigens could serve as the principal immunogens in infected humans. Monospecific antibodies against these antigens neutralized the virus and caused cytolysis of infected cells in vitro. Other workers prepared antibodies against VP 7/8, a major HSV-1 type specific glycoprotein.[117] Specificity was shown by neutralization of HSV-1 infectivity and by immunoprecipitation in vitro.

Hybridomas have been produced to viral antigens by fusing mouse myeloma cells with spleen cells from mice primed with HSV-1.[118] This procedure yielded five clones producing neutralizing antibody against homologous virus. There were two clones to glycoprotein C and two clones to type-common glycoprotein D. The four clones were independently derived and were not clonally related since antibody in each pair belonged to a different subclass of immunoglobulin. Work of this nature supports the diversity of the humoral responses to HSV. In vivo protection against herpesvirus has been reported using monoclonal antibodies.[119] It was found that monoclonal antibodies directed against HSV-1 type-common glycoproteins, gC and gD, protected mice from fatal HSV-1 infection if given 2 hr before virus challenge or 24 hr after virus inoculation. If antibodies were not given, mice died in 7 to 10 days. This provides evidence that monoclonal antibodies to type-common antigens can protect against HSV-1 infection.

Protection by immunization to HSV has been done in hairless mice immunized with infectious HSV, inactivated virus, or material enriched for viral glycoproteins.[120] Immunized mice were protected against primary facial and ganglionic infections. Live and inactivated virus induced neutralizing antibodies, but glycoprotein material did not. Glycoprotein material induced antibody largely directed against two glycopolypeptides with molecular weights of 120,000 to 130,000. Hairless mice immunized with glycoprotein responded faster than control mice in synthesizing neutralizing antibodies after challenge with infectious virus. Congenital athymic BALB/c (nu/nu) mice were protected against primary facial infections after immunization with glycoprotein materials, but glycoprotein-specific antibodies were not induced. It, therefore, appears that subunit components of virus such as glycoproteins can serve as immunogens to protect animals from virulent infections. Although antibodies are not synthesized as indicators of host immunization when glycoproteins are administered, nonimmune-mediated cellular protection may be a more important first line of defense. Perhaps immunization with glycoprotein material stimulates macrophages or NK cells directly.

A number of investigators have examined kinetics of antibody appearance correlated with host cell protection. The relationship of complement fixing antibody to an early HSV antigen (Ag-4) was investigated.[121] Sera from HSV-1 patients was devoid of complement-fixing antibody to Ag-4 antigen at the time of herpes lesion outbreak in 10 of 13 cases. One to four weeks after appearance of the lesion 28 of 30 patients acquired this complement-fixing antibody. This sera contained high levels of IgM antibody and the ability to precipitate Ag-4 antigen.

A second study which correlates antibody levels with protection to HSV indicates that higher-efficiency antibodies to HSV-1 have been found in subjects with oral herpes lesions than in subjects with no lesions at the time of sampling.[122] Those with lesions had greater antibody titers to structural components of HSV, and higher efficiency sera had a higher proportion of antibody to virus particles.

Additional studies have substantiated that antibody can inhibit the spread of HSV.[123,124] In one study young adult mice were infected with 3×10^6 PFU of HSV-2.[123] In 9 to 12 days virus had spread to the spinal cord and brain. Administration of HSV-2 neutralizing antisera from syngeneic mice or rabbits inhibited the spread of HSV-2 and prevented death if given 8 to 48 hr after infection, but not if given at 72 hr. HSV-2 antibody appeared responsible for protection. However, the question of latent infection was not addressed. If mice were irradiated 24 hr before antibody transfer, antibody resulted in no protection. Another component, probably cellular, was thought to be operating, but it was not identified. Speculatively, the mechanism could have been ADCC involving neutrophils in concert with the administered antibody. In another study BALB/c mice passively immunized with antibody to HSV-1 were protected when virally challenged, but control mice were not.[124] All control mice had HSV in dorsal root ganglia by 48 hr. Administration of antibody prevented acute neurological disease if it was given no later than 48 hr postinfection. In addition antibody restricted the extent of latent infection in the lumbrosacral ganglia as well. These data provide evidence that antibody is effective in preventing viral spread in peripheral and central nervous systems.

In order to determine antibody specificity and what types of antibodies are effective against HSV infections some workers have investigated antibodies in detail. Specificity of antibody was studied by administering three doses of one type of HSV to rabbits at 2-week intervals followed by the administration of another type of virus.[125] The rabbits had a homotypic antibody response to the primary virus which persisted for long intervals at a constant level. Heterotypic antibody developed after one or two injections of heterotypic virus. The titer tended to decline and was undetectable at 9 months. At 9 months a third injection of heterotypic virus was given, and antibody was rapidly produced. Titers exceeded or were near those of the homotypic antibody. The IgG fraction was found to contain mainly homotypic antibody and the IgM fraction heterotypic antibody. This suggests that the first virus (homotypic virus) latently infected animals and the second virus (heterotypic virus) was unable to infect animals for any length of time.

Antiviral antibodies were shown to affect the cellular immune response in herpesvirus infections by modulating the lymphoproliferative response to herpesvirus antigen.[126] Lymphocyte proliferation responses varied with the organ examined in rabbits with HSV infections of the cornea. Proliferation occured at day 5 in lymphoid cells from local lymph nodes, at day 11 in peripheral blood, and at day 14 in the spleen. Presence of autologous serum antibody in cultures suppressed lymphoproliferative responses of lymph node lymphocytes to HSV antigens at day 5; the time when antibody production was first noted. Peripheral blood and splenic lymphocytes were not affected early in infection, but at 7 months after infection the presence of autologous serum antibodies stimulated splenic lymphocytes from animals with recurrent disease. Early suppressive and late enhancing effects of specific antibodies suggested a regulating effect. Perhaps early in infection with excess antigen, antiviral antibody down regulates cells which have antigen bound to specific receptors. Late in infection with little antigen available antiviral antibody serves to stimulate antigen specific cells to proliferate.

Lastly, components of the immunoglobulin molecule necessary for protection against herpesvirus infections have been examined recently.[127] It was concluded that the Fc component of antibody is needed to resolve intracellular infection, and that the mechanism by which antibody-mediated recovery functions remains undefined. This mechanism does not appear to involve virus neutralization or complement activation. Rabbit anti-HSV-specific $F(ab')_2$ fragments were inactive in complement-mediated cytolysis while retaining the capacity to neutralize virus infectivity in mice in vivo. If $F(ab')_2$ fragments were passively transferred before or with inoculation, they were as efficient as intact IgG in protection, but if they were given 8 hr later, only intact IgG was protective. This did not seem to be due to the loss of $F(ab')_2$ fragments or the ability of $F(ab')_2$ to activate complement.

In summary, antibodies contribute to limiting the dissemination of BHV-1 in tissues. However, evidence from numerous workers suggests antibodies in natural infections do not protect animals from disease. Antibodies probably work in concert with cellular immunity to provide defensive protection.

B. Interferon

Several studies have shown that BHV-1 infection in vivo produces high levels of interferon in blood[128] and nasal secretions.[129,130] Large quantities of interferon were produced within 24 hr of in vitro stimulation by BHV-1 and purified protein derivative antigen from *Mycobacterium* sp.[131] T lymphocytes provided the antigen specific step for immune interferon production, and there was a two- to ten-fold increase when lymphocytes were combined with macrophages. These studies indicated that direct physical contact between lymphocytes and macrophages was required. Antibody-antigen complexes composed of irradiated virus-infected cells in the presence of antibody were as efficient or better at stimulating interferon than was free antigen. Because BHV-1 was inhibited by interferon levels stimulated in cultures by BHV-1 antigen, local production of interferon may play a role in control of virus dissemination in vivo. More recently supernatants of virus-infected bovine macrophages possessed interfering activity, probably due to interferon.[132] Others have also found bovine interferon to be inhibitory to BHV-1.[133,134]

In contrast other workers have found a negative correlation between the number of leukocytes and the amount of interferon available.[135] Leukocyte counts increased in all calves given hydrocortisone before virus inoculation. After BHV-1 inoculation the leukocyte count decreased more than 50% on the average. Interferon response of calves undergoing lymphocyte suppression due to hydrocortisone suppression was enhanced. Levels and types of interferon produced by different cell types were not assessed. It would have been interesting to determine the kinds of leukocytes remaining and their capability for interferon synthesis.

Two types of interferon have been produced by human peripheral blood mononuclear leukocytes exposed to HSV.[136] Alpha interferon was found in 2-day cultures in sero-positive and sero-negative donors and was not immunologically specific. Gamma interferon present in 5-day cultures was produced in significantly higher titers in sero-positive donors. This gamma interferon was immunospecific by virtue of its presence in cultures from only sero-positive individuals.

Human blood leukocytes exposed to HSV-1 produced high levels of antiviral activity identified as type-1 (alpha and beta) interferon.[137] Evidence that type-1 interferon was produced by Fc receptor-negative null cells was demonstrated. Since NK activity is known to require Fc receptor positive cells, this suggests that production of type-1 interferon represents one of the earliest functions in protection to HSV infection and is likely to develop before the appearance of Fc receptor bearing NK cell function.

An interactive effect of interferon, monocytes, and antibody in protecting against HSV infection has been reported.[138] Peritoneal macrophages elicited in mice by *Corynebacterium parvum* vaccine suppressed growth of HSV in infected cells by an interferon-dependent mechanism. The activation of cells capable of interferon synthesis during recurrent herpesvirus infections may function to limit the extent of infection. If specific or nonspecific immune modulators could be administered at critical points of infection perhaps the host could eliminate the recrudescence quickly.

In a recent study which examined the in vivo role of interferon and leukocytes, human interferon and cells were transferred for short intervals to newborn mice. Human interferon and mononuclear cells protected newborn mice given a lethal dose of HSV.[139] Thirty-five percent of the mice survived, and the active effector cell was identified as a lymphocyte or monocyte-macrophage. Neutrophils were incapable of serving as effector cells. Mice given nonleukocyte cells and interferon had a 55% survival rate. If given human interferon, mononuclear cells and subneutralizing doses of anti-HSV antibody before infecting with HSV, 92% of the mice survived, and did significantly better than those receiving mononuclear cells and interferon or mononuclear cells and antibody. Mononuclear cells from neonatal humans provided no protection.

Although interferon is present in animals challenged with BHV-1, interferon types are undefined and their roles not assessed in cattle. Of what hierarchical importance do different types of interferon have in protection of cattle from viral infection? In cattle do interferon types impact on other immune defense mechanisms or vice versa? Little information exists regarding these early soluble mediators in the bovine species.

VI. LATENCY

A. Proposed Mechanisms for BHV-1

Latency and recrudescence of BHV-1 and other herpesviruses is an area that is undergoing intensive research at present in order to better understand the disease chronicity and to find possible treatments or vaccines against the virus. Although latency has been observed in cattle with BHV-1, the mechanisms of this process are still uncertain. Presently, there are two models which describe herpesvirus latency. The models were developed by studies with HSV but probably apply to BHV-1 as well.

One of the most well accepted and documented theories of latency for HSV is that the virus travels up the trigeminal nerve and lies dormant after initial infection until a stimulus to the ganglion causes virus release. Virus then travels via the nerve to skin and causes clinical disease. Another theory which is compatible with the first was introduced in 1976.[140] The second theory proposes that if there is a stimulus on the skin which produces conditions favorable for virus growth the result is a recurrence of clinical disease. There may be a parallel condition in cattle involving BHV-1.

In support of this later theory, workers have shown that HSV could be isolated in vitro from skin epidermis of clinically normal but latently infected mice by cutting the skin.[141] This indicates that virus can be present in skin without associated disease.

Several strains of BHV-1 have been found capable of establishing latent infections in cattle.[142] Isolates were obtained from trigeminal ganglia following cocultivation with bovine cell lines and by superinfecting with a temperature-sensitive helper strain of BHV-1. Four isolates of latent BHV-1 were obtained from 44 pairs of trigeminal ganglia.

In mice, long-term persistent infections of BHV-1 were induced by intracerebral inoculation of athymic nude mice but not by intraperitoneal injection.[143] However, in vitro inoculation of outbred mouse spleen fragments with BHV-1 resulted in BHV-1 incorporation into host DNA and transformation of spleen macrophages. Presence of BHV-1 genetic information was confirmed by in situ hybridization. These BHV-1 transformed macrophages induced fibrosarcomas and cystic tumors in athymic nude mice. However, infectious virus could not be rescued from transformed cells by cocultivation with rabbit kidney cells, treatment with iododeoxyuridine, or UV irradiation.

These same workers have demonstrated persistent infection and transformation of BHV-1 infected mouse embryo fibroblasts.[144] Early detection of BHV-1 specific membrane and intracellular antigens was found in fibroblastoid cells. Transformed cells induced fibrosarcomas when implanted subcutaneously in athymic nude mice, but infectious virus was not rescued from transformed or tumor cells. Nontransformed control cells survived no more than ten in vitro passages and did not induce tumors in nude mice. These data point to the potential of BHV-1 in transforming mouse cells, and indicate the presence of viral genome or transforming portion of the genome residing in mouse cells.

B. Vaccination

Attempts to protect cattle by vaccination to BHV-1 have been performed by various workers. In one example, nonimmune calves exposed to BHV-1 rapidly cleared the virus such that 10^9 plaque-forming units (PFUs) of virus are removed from the nasal mucosa in less than 4 hr. An eclipse phase follows the clearance of viral inoculum, and replicating virus is first detected at 9 hr with maximal titers at 4 days. When these same animals were rechallenged 30 days after initial exposure 10^9 PFUs of virus are cleared in 1 hr and only an abortive reinfection occurs.[145] Although this study did not examine the mechanisms for protection upon viral rechallenge, evidence suggests that protection can be conferred by vaccination.

Attempts to develop a vaccine using temperature sensitive BHV-1 virus have been tested by several workers. A temperature sensitive virus vaccine produced latent infections as frequently as a nontemperature sensitive virus vaccine.[3] Treatment with dexamethasone resulted in reactivation of virus in cattle receiving either temperature sensitive BHV-1 vaccine or nontemperature sensitive BHV-1 vaccine. Reactivated virus was shown to be the same as the original virus used for vaccination. In contrast, others have found that vaccination with temperature sensitive intranasal vaccine or formalized BHV-1 in Freunds complete adjuvant protected calves from latent infection.[146]

Speculating that trigeminal ganglia may be one latent site for BHV-1, workers have studied two infections induced in 14 calves by dexamethasone at 2, 3, 5, 15, or 30 months after BHV-1 infection.[147] They found trigeminal ganglionitis in all dexamethasone-treated calves. Degeneration of ganglion cells and neuronophagia were prominent after the second infection. This study would support the model of BHV-1 latency as mentioned earlier. This virus, in infected cattle, spreads centripetally to sensory ganglia. Following reactivation by dexamethasone, the reactivated virus travels centrifugally through nerve fibers to the mucous membrane. This movement of virus between geographic locations indicates that after virus invades nerve fibers or ganglia, it is no longer accessible to circulating antibodies. Evidence for the inability of antibodies to protect was shown by the same workers when secondary recrudescence occured in infected calves following dexamethasone treatment.

Other workers have administered dexamethasone to initiate recrudescence of BHV-1.[148] In contrast, virus was not isolated from neuronal ganglia in the two animals examined in this experiment. Virus was isolated at the site of inoculation which was the muscosal surface of the nasal or vaginal regions in dexamethasone treated animals.

In other work all dexamethasone-treated calves (16 total) that were inoculated and challenged with virulent virus 49 days after initial vaccination developed nonsuppurative trigeminal ganglionitis and encephalitis.[149] Those treated with dexamethasone but not challenged with virus after initial vaccination did not have lesions in the peripheral nervous or central nervous system. BHV-1 virus antigens were not observed in tissues of any of the calves by fluorescent antibody techniques. Modified live BHV-1 virus neither produced lesions nor latent infection, and vaccination with modified live virus did not protect calves against establishment of latent infection after exposure to large doses ($10^{6.5}$ $TCID_{50}$) of virulent BHV-1. This would suggest that modified live BHV-1 vaccine may not be protective to cattle.

Calves infected with BHV-2, a related virus, and treated with dexamethasone 69 days postinfection were negative for BHV-2, but latent BHV-1 infections were induced in all calves.[150] There was no clinical evidence of disease, but BHV-1 was found in nasal and pharyngeal swabbings, and ganglionitis of trigeminal ganglia was observed. No reactivation of clinical conditions was associated with BHV-2 or shedding of BHV-2 virus was observed.

Similar findings of unexpected BHV-1 re-expression occured in work by others.[151] Cattle having antibodies to BHV-2 were examined for latent BHV-2. Virus was not isolated from sensory ganglia, epithelium, or lymph nodes. However, BHV-1 was recovered from one animal treated with corticosteroids, but BHV-2 was not detected.

C. HSV

Latency with HSV, like BHV-1, is one of the most problematic aspects of the disease. Recurrence and contagion potential are cause for concern in animal and human medicine. Many workers have investigated the mechanisms of latency in an effort to devise better methods of control and prevention.

1. Location of Latency

The manner in which herpesvirus spreads and where the virus latently resides has been studied by numerous workers. It has been shown that HSV is found in the peripheral nervous system (PNS) during latency as well as the central nervous system (CNS).[152] In the first week after infection, HSV was found in the CNS of mice inoculated in the cornea. Virus was found to progress from the PNS to the CNS and was capable of establishing a latent infection in the CNS. However, HSV could not be reactivated by explant techniques from the CNS.

HSV injected into nostrils of mice resulted in appearance of HSV antigens in neurons of trigeminal ganglia, main sensory and spinal tract trigeminal nuclei, and bilaterally in the locus ceruleus.[153] These results indicate that HSV spreads via axons, passes through a series of neurons, and then reaches the vital nuclei in the brainstem, including the monoaminergic neurons, from the primary replication area in the lip. In guinea pigs, it was shown that HSV-2 resides in footpads and lumbrosacral ganglia during latent infection.[154] Rat splenic lymphocytes and peritoneal exudate cells were found to be incapable of replicating HSV even if cells from sensitized animals are stimulated in vitro.[155] Both cell preparations offered some protection to rat dorsal root ganglia exposed to HSV. This system can be used to study latency since neurons of sensory ganglia are a natural site of latent herpesvirus.

Latent infections of HSV were established in cervical dorsal root ganglia of BALB/c mice following peripheral inoculation of mutants of HSV-1.[156] The neuron was identified unambiguously as the site of latent infection. The authors presented evidence showing that activation of virus infections in latently infected ganglia by neurosurgery results in a reduction of the number of latent foci in that ganglion. Their interpretation was that productive infection which follows activation of latent virus in the neuron leads to destruction of that cell.

Thirteen temperature sensitive mutants of HSV-1 were tested for their ability to establish latency in the brains of mice.[157] Eleven mutants established latency, and it was shown that the lesion caused by these temperature sensitive mutants was involved in establishing latency. No correlation of latency was found with the synthesis of various morphologically identifiable virus products in the brain. By comparing latency characteristics with previously established polypeptide phenotypes of mutants, the authors concluded that one immediate and one or more later virus functions are necessary for establishing and/or maintaining the latent state.

Persistent dynamic-state infections with HSV-1 have been maintained in human T lymphoblastoid (CEM9) cells for many months after initial infection with HSV wild type virus.[158] Persistently infected cells grew as well as uninfected cells, except during periods of crisis where increased viral replication and cytopathic effect were observed. When cells were maintained twice the usual time interval before subculture, they could survive the crisis and no interferon was detected in the cultures. Persistently infected cultures were "cured" with HSV antiserum or incubation at 39°C. These were resistant to reinfection with HSV but permissive for vesicular stomatitis virus replication. These results suggest that these treatments modulated a shift from the dynamic-state to the static-state, latent infection. Findings of work done by others are consistent with the hypothesis of static latency, provided that spontaneous activation of the silent genome occasionally occurs in a few ganglion cells.[159] A serum reacting with immediate early polypeptides, three early (beta) polypeptides, and the capsid polypeptide of HSV was used to identify sections from trigeminal ganglia of rabbits with established latent HSV infections. Nonstructured polypeptides were detected in ganglion cells before culturing in only 1 of 12 rabbits. Good correlation was found between occurrence of nonstructural and structural antigens in explanted fragments starting from 72 hr in culture.

After examining human paravertebral ganglia from 40 autopsies for the presence of specific sequences of viral RNA by *in situ* cytological hybridization, it has been convincingly argued that specific transcription of the HSV genome occurs in latently infected human ganglion cells.[160] A hypothesis has been proposed that a block in virus replication can be promoted by an inhibitor of an HSV specific regulatory protein, and can be overcome by the addition of HSV DNA copies in the infected cell.[161] In this work, it was found that the difference of HSV concentrations did not account for the variation in virus yields and number of infectious centers of HSV infected mouse neuroblastoma cells (c1300) infected at different multiplicities of infections (MOI). The authors suggest that a c1300 cell had to be infected with more than one HSV particle in order to produce progeny virus-multiplicity activation. This greater than expected enhancement of virus production of c1300 cell cultures receiving MOI of HSV was probably not due to improved virus adsorption, or influenced by nonvirus factors in the virus inoculum stimulatory of HSV replication.

Continuous production of infectious extracellular virus in a human lymphoblastoid cell line infected with HSV was observed for 15 months.[162] By the 5th day postinfection 75% of the cells produced HSV antigen and 90% did so by the 10th day as detected by fluorescent antibody. Only 11% of the isolated cells produced detectable infectious virus.

2. Role of Latent Infections vs. Virulent Infections

One avenue of investigation in finding ways to prevent serious HSV disease has been to test the results of inoculation with a milder strain of HSV. Ganglia of rabbits infected with a relatively benign strain of HSV and challenged with one or two more virulent strains, resulted in milder disease.[163] Ganglia from these rabbits were colonized only by the initial virus strain. A similar study by these same authors showed that HSV induced ocular disease was minimal when animals were first challenged with a benign strain of HSV-1 and subsequently challenged with a more virulent strain.[164]

Mice inoculated intraperitoneally with one of two strains of HSV were protected against more lethal strains.[165] This occurred if the less pathogenic strain was given 4 to 27 hr before the more lethal strain, or if it was given simultaneously. Mice found to survive HSV infection were protected from a second more lethal dose of HSV-1 or 2.[166] The genome recovered from the spinal cord of these mice was found unchanged. Juvenile mice were able to survive infection with a temperature sensitive mutant of HSV-2 which protected them from infection with more lethal doses of HSV-1 or 2.

Results from the above experiments indicate that immunization with milder forms of HSV may offer protection to animals against lethal or more virulent herpesvirus infections. Appreciation of the mechanisms of benign virus genome expression and host immune response may offer avenues for immunization with segments of viral DNA which protect the host but do not result in disease.

As an example, superinfection of partially homogenized HSV-2 resulted in activation of HSV-1, which demonstrated that HSV-2 DNA synthesis is not required for HSV-1 activation.[167] HSV was maintained in a repressed form in human embryo lung cells. Other workers have found that reducing incubation temperature or superinfecting with a heterologous virus (cytomegalovirus) results in activation of HSV-1 replication.

Other studies involving immunization have shown that this treatment may help in reducing active and latent HSV infection. As an example, hairless mice were immunized with an envelope (HSV-1) antigen and challenged with HSV-1 after 2 weeks.[168] A consistent cell-mediated immune response in all immunized mice was found following immunization. Humoral response was low or not detectable. After challenge a marked secondary humoral and cell-mediated response was seen in all immunized mice. Animals were protected against development of skin lesions and fatal outcome of infection. However, establishment of latent infections in sensory ganglia was not prevented by the immunization procedure. In contrast, cytotoxic T cell activity has been observed during active HSV infections, but it has been reported that after day 12 and during the latent period no cytotoxic T cell activity was evident.[169] Cytotoxic T cell activity appeared again only after in vitro culturing of lymphocytes for 3 days.

Mice inoculated with HSV-1 by lip or cornea and passively immunized with rabbit antibody to HSV developed latent infection in the trigeminal ganglia by 96 hr.[170] Neutralizing antibody was cleared from the system and not detected after 2 months in most mice. Latent infections in ganglia from antibody negative mice revealed that latent virus was present in over 90% of the mice. This would indicate that serum neutralizing antibody is not necessary to maintain the latent state. When lips and cornea were traumatized, viral reactivation occured in 90% of the mice as demonstrated by appearance of neutralizing antibody.

The drug, Acyclovir, inhibited latent herpes growth in vitro with explanted cultures from hairless mice as long as the drug was continuously present.[171] Administration of the drug to mice with latent herpes did not eliminate latent virus from the trigeminal ganglia.

Another example using immunization to prevent virulent infection is with segments of viral DNA. Three doses of subunit viral vaccine were used to immunize albino rabbits who were then challenged with HSV in the cornea.[172] Latency in the homolateral Gasserian ganglion was found in 18 of 22 immunized animals. The proportion of HSV yielding fragments was higher in nonimmunized animals (42.6%), compared to those receiving whole virion vaccine (14.6%), but more importantly fewer Gasserian ganglion had HSV fragments when animals received subunit viral vaccine (5.4%). Immunity from vaccination restricted five times the number of ganglion cells which became virus carriers.

Host age was found to be the most important variable in influencing viral latency in female monkeys.[173] Sixteen adolescents and nineteen adults were infected vaginally with HSV-2. Sixty-nine percent of adolescents and 42% of adults had HSV-2 latent infections. Those who developed latent infections were found to be significantly easier to infect, shed virus during primary infection significantly longer, and had more severe primary disease than those not latently infected.

D. Reactivation of Latent Infection

Various ways of reactivating HSV have been investigated. Latent HSV infection of the cervical autonomic ganglion was reactivated by in vivo postganglionic neurectomy.[174] Cell-mediated immune defenses, specifically T cells, were shown to be highly efficient in eliminating reactivated virus from ganglion in vivo. Antithymocyte serum prolonged the time that virus could be detected. Cyclophosphamide increased the percentage of mice exhibiting reactivated ganglion infection and viral titers. Neither antithymocyte serum nor cyclophosphamide could reactivate HSV in the absence of neurectomy.

In contrast to neurectomy studies reactivation of latent infections has been accomplished using electrical stimulation. Physiological levels of electrical current were delivered via an electrode implanted over the trigeminal ganglion of latently infected rabbits. With this treatment it was possible to modify and synchronize virus shedding in preocular tear film and to cause multiple episodes or reactivation in a single animal.[175]

In another model of latent reactivation, human embryonic fibroblasts were infected with UV-irradiated HSV-2.[176] Synthesis of virus was induced at 35.5°C, but in contrast grew poorly and was inactivated at 40.5°C. Enhanced reactivation occurred at 36.5°C when HSV-2 latently infected cultures were superinfected with human cytomegalovirus or irradiated with a small dose of UV light. Cytomegalovirus did not enhance synthesis of HSV-2 during the normal growth cycle, but did enhance synthesis of UV irradiated HSV-2. These observations suggest that in this in vitro latency system, some HSV genomes were damaged by UV irradiation and were maintained in a nonreplicating state without being destroyed or significantly impaired. Others[177] have found that reactivation was enhanced when cells were treated with UV light or mitomycin C prior to infection with HSV.

VII. CONCLUSIONS

BHV-1 is a virus of considerable complexity in eluding the host's immune response and maintaining a state of latency. By better understanding basic concepts of the host's immune response, mechanisms of viral recognition and methods central to destruction of virally infected cells will be appreciated. The time when immune parameters can offer maximal protection and how these parameters might be improved through modulatory therapy with immunoregulatory cells or molecules may lead to improved immunization and treatment of infected animals.

Secondly, there is a continued need to better understand how the virus interacts with host cells. Receptor sites on immune and nonimmune cells which allow protection or infection to develop need to be appreciated. Genetic or receptor mechanisms which allow reinfection with the same virus or prevent infection with different strains of the same virus need further clarification.

Lastly, the development of viral components which induce the host's immune response need considerable exploration with BHV-1. Development of components as vaccines which offer protection from virulent infection has received little attention with BHV-1. Understanding which viral components produce maximal immune responses and ultimately maximal protection would greatly aid in understanding this elusive pathogen.

REFERENCES

1. **Gibbs, E. P. J. and Riweyemamu, M. M.**, Bovine herpesvirus. I. Bovine herpesvirus-1, *Vet. Bull. (London)*, 47, 317, 1977.
2. **Homan, E. J. and Easterday, B. C.**, Isolation of bovine herpesvirus-1 from trigeminal ganglia of clinically normal cattle, *Am. J. Vet. Res.*, 41, 1212, 1980.
3. **Pastoret, P. P., Babiuk, L. A., Misra, V., and Grisbel, P.**, Reactivity of temperature-sensitive and non-temperature-sensitive infectious bovine rhinotracheitis vaccine virus with dexamethasone, *Infect. Immun.*, 29, 483, 1980.
4. **Rouse, B. T. and Babiuk, L. A.**, Host defense mechanisms against infectious bovine rhinotracheitis virus: *in vitro* stimulation of sensitized lymphocytes by virus antigen, *Infect. Immun.*, 10, 681, 1974.
5. **Zair, J. A., Leary, P. L., and Levin, M. J.**, Specificity of the blastogenic response of human mononuclear cells to herpesvirus antigens, *Infect. Immun.*, 20, 646, 1978.
6. **Forman, A. J. and Babiuk, L. A.**, Effect of infectious bovine rhinotracheitis virus infection on bovine alveolar macrophage function, *Infect. Immun.*, 35, 1041, 1982.
7. **Forman, A. J., Babiuk, L. K., Baldwin, F., and Friend, S. C. E.**, Effect of infectious bovine rhinotracheitis virus infection of calves on cell populations recovered by lung lavage, *Am. J. Vet. Res.*, 43, 1174, 1982.
8. **McGeorge, M. B. and Morahan, P. S.**, Comparison of various macrophage-inhibitory agents on vaginal and systemic herpes simplex virus type 2 infections, *Infect. Immun.*, 22, 623, 1978.
9. **Mogensen, S. C. and Anderson, H. K.**, Role of activated macrophages in resistance of congenitally athymic nude mice to hepatitis induced by herpes simplex virus type 2, *Infect. Immun.*, 19, 792, 1978.

10. Babiuk, L. A. and Misra, V., Levamisole and bovine immunity: *in vitro* and *in vivo* effects on immune responses to herpesvirus immunization, *Can. J. Microbiol.*, 27, 1312, 1981.
11. Anon., Levamisole, *Lancet*, 1, 151, 1975.
12. Babiuk, L. A. and Misra, V., Effect of levamisole in immune responses to bovine herpesvirus-1, *Am. J. Vet. Res.*, 43, 1349, 1982.
13. Braathan, L. R., Berle, E., Mobech-Hanssen, U., and Thorsby, E., Studies on human epidermal Langerhans cells. II. Activation of human T lymphocytes to herpes simplex virus, *Acta Derm. Venereol.*, 60, 381, 1980.
14. Berle, E. J. and Thorsby, E., Human T cell response to herpes simplex virus antigen *in vitro*, *Acta Pathol. Microbiol. Scand.*, 88, 31, 1980a.
15. Berle, E. J. and Thorsby, E., The proliferative T cell response to herpes simplex virus (HSV) antigen is restricted by self HLA-D, *Clin. Exp. Immunol.*, 39, 668, 1980b.
16. Usinger, W. R. and Splitter, G. A., A simple method for specific depletion of B lymphocytes from bovine blood mononuclear cells, *Immunol. Commun.*, 9, 693, 1980.
17. Usinger, W. R. and Splitter, G. A., Two molecularly independent surface receptors identify bovine T lymphocytes, *J. Immunol. Meth.*, 45, 209, 1981.
18. Schrader, J. W., Chen, W-F., and Scollay, R., The aquisition of receptors for peanut agglutinin by peanut agglutinin-negative thymocytes abd peripheral T cells, *J. Immunol.*, 129, 545, 1982.
19. LeJan, C. and Asso, J., The local and systemic immune response of calves following experimental infection with I.B.R. virus, in *The Ruminant Immune System*, Butler, J. E., Ed., Plenum Press, New York, 1981, 677.
20. Nyaga, P. N. and McKercher, D. G., Pathogenesis of bovine herpesvirus-1 (BHV-1) infections: interactions of the virus with peripheral blood cellular components, *Comp. Immun. Microbiol. Infect. Dis.*, 2, 587, 1980.
21. Rouse, B. T. and Babiuk, L. A., Host defense mechanisms against infectious bovine rhinotracheitis virus. II. Inhibition of viral plaque formation by immune peripheral blood lymphocytes, *Cell. Immunol.*, 17, 43, 1975.
22. Kapoor, A. K., Ling, N. R., Nash, A. A., Bachan, A., and Wildy, P., *In vitro* stimulation of rabbit T lymphocytes by cells expressing herpes simplex antigens, *J. Gen. Virol.*, 59, 415, 1982.
23. Hilfenhaus, J., Christ, H., Kohler, R., Moser, H., Kirchner, H., and Levy, H. B., Protectivity of herpes simplex antigens: studies in mice on the adjuvant effect of PICLC and on the dependence of protection on T cell competence, *Med. Microbiol. Immunol. (Berlin)*, 169, 225, 1981.
24. Katz, E., Mitrani-Rosenbaum, S., Margalith, E., and Ben-Bassat, H., Interaction of herpes simplex virus with human cell lines at various stages of lymphoid differentiation, *Intervirology*, 16, 33, 1981.
25. Mahan, K. B., Hoffmann, P. J., and Hoffmann, C. C., A reproducible method for the optimum infection of human mononuclear cells and lymphoid cell lines by herpes simplex virus type 1, *J. Virol. Meth.*, 4, 63, 1982.
26. Rinaldo, C. R., Richter, B. S., Black, P. H., and Hirsch, M. S., Persistent infection of human lymphoid and myeloid cell lines with herpes simplex virus, *Infect. Immun.*, 25, 521, 1979.
27. Rosenberg, G. L., Farber, P. A., and Notkins, A. L., *In vitro* stimulation of sensitized lymphocytes by herpes simplex virus and vaccinia virus, *Proc. Natl. Acad. Sci. U.S.A.*, 69, 756, 1972.
28. Rosenberg, G. L. and Notkins, A. L., Induction of cellular immunity to herpes simplex virus: relationship to humoral immune response, *J. Immunol.*, 112, 1019, 1974.
29. Letarte, M., Teh, H. S., and Meghji, G., Increased expression if Ia and Thy-1 antigens on mitogen-activated murine spleen lymphocytes, *J. Immunol.*, 125, 370, 1980.
30. Koide, Y., Awashima, F., Yoshida, T. O., Takenouchi, T., Wakisaka, A., Moriuchi, J., and Aizawa, M., The role of three distinct Ia-like antigen molecules in human T cell proliferative responses: effect of monoclonal anti-Ia-like antibodies, *J. Immunol.*, 129, 1061, 1982.
31. Yachie, A., Miczawaki, T., Yokoi, T., Nagaoki, T., and Taniguchi, N., Ia positive cells generated by PWM-stimulation within OKT-4$^+$ subset interact with OKT-8$^+$ cells for inducing active suppression on B cell differentiation, *J. Immunol.*, 129, 103, 1982.
32. Indiveri, F., Wilson, B. S., Russo, C., Quaranta, V., Pellegrino, M. A., and Ferrone, S., Ia-like antigens on human T lymphocytes: relationship to other surface markers, role in mixed lymphocyte reactions, and structural profile, *J. Immunol.*, 125, 2673, 1980.
33. Parish, C. R., Freeman, R. R., McKenzie, I. F. C., Cheers, C., and Cole, G. A., Ia antigens in serum during different murine infections, *Infect. Immun.*, 26, 422, 1979.
34. Para, M. F., Baucke, R. B., and Spear, P. G., Immunoglobulin G (Fc)-binding receptors on virions of herpes simplex virus type 1 and transfer of these receptors to the cell surface by infection, *J. Virol.*, 34, 512, 1980.
35. Adler, R., Glorisso, J. C., Cassman, J., and Levine, M., Possible role of Fc receptors on cells infected and transformed by herpesvirus: escape from immune cytolysis, *Infect. Immun.*, 21, 442, 1978.
36. Smith, K. A., Lachman, L. B., Oppenheim, J. J., and Favata, M. A., The functional relationship of the interleukins, *J. Exp. Med.*, 151, 1551, 1980.

37. Varesio, L., Holden, H. T., and Taramelli, D., Mechanism of lymphocyte activation. II. Requirements for macromolecular synthesis in the production of lymphokines, *J. Immunol.*, 125, 2810, 1980.
38. Marcucci, F., Waller, M., Kirchner, H., and Krammer, P., Production of immune interferon by murine T-cell clones from long-term cultures, *Nature (London)*, 291, 79, 1981.
39. Duncan, M. R., George, F. W., IV, and Hadden, J. W., Concanavalin A-induced human lymphocyte mitogenic factor: activity distinct from interleukin 1 and 2, *J. Immunol.*, 129, 56, 1982.
40. Baker, P. E., Gillis, S., and Smith, K. A., Monoclonal cytolytic T-cell lines, *J. Exp. Med.*, 149, 273, 1979.
41. Watson, J., Continous proliferation of murine antigen-specific helper T lymphocytes in culture, *J. Exp. Med.*, 150, 1510, 1979.
42. Pfizenmaier, K., Delzeit, R., Rollinghoff, M., and Wagner, H., T-T cell interactions during *in vitro* cytotoxic T lymphocyte responses, *Eur. J. Immunol.*, 10, 577, 1980.
43. Dubey, D. P., Stux, S., Tsong, Y. K., and Yunis, E., Crossreactive cytotoxic T-cell line expanded by TCGF, *Transplant. Proc.*, 13, 1153, 1981.
44. Lutz, C. T., Glasebrook, A. L., and Fitch, F. W., Alloreactive cloned T cell line. IV. Interaction of alloantigen and TCGF to stimulate cloned cytolytic T lymphocytes, *J. Immunol.*, 127, 391, 1981.
45. Stotter, H., Rude, E., and Wagner, H., T cell factor (interleukin 2) allows *in vivo* induction of T helper cells against heterologous erythrocytes in athymic (nu/nu) mice, *Eur. J. Immunol.*, 10, 719, 1980.
46. Wagner, H., Hardt, C., Heeg, K., Pfizenmaier, K., Stotter, H., and Rollinghoff, M., The *in vivo* effects of interleukin 2 (TCGF), *Immunobiology*, 161, 139, 1982.
47. DeVries, J. E., Vyth, F. A., and Mendelsohn, J., T-cell growth factor-mediated proliferation of lymphocytes from a T-chronic lymphocytic leukemia patient lacking mitogen and alloantigen responsiveness, *Clin. Exp. Immunol.*, 43, 302, 1981.
48. Gillis, S., Kozak, R., Durante, M., and Weksler, M. E., Immunological studies of aging. Decreased production of and response to T cell growth factor by lymphocytes from aged humans, *J. Clin. Invest.*, 67, 937, 1981.
49. Wofsy, D., Murphy, E. D., Roths, J. B., Dauphinee, M. J., Kipper, S. B., and Talal, N., Deficient interleukin 2 activity in MRL/M and C57BL/6J mice bearing the 1pr gene, *J. Exp. Med.*, 154, 1671, 1981.
50. Alcocer-Varela, L. and Alarcon-Segovia, D., Decreased production of and response to interleukin-2 by cultured lymphocytes from patients with systemic lupus erythematosus, *J. Clin. Invest.*, 69, 1388, 1982.
51. Gilman, S. C., Rosenberg, J. S., and Feldman, J. D., T lymphocytes of young and aged rats. II. Functional defects and the role of IL-2, *J. Immunol.*, 128, 644, 1982.
52. Lopez-Botet, M., Fontan, G., Rodriguez, M. C. G., and de Landazuri, M. O., Relationship between IL-2 synthesis and proliferative response to PHA in different primary immunodeficiencies, *J. Immunol.*, 128, 679, 1982.
53. Prowse, S. J., Lymphokine (interleukin 2) secretion as a measure of T cell recognition of parasite antigens, *AJEBAK*, 59, 695, 1981.
54. Warren, H. S. and Pembrey, R. G., Lymphokine production by peripheral blood leukocytes: quantitation of T-cell growth factor activity for assessment of immune response capability, *Aust. N.Z. J. Med.*, 11, 475, 1981.
55. Ilonen, J. and Salmi, A., Detection of antigen-specific cellular immune response by the *in vitro* production of T-cell growth factor, *Scand. J. Immunol.*, 15, 521, 1982.
56. MacSween, J. M., Cohen, A. D., Rajaraman, K., and Fox, R. A., Lymphokine responses to mitogenic and antigenic stimulation. Predictive value in renal transplantation, *Transplantation*, 34, 196, 1982.
57. Watson, J., Gillis, S., Marbrook, J., Mochizuki, D., and Smith, K. A., Biochemical and biological characterization of lymphocyte regulatory molecules. I. Purification of a class of murine lymphokines, *J. Exp. Med.*, 150, 849, 1979.
58. Shore, S. L., Nahmias, A. J., Starr, S. E., Wood, P. A., and McFarlin, D. E., Detection of cell-dependent cytotoxicity antibody to cells infected with herpes simplex virus, *Nature (London)*, 251, 350, 1974.
59. Rouse, B. T., Wardley, R. C., and Babiuk, L. A., Antibody dependent cell-mediated cytotoxicity in cows: comparison of effector cell activity against heterologous erthrocyte and herpesvirus-infected bovine target cells, *Infect. Immun.*, 13, 1433, 1976.
60. Grewal, A. S., Rouse, B. T., and Babiuk, L. A., Mechanisms of resistance to herpesvirus: comparison of effectiveness of different cell types in mediating antibody-dependent cell-mediated cytotoxicity, *Infect. Immun.*, 15, 698, 1977.
61. Rouse, B. T. and Babiuk, L. A., Mechanisms of viral immunopathology, *Adv. Vet. Sci. Comp. Med.*, 23, 103, 1979.

62. Kohl, S. and Loo, L. S., Protection of neonatal mice against herpes simplex virus infection: probable *in vivo* antibody-dependent cellular cytotoxicity, *J. Immunol.*, 129, 370, 1982.
63. Kohl, S. and Loo, L. S., Ontogeny of murine cellular cytotoxicity to herpes simplex virus-infected cells, *Infect. Immun.*, 30, 847, 1980.
64. Rager-Zisman, B. and Allison, A. C., Mechanism of immunological resistance to herpes simplex virus 1 (HSV-1) infection, *J. Immunol.*, 116, 35, 1976.
65. Wardley, R. C., Rouse, B. T., and Babiuk, L. A., Antibody dependent cytotoxicity mediated by neutrophils: a possible mechanism of antiviral defense, *J. Reticuloendothel. Soc.*, 19, 323, 1976.
66. Grewal, A. S., Carpio, M., and Babiuk, L. A., Polymorphonuclear neutrophil-mediated antibody-dependent cell cytotoxicity of herpesvirus-infected cells: ultrastructural studies, *Can. J. Microbiol.*, 26, 427, 1980.
67. Rouse, B. T., Wardley, R. C., and Babiuk, L. A., The role of antibody dependent cytotoxicity in recovery from herpesvirus infections, *Cell. Immunol.*, 22, 182, 1976.
68. Sheppard, A. M. and Smith, J. W., Antibody mediated destruction of keratocytes infected with herpes simplex virus, *Curr. Eye Res.*, 1, 397, 1981.
69. Norrild, B., Shore, S. L., Cromeans, T. L., and Nahmias, A. J., Participation of three major glycoprotein antigens of herpes simplex virus type 1 early in the infectious cycle as determined by antibody-dependent cell-mediated cytotoxicity, *Infect. Immun.*, 28, 38, 1980.
70. Shore, S. L., Nahmias, A. J., Starr, S. E., Wood, P. A., and McFarlin, D. E., Detection of cell-dependent cytotoxic antibody to cells infected with herpes simplex virus, *Nature (London)*, 251, 350, 1974.
71. Romano, T. J. and Shore, S. L., Analysis of the adhesion step in the herpes simplex virus antibody-dependent cellular cytotoxicity system, *Infect. Immun.*, 26, 163, 1979.
72. Rouse, B. T., Grewal, A. S., Babiuk, L. A., and Fujimiya, Y., Enhancement of antibody dependent cell-mediated cytotoxicity of herpesvirus-infected cells by complement, *Infect. Immun.*, 18, 660, 1977.
73. Rouse, B. T., Grewal, A. S., and Babiuk, L. A., Complement enhances antiviral antibody-dependent cell cytotoxicity, *Nature (London)*, 266, 456, 1977.
74. Grewal, A. S., Rouse, B. T., and Babiuk, L. A., Mechanisms of recovery from viral infections: destruction of infected cells by neutrophils and complement, *J. Immunol.*, 124, 312, 1980.
75. Grewal, A. A. and Rouse, B. T., Destruction of virus infected cells by neutrophils and complement, *Experientia*, 36, 352, 1980.
76. Santoli, D. and Koprowski, H., Mechanisms of activation of human natural killer cells against tumor and virus-infected cells, *Immunol. Rev.*, 44, 125, 1979.
77. Minato, N., Reid, L., Cantor, H., Lengyel, P., and Bloom, B. R., Mode of regulation of natural killer cell activity by interferon, *J. Exp. Med.*, 152, 124, 1980.
78. Piontek, G. E., Weltzin, R., and Tompkins, W. A. F., Enhanced cytotoxicity of mouse natural killer cells for vaccinia and herpes virus-infected targets, *J. Reticuloendothel. Soc.*, 27, 175, 1980.
79. Trinchieri, G. and Santoli, D., Antiviral activity induced by culturing lymphocytes with tumor-derived or virus-transformed cells. Enhancement of human natural killer cell activity by interferon and antigonistic inhibition of susceptibility of target cells to lysis, *J. Exp. Med.*, 147, 1314, 1978.
80. Fujimiya, Y., Babiuk, L. A., and Rouse, B. T., Direct lymphocytotoxicity against herpes simplex virus infected cells, *Can. J. Microbiol.*, 24, 1076, 1978.
81. Rola-Pleszczynski, M., *In vitro* induction of human cell-mediated cytotoxicity directed against herpes simplex virus-infected cells: characterization of the effector lymphocyte, *J. Immunol.*, 125, 1475, 1980.
82. Johnson, D. R. and Jondal, M., Herpesvirus-transformed cytotoxic T-cell lines, *Nature (London)*, 291, 81, 1981.
83. Armerding, D., Simon, M. M., Hammerling, U., Hammerling, G. J., and Rossiter, H., Function, target-cell preference and cell-surface characteristics of herpes simplex virus 2-induced nonantigen-specific killer cells, *Immunobiology*, 158, 347, 1981.
84. Quinnan, G. V. and Manischewitz, J. E., The role of natural killer cells and antibody-dependent cell-mediated cytotoxicity during murine cytomegalovirus infection, *J. Exp. Med.*, 150, 1549, 1979.
85. Welsh, R. M., Jr., Mouse natural killer cells: induction, specificity, and function, *J. Immunol.*, 121, 1631, 1978.
86. Herberman, R. B., Djien, J. Y., Kay, H. D., Ortaldo, R. J., Riccardi, C., Bonnard, G. D., Holden, H. T., Fagnani, R., Santoli, A., and Puccetti, P., Natural killer cells: characteristics and regulation of activity, *Immunol. Rev.*, 44, 42, 1979.
87. Gildlund, M., Orn, A., Wigzell, H., Senik, A., and Gresser, I., Enhanced NK cell activity in mice injected with interferon and interferon inducers, *Nature (London)*, 273, 759, 1978.
88. Kirchner, H., Peter, H. H., and Hilfenhaus, J., Interactions between herpes simplex virus and the cells of the immune system, in *Virus Lymphocyte Interactions: Implications for Disease*, Proffitt, A., Ed., Elsevier/North Holland, Amsterdam, 1981, 259.

89. Herberman, R. B. and Ortaldo, J. R., Natural killer cells: their role in defenses against diseases. Characteristics and regulation of activity, *Science*, 214, 24, 1981.
90. Haller, O., Arnheiter, H., Gresser, I., and Lindermann, J., Genetically determined interferon-dependent resistance to influenze virus in mice, *J. Exp. Med.*, 149, 601, 1979.
91. Sethi, K. K. and Brandis, H., *In vitro* acquisition of resistance against herpes simplex virus by permissive murine macrophages, *Arch. Virol.*, 59, 157, 1979.
92. Sutton, A. L., Smithwick, E. M., Seligman, S. L., and Kim, D. S., Fatal disseminated herpesvirus hominis type 2 infection in an adult with associated thymic dysplasia, *Am. J. Med.*, 56, 545, 1974.
93. Ching, C. and Lopez, C., Natural killing of herpes simplex virus type 1-infected target cells: normal human responses and influence of antiviral antibody, *Infect. Immun.*, 26, 49, 1979.
94. Rouse, B. T. and Babiuk, L. A., Host responses to infectious bovine rhinotracheitis virus. III. Isolation and immunologic activities of bovine T lymphocytes, *J. Immunol.*, 113, 1391, 1974.
95. Rouse, B. T. and Babiuk, L. A., The direct antiviral cytotoxicity by bovine lymphocytes is not restricted by genetic incompatability of lymphocytes and target cells, *J. Immunol.*, 118, 618, 1977.
96. Schlabach, A. J., Martinez, D., Field, A. K., and Tytell, A. A., Resistance of C58 mice to primary systemic herpes simplex virus infection: macrophage dependence and T-cell independence, *Infect. Immun.*, 26, 615, 1979.
97. Kutinov'a, L., Vonka, V., and Slichtov'a, V., Immunogenicity of subviral herpes simplex virus preparations: protection of mice against intraperitoneal infection with live virus, *Acta. Virol. (Praha)*, 24, 391, 1980.
98. Donnenberg, A. D., Chaikof, E., and Aurelian, L., Immunity to herpes simplex virus type 2: cell-mediated immunity in latently infected guinea pigs, *Infect. Immun.*, 30, 99, 1980.
99. Pfizenmaier, K., Jung, H., Starzinski-Powitz, A., Rollinghoff, M., and Wagner, H., The role of T cells in anti-herpes simplex virus immunity. I. Induction of antigen-specific cytotoxic T lymphocytes, *J. Immunol.*, 119, 939, 1977.
100. Sethi, K. K. and Brandis, H., Specifically immune mouse T cells can destroy H-2 compatible murine target cells infected with herpes simplex virus type 1 or 2, *Z. Immun. Forsch.*, 150, 162, 1977.
101. Lawman, M. J., Rouse, B. T., Courtney, R. J., and Walker, R. D., Cell-mediated immunity against herpes simplex induction of cytotoxic T lymphocytes, *Infect. Immun.*, 27, 133, 1980.
102. Nash, A. A., Field, H. J., and Quartey-Papafio, R., Cell-mediated immunity in herpes simplex virus-infected mice: induction, characterization and antiviral effects of delayed type hypersensitivity, *J. Gen. Virol.*, 48, 351, 1980.
103. Rouse, B. T. and Lawman, M. J., Induction of cytotoxic T lymphocytes against herpes simplex virus type 1: role of accessory cells and amplifying factor, *J. Immunol.*, 124, 2341, 1980.
104. Carter, V. C., Rice, P. L., and Tevethia, S. S., Intratypic and intertypic specificity of lymphocytes involved in the recognition of herpes simplex virus glycoproteins, *Infect. Immun.*, 37, 116, 1982.
105. Eberle, R., Russell, R. G., and Rouse, B. T., Cell-mediated immunity to herpes simplex virus: recognition of type-specific and type-common surface antigens by cytotoxic T cell populations, *Infect. Immun.*, 34, 795, 1981.
106. Sethi, K. K. and Wolff, M. H., The nature of host-cell herpes-simplex virus interaction(s) that renders cells susceptible to virus-specific cytotoxic T cells, *Immunobiology*, 157, 365, 1980.
107. Sethi, K. K., Stroehmann, I., and Brandis, H., Human T-cell cultures from virus-sensitized donors can mediate virus-specific and HLA-restricted cell lysis, *Nature (London)*, 286, 718, 1980.
108. Schmid, D. S., Larsen, H. S., and Rouse, B. T., The role of accessory cells and T cell-growth factor in induction of cytotoxic T lymphocytes against herpes simplex virus antigens, *Immunology*, 44, 755, 1981.
109. Sheridan, J. F., Donnerberg, A. D., Aurelian, L., and Elpern, D. J., Immunity to herpes simplex virus type 2, *J. Immunol.*, 129, 326, 1982.
110. Babiuk, L. A. and Rouse, B. T., Interactions between effector cell activity and lymphokines: implications for recovery from herpes virus infections, *Int. Arch. Allergy Appl. Immun.*, 57, 62, 1978.
111. Moller-Laren, A., Haahr, S., and Black, F. T., Cellular and humoral immune responses to herpes simplex virus during and after primary gingivostomatitis, *Infect. Immun.*, 22, 445, 1978.
112. Babiuk, L. A., Wardley, R. C., and Rouse, B. T., Defense mechanisms against bovine herpesvirus: relationship of virus-host cell events to susceptibility to antibody-complement cell lysis, *Infect. Immun.*, 12, 958, 1975.
113. Hyland, S. J., Easterday, B. C., and Pawlisch, R., Antibody levels and immunity to infectious bovine rhinotracheitis virus (IBR) infections in Wisconsin Dairy cattle, *Dev. Biol. Stand.*, 28, 510, 1975.
114. Hochman, N., Rones, Y., Ehrlich, J., Levy, R., and Zakay-Rones, Z., Antibodies to herpes simplex virus in human gingival fluid, *J. Periodontol.*, 52, 324, 1981.
115. Darcel, C. L. Q., Bradley, J. A., and Mitchell, D., Immune responses of cattle to antigens obtained from bovine herpesvirus 1-infected tissue culture, *Can. J. Comp. Med.*, 43, 288, 1979.
116. Vestergaard, B. F., Herpes simplex virus antigens and antibodies: survey of studies based on quantitative immunoelectrophoresis, *Rev. Infect. Dis.*, 2, 899, 1980.

117. Ching, C. Y. and Lopez, C., A type-specific antiserum induced by a major herpesvirus type 1 glycoprotein, *J. Immunol. Meth.*, 32, 383, 1980.
118. Pereira, L., Klassen, T., and Baringer, J. R., Type-common and type-specific monoclonal antibody to herpes simplex virus type 1, *Infect. Immun.*, 29, 724, 1980.
119. Dix, R. D., Pereira, L., and Baringer, J. R., Use of monoclonal antibody directed against herpes simplex virus glycoproteins to protect mice against acute virus-induced neurological disease, *Infect. Immun.*, 34, 192, 1981.
120. Zweerink, H. J., Martinez, D., Lynch, R. J., and Stanton, L. W., Immune responses in mice against herpes simplex virus: mechanisms of protection against facial and ganglionic infections, *Infect. Immun.*, 31, 267, 1981.
121. Arsenakis, M. and May, J. T., Complement-fixing antibody to the AG-4 antigen in herpes simplex virus type 2-infected patients, *Infect. Immun.*, 33, 22, 1981.
122. Ratner, J. J. and Smith, K. O., Serum antibodies to herpes simplex virus type 1 during active oral herpes infection, *Infect. Immun.*, 27, 113, 1980.
123. Oakes, J. E. and Rosemond-Hornbeak, H., Antibody-mediated recovery from subcutaneous herpes simplex virus type 2 infection, *Infect. Immun.*, 21, 489, 1978.
124. McKendall, R. R., Klassen, T., and Baringer, J. R., Host defenses in herpes simplex infections of the nervous system: effect of antibody on disease and viral spread, *Infect. Immun.*, 23, 305, 1979.
125. Ohashi, M. and Ozaki, Y., Studies of the neutralizing antibody to herpes simplex virus. I. Effect of heterotypic antigenic stimulus on the type specificity of neutralizing antibody in rabbits, *Arch. Virol.*, 67, 57, 1981.
126. Meters-Elliott, R. H. and Chitjian, P. A., Induction of cell-mediated immunity in herpes simplex virus keratitis. Kinetics of lymphocyte transformation and the effect of antiviral antibody, *Invest. Ophthalmol. Vis. Sci.*, 19, 920, 1980.
127. Oakes, J. E. and Lausch, R. N., Role of Fc fragments in antibody-mediated recovery from ocular and subcutaneous herpes simplex virus infections, *Infect. Immun.*, 33, 109, 1981.
128. Rosenquist, B. D. and Loan, R. W., Interferon induction in the bovine species by infectious rhinotracheitis virus, *Am. J. Vet. Res.*, 30, 1305, 1969.
129. MacLaughlin, N. J. and Rosenquist, B. D., Duration of protection of calves against rhinovirus challenge exposure by infectious bovine rhinotracheitis virus-induced interferon in nasal secretions, *Am. J. Vet. Res.*, 43, 289, 1982.
130. Cummins, J. M. and Rosenquist, B. D., Temporary protection of calves against adenovirus infection by nasal secretion interferon induced by infectious bovine rhinotracheitis virus, *Am. J. Vet. Res.*, 43, 955, 1982.
131. Babiuk, L. A. and Rouse, B. T., Immune interferon production by lymphoid cells: role in the inhibition of herpesviruses, *Infect. Immun.*, 13, 1567, 1976.
132. Forman, A. J., Babiuk, L. A., Misra, V., and Baldwin, F., Susceptibility of bovine macrophages to infectious bovine rhinotracheitis virus infection, *Infect. Immun.*, 35, 1048, 1982.
133. Rosenquist, B. D. and Loan, R. W., Production of circulating interferon in the bovine species, *Am. J. Vet. Res.*, 30, 1293, 1969.
134. Fulton, R. W. and Root, S. K., Antiviral activity in interferon-treated bovine tracheal organ cultures, *Infect. Immun.*, 21, 672, 1978.
135. Cummins, J. M. and Rosenquist, B. D., Leukocyte changes and interferon production in calves infected with hydrocortisone and infected with infectious bovine rhinotracheitis virus, *Am. J. Vet. Res.*, 40, 238, 1979.
136. Green, J. A., Yeh, T. J., and Overall, J. C., Jr., Sequential production of IFN alpha and immune-specific IFN gamma by human mononuclear leukocytes exposed to herpes simplex virus, *J. Immunol.*, 127, 1192, 1981.
137. Peter, H. H., Dallugge, H., Zwatzky, Leibold, W., and Kirchner, H., Human peripheral null lymphocytes. II. Producers of type-1 interferon upon stimulation with tumor cells, herpes simplex virus and *Cornybacterium parvum*, *Eur. J. Immunol.*, 10, 547, 1980.
138. Morse, S. S. and Morahan, P. S., Activated macrophages mediate interferon-independent inhibition of herpes simplex virus, *Cell. Immunol.*, 58, 72, 1981.
139. Kohl, S., Loos, L. S., and Greenberg, S. B., Protection of newborn mice from a lethal herpes simplex virus infection by human interferon, antibody and leukocytes, *J. Immunol.*, 128, 1107, 1982.
140. Hill, T. J. and Blyth, W. A., An alternative theory of herpes-simplex recurrence and a possible role for prostaglandins, *Lancet*, 1, 397, 1976.
141. Hill, T. J., Harbour, D. A., and Blyth, W. A., Isolation of herpes simplex virus from the skin of clinically normal mice during latent infection, *J. Gen. Virol.*, 47, 205, 1980.
142. Homan, E. J. and Easterday, B. C., Further studies of naturally occurring latent bovine herpesvirus infections, *Am. J. Vet. Res.*, 42, 1811, 1981.
143. Geder, L., Lee, K. J., Dawson, M. S., Engle, R., Maliniak, R. M., and Lans, C. M., Induction of persistent infection in mice and oncogenic transformation of mouse macrophages with infectious bovine rhinotracheitis virus, *Am. J. Vet. Res.*, 42, 300, 1981.

144. Geder, L., Lee, K. J., Dawson, M. S., Hyman, R. W., Maliniak, R. M., and Rapp, F., Properties of mouse embryo fibroblasts transformed *in vitro* by infectious bovine rhinotracheitis virus, *J. Natl. Cancer Inst.*, 65, 441, 1980.
145. Lupton, H. W. and Reed, D. E., Clearance and shedding of infectious bovine rhinotracheitis virus from the nasal mucosa of immune and nonimmune calves, *Am. J. Vet. Res.*, 41, 117, 1980.
146. Rossi, C. R. and Kiesel, G. K., Effect of infectious bovine rhinotracheitis virus immunization on viral shedding in challenge-exposed calves treated with dexamethasone, *Am. J. Vet. Res.*, 43, 1576, 1982.
147. Narita, M., Innui, S., Nanba, K., and Shimizu, Y., Recrudescence of infectious bovine rhinotracheitis virus and associated neural changes in calves treated with dexamethasone, *Am. J. Vet. Res.*, 42, 1192, 1981.
148. Rossi, C. R., Kiesel, G. K., and Rumph, P. F., Association between route of inoculation with infectious bovine rhinotracheitis virus and site of recrudescence after dexamethasone treatment, *Am. J. Vet. Res.*, 43, 1440, 1982.
149. Narita, M., Inui, S., Nanba, K., and Shimizu, Y., Neural changes in vaccinated calves calves challenge exposed with virulent infectious bovine rhinotracheitis virus, *Am. J. Vet. Res.*, 41, 1995, 1980.
150. Castrucci, G., Frigeri, F., Cilli, V., Tesei, V., Arush, A. M., Pedini, B., Ranucci, S., and Rampichini, L., Attempts to reactivate bovid herpesvirus-2 in experimentally infected calves, *Am. J. Vet. Res.*, 41, 1890, 1980.
151. Letchworth, G. J. and Carmichael, L. E., Bovid herpesvirus 2 latency: failure to recover virus from central sensory nerve ganglia, *Can. J. Comp. Med.*, 46, 76, 1982.
152. Cabrera, C. V., Wohlenberg, C., Openshaw, H., Rey-Mendez, M., Puga, A., and Notkins, A. L., Herpes simplex virus DNA sequences in the CNS of latently infected mice, *Nature (London)*, 288, 288, 1980.
153. Kristensson, K., Nennesmo, L., Persson, L., and Lycke, E., Neuron to neuron transmission of herpes simplex virus. Transport of virus from skin to brainstem nuclei, *J. Neurol. Sci.*, 54, 149, 1982.
154. Fong, B. S. and Scriba, M., Use of (^{125}I) deoxycytidine to detect herpes simplex virus-specific thymidine kinase in tissues of latently infected guinea pigs, *J. Virol.*, 34, 644, 1980.
155. Hartman, M. and Ziegler, R. J., Protective effects of rat splenic lymphocytes and peritoneal exudate cells on herpes simplex virus infection of rat dorsal root ganglia in culture, *J. Neuropathol. Exp. Neurol.*, 38, 165, 1979.
156. McLenna, J. L. and Darby, G., Herpes simplex virus latency: the cellular location of virus in dorsal root ganglia and the fate of the infected cell following virus activation, *J. Gen. Virol.*, 51, 233, 1980.
157. Watson, K., Stevens, J. G., Cook, M. L., and Subak-Sharpe, J. H., Latency competence of thirteen HSV-1 temperature-sensitive mutants, *J. Gen. Virol.*, 49, 149, 1980.
158. Cummings, P. J., Lakomy, R. J., and Rinaldo, C. R., Jr., Characterization of herpes simplex virus persistence in a human T lymphoblastoid cell line, *Infect. Immun.*, 34, 817, 1981.
159. Rajcani, J. and Matis, J., Immediate early and early polypeptides in herpesvirus latency, *Acta Virol. (Praha)*, 25, 371, 1981.
160. Galloway, D. A., Fenoglio, C. M., and McDougall, J. K., Limited transcription of the herpes simplex virus genome when latent in human sensory ganglia, *J. Virol.*, 41, 686, 1982.
161. Vahine, A., Nilheden, E., and Svennerholm, B., Multiplicity activation of herpes simplex virus in mouse neuroblastoma (C1300) cells, *Arch. Virol*, 70, 345, 1981.
162. Roumillat, L. F., Feorino, P. M., Caplan, D. D., and Luker, P. D., Analysis and characterization of herpes simplex virus after its persistence in a lymphoblastoid cell lines for 15 months, *Infect. Immun.*, 29, 671, 1980.
163. Centifanto-Fitzgerald, Y. M., Varnall, E. D., and Kaufman, H. E., Initial herpes simplex virus type 1 infection prevents ganglionic superinfection by other strains, *Infect. Immun.*, 35, 1125, 1982.
164. Varnell, E. D., Centifanto-Fitzgerald, Y. M., and Kaufman, H. E., Herpesvirus infection and its effects on virulent superinfection, ganglionic colonization, and shedding, in *Herpesvirus: Clinical Pharmacological and Basic Aspects*, Shiota, H., Cheng, Y. C., and Prusoff, W. H., Eds., Excerpta Medica, Amsterdam, 1982, 21.
165. Schroder, C. H., Engler, H., and Kirchner, H., Protection of mice by an apathogenic strain HSV-1 against lethal infection by a pathogenic strains of HSV-1, *J. Gen. Virol.*, 52, 159, 1981.
166. Darai, G., Zoller, L., Matz, B., Schwaier, A., Flugel, R. M., and Munk, K., Experimental infection and the state of viral latency of adult tupaia with herpes simplex virus type 1 and 2 and infection of juvenile tapaia with temperature-sensitive mutants of HSV type 2, *Arch Virol.*, 65, 311, 1980.
167. Wigdahl, B. L., Isom, H. C., DeClercq, E., and Rapp, F., Activation of herpes simplex virus (HSV) types 1 genome by temperature-sensitive mutants of HSV type 2, *Virology*, 116, 468, 1982.
168. Klein, R. J., Buimovici-Klein, E., Moser, H., Moucha, R., and Hilfenhaus, J., Efficacy of a virion envelope herpes simplex virus vaccine against experimental skin infections in hairless mice, *Arch. Virol.*, 68, 73, 1981.

169. Nash, A. A., Quartey-Papafio, R. and Wildy, R., The functional characteristics of lymphoid cell-mediated immunity in herpes simplex virus infected mice: functional analysis of lymph node cells during periods of acute and latent infection with reference to cytotoxic and memory cells, *J. Gen. Virol.*, 49, 309, 1980.
170. Sekizawa, T., Openshaw, H., Wohlenberg, C., and Notkins, A. L., Latency of herpes simplex virus in absence of neutralizing antibody: model for reactivation, *Science*, 210, 1026, 1980.
171. Klein, R. J., DeStefano, E., Friedman-Kien, A. E., and Brady, E., Effect of acyclovir on latent herpes simplex virus infections in trigeminal ganglia of mice, *Antimicrob. Agents Chemother.*, 19, 937, 1981.
172. Rajcani, J., Kutinova, L., and Vonka, V., Restriction of latent herpes virus infection in rabbits immunized with subviral herpes simplex virus vaccine, *Acta Virol. (Praha)*, 24, 183, 1980.
173. Reeves, W. C., DiGiacomo, R., and Alexander, E. R., A primate model for age and host response to genital herpetic infection: determinants of latency, *J. Infect. Dis.*, 143, 554, 1981.
174. Price, R. W. and Schmitz, J., Reactivation of latent herpes simplex virus infection of the autonomic nervous system by postganglionic neurectomy, *Infect. Immun.*, 19, 523, 1978.
175. Green, M. T., Rosborough, J. P., and Dunkel, E. G., *In vivo* reactivation of herpes simplex virus in rabbit trigeminal ganglia: electrode model, *Infect. Immun.*, 34, 69, 1981.
176. Nishiyama, J. and Rapp, R., Latency *in vitro* using irradiated herpes simplex virus, *J. Gen. Virol.*, 52, 113, 1981.
177. Zamansky, G. B., Kleinman, L. F., Black, P. H., and Kaplan, J. C., Reactivation of herpes simplex virus in a cell line inducible for simian virus 40 synthesis, *Mutat. Res.*, 70, 1, 1980.

Chapter 6

PORCINE HERPESVIRUS 1

Peter L. Nara

TABLE OF CONTENTS

I. INTRODUCTION

A. Historical Perspective

Pseudorabies is a natural disease of some domestic and wild animals. Experimentally, the disease infects many classes of mammals and birds but is noninfectious for apes, chimpanzees, reptiles, and insects. It is caused by a virus which is a member of the family *Herpesviridae.* Historically, the disease has been traced back in the literature by Hanson[1] and found to be present in the U.S. as early as 1813. This, however, was not a scientific publication and, as such, can only be speculation. The first reported and documented case in the scientific literature occurred in 1902. Aujeszky[2] described the disease in its fatal form when it occurred in a cow, subsequently in a dog, and then a cat. He used the brain tissues from these animals for successful serial transmission studies in laboratory animals. Two years later Marek[3] described the disease in laboratory white rabbits coining the term "infectious bulbar paralysis". It was not until 1910 that the agent was shown to pass through bacteria-retaining filters and still cause disease.[4] "Mad itch" was an adopted term from the clinical signs in cattle described by North American clinicians years later. Occasionally the signs of drooling and chomping of the jaws were noted in natural cases involving cattle and pigs, hence the inappropriate comparison to rabies and the term "pseudorabies". In 1931 Shope[5] established the serologic identity of "mad itch" to Aujeszky's Disease and opened the whole area of research.

This disease has been reported in all countries of central and eastern Europe,[6-11] Russia,[12] South America[13,14] British Isles,[15] Northern Ireland,[16,17] Republic of Ireland,[18] England,[19,20] North America,[5,20,22] New Zealand,[23,24] China and Taiwan.[25] The disease, however, has not been reported in Japan or Australia and only unconfirmed reports of the disease in the northern nations of Africa have been noted.

B. Host Range

Natural infections (Table 1) seem to occur most often in swine, cattle, sheep, dogs, and cats. Wild animals reported to be naturally infected include wild rats and mice,[26] polar and silver foxes,[27] mink and silver blue foxes,[28] red foxes,[29] roe deer,[30] and field hares[31] in Czechoslovakia. Outbreaks of the disease have also been reported in European fur farms, such as silver and blue foxes in Denmark,[32] mink in Belgium,[33] Holland,[34] Greece,[35] Yugoslavia, and Russia.[36] These infections were found to be in association with the feeding of virus-contaminated food.

Experimentally, infections occur in many species of animals and birds.[36] These have also been listed in Table 1. Wright,[37] in 1980, evaluated the role of the raccoon in a pseudorabies enzootic area of Missouri. The raccoons were found to be negative for neutralizing antibody to pseudorabies virus. However, raccoons that were challenged with 10^4 to 10^5 $TCID_{50}$ of virus either died or showed clinical signs, indicating that the raccoon may serve as a short-term reservoir for pseudorabies virus. Man does not appear to be a susceptible species as indicated by Gustafson.[36] Laboratory workers involved with the handling of infected pigs and virus aerosols have remained asymptomatic and without the development of virus-neutralizing antibodies in their serums. Also, in the same report, a disease control officer was bitten by an infected pig with virus in its tonsillar tissue and did not develop virus-neutralizing antibodies or clinical signs. However, a report by Tuncman[38] leaves some doubt. Two laboratory personnel working with infectious pseudorabies virus exhibited pruritus which extended from the hand to the scapular region and lasted approximately 48 hr. Headache, weakness, and an apthous eruption remained for 3 to 4 days following the episode of pruritus. Two rabbits were inoculated with blood samples taken from the workers at the height of the pruritus; both of the rabbits became ill and died within 18 hr postinjection. A second intracerebral injection was performed on another rabbit with brain tissue from the first

Table 1
LIST OF ANIMAL SPECIES INFECTED BY PORCINE HERPESVIRUS 1 (AUJESZKY'S)

Naturally Infectable Species

Domesticated large animals	Companion small animals	Wild animals	Other
Swine	Canine	Rats	Man?
Bovine	Feline	Mice (field)	
Equine		Polar foxes	
Donkey		Silver foxes	
Sheep		Blue foxes	
Goat		Red foxes	
		Mink	
		Roe deer	
		Rabbits	
		Coyote	
		Raccoon	

Experimentally Infectable Species

Laboratory animals	Wild animals	Birds	Other
Mouse (field, gray, white)	Deer	Pigeon	Man?
Guinea pig	Groundhog	Goose	
Gerbil	Jackal	Mallard duck	
Norwegian rat (both white and gray)	Muskrat	Turkey buzzard	
Ferret	Opossum	Sparrow hawk	
Marmoset monkey	Porcupine	Chickens	
Rhesus monkey	Brown bat	Turkey	
	Raccoon		
	Rats		
	Mice		
	Foxes (blue, silver, polar, red)		
	Mink		
	Rabbits		
	Coyote		

Experimentally Refractive Species

Mammals	Poikilotherms	Insects
Ape (Barbary)	Frog	Lice (swine)
Chimpanzee	Toad	Tick (swine)
	Tortoise	
	Snake	

ones and the results were identical. The two workers both recovered uneventfully in a week's time. Therefore, it appears that although chimpanzees and Barbary apes are resistant, at least two other subhuman primates, rhesus monkeys and marmosets, are susceptible. It is, therefore, wise to regard man as having a fairly high threshold for infection via the respiratory, oral mucosa, and skin routes; however, a genuine respect should be maintained when working with the virus.

The disease is of economic importance in the U.S. In 1962 pseudorabies outbreaks occurred in a virulent form in Indiana and appeared to be the focus of infection for the nation. California[22] experienced a large outbreak in 1968 which appeared to be related to the practice of feeding uncooked garbage. An increasing number of cases were noted in 1973 and 1974[36] and have continued to cause problems for the swine industry ever since.

II. CHARACTERISTICS OF THE PORCINE HERPESVIRUS 1 (PHV-1)

A. Evolutionary Origin and Association of PHV-1 to the Family Herpesviridae

Based on the premise that viruses require other living forms to multiply, the current concept of viruses being derived from genetically specific components of eukaryotic cells remains the most plausible.[37] The first eukaryotic cells originated approximately 700 million years ago. By definition, the capability to acquire a nuclear membrane envelope, a characteristic feature of all herpesviruses, could not have existed prior to this cellular modification. The presence of the various herpesviruses in today's species tells us only that a herpesvirus found in a particular species could not have occurred in its present form earlier than when the species evolved.

An early origin of the pig herpesviruses is also suggested by the result of genotypic analysis. The quanine + cytosine content of the DNA of several vertebrate viruses have been examined.[39-41] A wide range of 45 to 74 mol% is indicated by these data and imply a substantial divergence in DNA base composition, thus indicating a wide evolutionary separation among herpesviruses. It is interesting in this regard to note that members of the order Rodentia (mice and rats) can become infected naturally with porcine herpesvirus[26,42,43] and show remarkable similarities to Aujeszky's in pigs with regard to clinical signs, carrier status, localization of virus, and susceptibility.[44] Evolutionary association can be made between the giant rodents of the suborder Simplicidentata (which includes mice and rats), family Dasyproctidae, genus *Coelogenys* and *Agoutis,* and the forerunner of the wild pig, suborder Artiodactyla, families Suidae and Tragulidae (Indian Chevrotain).[45] It is quite possible that porcine herpesvirus 1 has had a long evolutionary association with early members of the rodent family, allowing these viruses to diverge with the subsequent development of the modern porcine species.

To complete the available information concerning the evolution of the porcine herpesvirus, one other study must be mentioned. Subak-Sharpe et al.[46] analyzed the pattern of nearest-neighbor base sequences for herpes simplex virus-1, porcine pseudorabies virus, and equine rhinopneumonitis (type 1) viruses. The patterns differed widely among the three herpes-viruses themselves and also with that found in mammalian cells. These workers concluded that it was unlikely that these herpesviruses originated from mammalian cells.

B. Classification, Biochemical, Biophysical Properties

Aujeszky's disease virus is listed by the Herpesvirus Study Group of the International Committee for the Nomenclature of Viruses as family Herpesviridae, subfamily Alphaherpesviridae, host subfamily porcine (alpha) herpesvirus 1 (PHV-1) — the prototype of which is human herpes-virus 1 (herpes simplex 1). The principle objective of this system is to assign the virus based primarily on its biological properties and, to a lesser degree, on the structure of viral DNAs.

The complete virus particle measures 150 to 180 nm when visualized by negative contrast methods.[47] Component parts include (1) core-containing genome measuring approximately 77.5 nm in diameter, (2) the capsid surrounding the core, consisting of an icosohedron measuring 105 to 110 nm and containing 162 capsomers which measure approximately 12 to 13 nm long by 9 to 10 nm wide with a central hole of 4 nm, and (3) outer membrane (envelope) enclosing the nucleocapsid, measuring approximately 180 nm in diameter with 8 to 10 nm projections. The genome contains a linear, double-stranded DNA molecule of approximately 90×10^6 daltons[48] and a high (73 mol%

quanine/cytosine) base composition.[49] It consists of a terminal sequence (mol wt ≅ 9.9 × 10^6) which is inverted internally at the other side of a short sequence (mol wt ≅ 6.0 × 10^6). The remainder of the molecule forms the so-called "long unique sequence".[50,51] Based on this information the molecular weight for the complete virion is established at 70 × 10^6 daltons or containing approximately 0.12 × 10^{-15} g of DNA. Porcine herpesvirus 1 has a buoyant density of 1.278 g/cm^3.[5] Amino acid composition has not been analyzed. The analysis of viral proteins requires much effort to purify them. Their inherent instability and tendency to aggregate limit a pure quantity for manipulation along with the potential for contamination by cell source proteins. The most effective means of protein analysis has been by polyacrylamide gel electrophoresis (PAGE). The porcine herpesvirus virion contains core, capsid, and envelope proteins. Conclusions reached indicate that there is a major capsid protein (120 × 10^3 daltons or more) and that the envelope proteins include most of the glucosamine residues and mostly lipid types and total about 40% of the proteins of the virion.[52] Recently, however, Ben-Porat et al.[53] have shown the major nonstructural DNA binding proteins (136 K) to be critical for completion of first round DNA synthesis and provide for "stabilization" of progeny viral DNA in the infected cell, thus protecting it from nucleolytic attack.

Since the envelope contains significant lipids, porcine herpesviruses are sensitive to lipid solvents. As examples, ether[54] and fluorocarbons[55] have been found to affect them as well as enzymes such as trypsin, pronase, phospholipase C, and acid and alkaline phosphatases.[56,57] Chemical inactivation occurs with ethidium bromide,[58] nitrous acid,[59] 5-fluorovracil (producing noninfective particles),[60] dithiothicitol (under mild alkaline conditions),[61] urea, and detergents.[36]

Thermostability factors have been investigated for PHV-1. The virus can be stabilized against heat by a variety of agents, including 1 *M* $NaSO_4$ or Na_2PO_4.[62] As examples of heat lability, Kapland and Vatter[54] found that only 28% of the infectivity of the virus in cell culture fluids survived after heating aliquots to 44°C for 5 hr and Huang and Cheng,[62] using cell culture fluids, inactivated the virus in 15 min at 56°C, in 5 min at 70°C, and in 1 min at 100°C. Other biophysical phenomena investigated include a study by Wallis et al.[64] demonstrating that enveloped viruses such as herpesvirus, measles, sindbis, and vesicular stomatitis viruses are photosensitive when the virions are dissociated from the protective effects of organic compounds contained in the virus harvest, while under the same conditions nonenveloped viruses are photoresistant. In this study, use of a 425-nm wavelength monochromatic light was found to be most effective for virus inactivations. It was also reported that PHV-1 becomes markedly photosensitive when replicated in cells pretreated with nontoxic neutral red.[36] Pfefferkorn et al.[65] reported that PHV-1 in cell cultures are killed by exposure to a 15-W germicidal lamp at a distance of 27 cm and can undergo reinactivation by irradiation with a "cool white" fluorescent discharge lamp at a 24-cm distance, at 37°C for a day, and an additional 2 days of incubation in the dark. Radiation sources[66] have been reported to be viricidal to PHV-1. Inactivation of virus in frozen cell culture fluids containing approximately 1.0 × 10^8 $TCID_{50}$ of PHV-1 using a ^{6}Co source required a total dose of 3.6511 x 10^6 rad. A variety of effects of UV radiation on herpesviruses have been adequately reviewed by Gentry and Randall[67] and is suggested for further information regarding this subject.

Agents have been reported to block the adsorption of herpesviruses to cells both in vitro and in vivo. Agar, heparin, and other sulfonated polyanions will block the adsorption of virus to cells, especially in blood or other tissues to be used for virus isolation.[36] Cyaurate or labile carbonyl compounds[68] given orally to mice, rabbits, and rats prevented the initiation of infection. However, infection was successful in these animals at a later time.

C. Tissue Culture Characteristics

The virus was first grown in tissue culture by Traub,[69] who obtained growth in the Maitland-type tissue culture system of rabbit testis, guinea pig testis, and chick embryo. The PHV-1 has been grown in a variety of monolayer cultures of cells: chick embryo fibroblasts,[70] pig kidney,[72,73] lamb kidney,[74] calf kidney,[75] rabbit kidney,[76] dog kidney,[79] cat kidney,[78] monkey kidney,[79,80] human tissue,[81] foal kidney,[82] calf testis,[83] Madison Darby bovine kidney (MDBK) ferret kidney,[84] and in a number of continuous cell lines such as HeLa,[85] BHK 21,[86] Hep 2, and Vero Pk.[87] Not all the tissue cultures exhibited the same sensitivity to the virus. Burrows[84] showed that rabbit, pig, dog, sheep, and ferret kidney cells and the "Stice 2a" line of pig kidney cells were the most sensitive. McFerran[87] found that pig kidney, PK 15, Vero, lamb kidney, and calf testes were also very sensitive.

The time required for absorption seems to vary with the cells used. PHV-1 absorbed 50% in 30 min using rabbit kidney cells, but in chick embryo fibroblasts 86% absorption was obtained by 2 hr and required 5 hr before maximum absorption occurred.[88] In rabbit kidney cells, the eclipse phase of 3 hr is followed by a latent period of 5 hr and an exponential increase phase of 5 hr.[88] In monkey cells, however, the latent and exponential phases of both increased to 12 hr.[79] The virus forms plaques in both chick embryo fibroblasts[70] and in pig kidney cells[72] where a direct linear relationship exists between size of inoculum and the number of plaques produced. Plaque size, however, can be a function of the virus strain and will be discussed in more detail later.

In a number of cell lines infected by PHV-1 a persistent infection has resulted in that line; for example, Beladi[89] found no cytopathic effect following inoculation of the virus onto calf kidney cells, but was able to demonstrate one infected cell per 500 to 1000 was infected. Another report has indicated viral persistence in a kidney pig cell line.[90]

Two types of cytopathic effects occur in tissue culture cells. One type is composed mainly of syncytia and the other of rounded, highly refractile (balloon) cells, with occasional giant cells[88] being formed by the dissolution of opposing cell membranes prior to the rounding of cells as the foci of infection widens. The cytopathic effect may not progress to complete destruction of the cell sheet. Foci of infection with typical cytopathic effect may appear to stop, leaving cell sheets with holes surrounded by infected cells — the margins of infected cells extending only a few cells into the sheet. Infected cells have an eosinophilic homogenous Feulgen-positive intranuclear inclusion body.[36] The inclusion body later shrinks and becomes surrounded by a halo, forming a typical Cowdry type A inclusion. At this stage the inclusion body becomes Feulgen-negative and eosinophilic.

D. Virus Strain Variation

Various viral strains have been isolated and maintained in laboratories based on peculiar characteristics of the virion either in a natural setting or in artificial laboratory conditions. The following is a short discussion of the more common strains being used today. Northern Ireland has four strains characterized: strain NIA-1 is neurotropic, normally causing 5 to 20% deaths in 7-week-old pigs;[91] strain NIA-2 causes a similar mortality but differs from NIA-1 causing striking lung lesions and a severe rhinitis;[92] strain NIA-3 was isolated from a farm where it displayed 13 to 20% mortality in 14 to 20-week-old pigs and 80 to 100% mortality in 7-week-old pigs,[93] and strain NIA-4 which was isolated from the lymph node of a cow during investigation of an outbreak of bovine malignant catarrhal fever.[93] This strain is not pathogenic for any of the animal species so far tested. Other strains of reduced virulence include the K,[56] SUCH-1,[94] and avirulent live virus strains (ALVA).[95] Other virulent strains include: Shope strain, Dekkinge strain (both moderately infectious), and the highly infectious strains (Indiana and Iowa).[95]

Recently, various strains have been passed at high multiplicities of infection (in vitro) to yield populations of virions containing defective DNA, these have been termed Pr 1 and Pr 2 defective-interfering particles (DI). Rixon and Ben-Porat[96] have derived two separate defective virions (based on the degree of sequence reiteration and guanine/ cytosine content) from a single parental stock. These DI particles interfere equally well with the growth of standard virus strains in tissue culture but neither significantly interferes with the absorption or the replication of standard virus, also, neither one causes a significant increase in the degree of breakage and reunion of standard parental DNA strands. However, in cells co-infected with standard virus and either population of defective virions, there is a delay in the maturation of concatemeric nascent forms of virus DNA to unit size molecules, as well as a reduction in the number of virus particles produced compared to cells infected with standard virus alone. It is hoped that these defective particles can help explain the intricate mechanisms controlling the expression of the herpesvirus genome during the normal course of infection in eukaryotic cells.

One last variation of the normal PHV-1 is the development of temperature-sensitive (ts) mutants.[97] These virions are grown in various temperatures and selected based on their partial mechanism of assembly. PHV-1, when grown at nonpermissive temperatures, are blocked in the nucleocapsid formation stage. These empty capsids (devoid of DNA) are composed of three major proteins and are the precursors of normal PHV-1 DNA-containing nucleocapsids. It is again hoped that these ts mutants can shed some light on the intercellular pathways utilized by the herpesvirus in general in the processes of virion production within the host.

III. VIRUS AND HOST'S INTERACTIONS

A. Epizootiology of PHV-1

The epizootiology of PHV-1 is not completely understood. Although swine are considered natural hosts and the principal reservoir of the virus, the mechanism of transmission is not always clear. A number of investigators have suggested the carrier pig as the major source of infection for swine.[36,93,98] The introduction of carriers is probably the most frequent source of infection in swine herds. However, epizootics have occurred on farms closed against the introduction of new stock and have implicated wild mammals or birds as a possible source.

Only limited information exists concerning the pathogenesis of PHV-1 in wild animals and their role in the spread of the virus. The Norway rat (*Rattus norvegicus*) has been suggested as a possible reservoir involving a rat-pig-cattle-rat cycle of infection.[99,100] Carriers within the rat species have been demonstrated[100] and survivors of experimental infection harbored the virus up to 131 days after infection. Also, naturally infected rats can remain carriers for up to 100 days and could transmit the disease to pigs if the infected carcasses were consumed. Laboratory mice were found more resistant than laboratory rats[101] to experimental PHV-1 infection. Although rats and mice (*Mus musculus*) may play some role in the dissemination of the virus, their resistance to infection with PHV-1 would seemingly tend to minimize their importance.

Raccoons (*Procyon lotor*) and opossums (*Didelphis marsupialis*) also have been suggested as possible vectors.[102] Transmission experiments[102,103] dealing with PHV-1 in raccoons indicate that natural infections occur in this species; the disease can be transmitted reciprocally between raccoons and swine by contact and when either consumed infected carrion of the other, and lastly, that horizontal transmission in raccoons does not exist. Thus, it appears from a limited number of investigators and a limited number of wild mammals that virus is present and is released in the saliva and nasal discharges, possibly resulting in a low measure of aerosolation of virus. The contamination of feed by saliva from infected animals and the possibility of swine or other animals consuming their carcasses constitute a potential hazard. Infectability of the virus has been shown to occur from skeletal muscle of a pig carcass after 30 days of storage at 1.5 to 2.0°C.[104] It has also been reported to survive on fomites such as hay, wood, and food for up to 46 days at −8°C and 10 to 30 days at 24°C.[104] In experimental investigations, urine and feces have not been found to contain virus from feral animals live-trapped on farms where pseudorabies has been present in swine.

The relationship of PHV-1 in swine and other domesticated farm species has been investigated. Sheep, goats, cats, and dogs all can be infected by swine shedding virus. All of these species may suffer fatal infections. Reports of recovery, with antibody present that is the result of infection, have not been reported. The exception is cattle, however, where a couple of reported[105,106] recoveries in beef cattle are recognized. In a study by Crandell et al.,[107] horizontal transmission in dairy calves was not established following experimental infection, however, virus could be isolated from the nasal secretions of these animals and one of two pigs co-mingled with these calves did seroconvert. Infections of dogs and cats in all instances have been reported to be caused by close association with infected pigs.[108,109] Exposure occurs by ingestion with infected dead pigs, consumption of infected swine offal,[110] and/or skin lacerations from fighting with infected pigs.[111] Lateral spread among dogs and cats has not been noted, although it seems quite possible because of the presence of the virus in the nasal and oral secretions of infected individuals and the excessive salivation that is part of the syndrome. Lateral spread has been observed in herds of cattle and flocks of sheep, presumably through aerosols and ingestion of contaminated feed. Horses seem to be infected only rarely and appear to be rather resistant to the virus. Their role in the spread of PHV-1 is unknown. It appears that infection among the domestic animals occurs in conjunction with an association to swine. However, examples can be found of infections occurring in which a connection cannot be identified, suggesting the existence of a reservoir other than swine.

Many kinds of birds have been found to be susceptible under experimental conditions (Table 1), but reports of natural infections are lacking. Producers have testified on occasion to finding dead birds during an episode of pseudorabies in their swine, however, scientific data are lacking to support this. Day-old chicks are susceptible to the virus on oral exposure.[111] Seed-eating birds and birds of prey such as the hawk have not been sufficiently investigated to evaluate them as infected shedders or passive vectors of the virus.

B. Epidemiology of PHV-1

The epidemiology aspects of PHV-1 in swine is based on the fact that pigs are considered the natural host and reservoir of the virus. Mortality due to PHV-1 infection is greatest in baby pigs and least in mature swine. The severity of the syndrome is dependent upon the virulence of the viral strain, age of the swine, dose of virus received, and route of exposure. Both lateral and vertical[112] spread of the infection occurs in swine. This fact is quite the exception to most members of the Herpesviridae family.

PHV-1 has become a disease of increasing importance to the swine industry. In 1977, 1256 confirmed cases of the disease were reported in the U.S. by animal health officials. This contrasts with the 714 cases reported 1 year earlier.[113] More recently, Pirtle[114] showed that approximately 10% of the 1246 slaughtered 6-month-old hogs in the U.S. had PHV-1 serum antibody and most of the exposures resulted in subclinical infections.

Commercial production methods influence incidence and severity of disease episodes. Swine assembled from several sources, sorted by size, and redistributed over a period of a few days while some are in incipient stages of the disease, have caused wide geographic distribution of the disease throughout the world. A continuous, open-end operation where the adding of new animals without appropriate isolation measures to replace those sold assures a constant supply of various virus strains onto the premises. During these times swine of all ages can be clinically affected and losses are especially high when nonimmune gilts or sows are exposed to the virus during gestation. Episodes of the disease in these herds, especially those in the close confinement situation of "farrow to finish" operations, are uniformly greater with regard to losses than episodes where pigs are farrowed on pasture and raised in pens separated from each other. Stresses associated with severe weather, transportation, and starvation are believed to be factors adversely affecting the course of disease and resistance to PHV-1.

The carrier state of swine is considered the main source of the virus in the infected herd.[36,93,98] Virus has been recovered in chronic infections in swine over long periods of time. In one study,[115] vaccinated swine surviving natural challenge yielded the virulent strain for as long as 11 months in 2 to 24% of animals tested. In another study,[116] virus was recovered from tonsil explants as long as 6 months after infection was observed. Latent infections in swine remain largely a mystery as to conditions of establishment, rate of occurrence, and conditions for and duration of recrudescence. Through DNA hybridization studies,[117] it was demonstrated that PHV-1 DNA sequences were present in the trigeminal nerve ganglion in 8 of 12 recovered swine from which virus could not be recovered by conventional means. Also, swine vaccinated with modified live virus and exposed to virulent virus at 3 weeks, yielded virus 3 months later after treatment with dexamethasone.[118] To add to the already complex picture of virus-host interaction, PHV-1 can be transmitted in the milk of infected sows,[119] transplacentally, and, in the semen of boars.[112] Finally, Bolin[120] reported that 48 to 72 hr-old embryos exposed to virus (in vitro) could transfer the infection to susceptible sows. Thus, it appears that PHV-1 does have a complicated epidemiology that will have to await further definitive investigations.

C. Immunobiology and Immunopathogenesis

The immunobiology and immunopathogenesis for any viral disease is a very dynamic relationship between many variables of the virion and the host. In this section, two somewhat idealized hypothetical examples will be discussed to permit the reader a greater appreciation of those variables and their consequences. The first hypothetical case will be that of virus-host interaction in a pig having received maternal antibody for PHV-1 previously, and interacting with the virus for the first time. The second example will be that of the neonate, with discussions of its own unique and complex problems in dealing with the virus.

The PHV-1 is generally introduced into the host via the oral-nasal route either by aerosolation (inhalation) or rooting behavior during feeding in virus-contaminated areas. Initially, the success of the virus will be dependent on many factors. Firstly, virulence of the virus for its host is obvious for without this characteristic the hosts response is academic. The concentration of the virulent virus will be dependent for the most part by the host. Virus introduced to the nasopharynx of a normal healthy pig will encounter a complement of respiratory defense mechanisms. The mucociliary escalator is a complex made up of mucoserous glands, goblet cells, ciliated pseudostratified columnar epithelium along with the associated lymphoid cells of the submucosa, lamina propria, and pharyngeal and tubal tonsilar tissue. Lowering of humidity and temperature tend to inhibit this mechanism and, when combined with dehydration of the animal due to lack of water intake, a significant decrease in mucoserous production occurs. When the animal is compromised in such fashion the viruses' chances of undergoing adsorption on to susceptible respiratory epithelial cells is greatly enhanced. It appears that adsorption is dependent on electrostatic function of ionic groups on the exterior of the PHV-1 virions and susceptible host cells.[121,122] This microenvironment is also demonstrated to be dependent on various electrolyte concentrations[121,122] and not so dependent on temperature. However, variations are found with all these parameters depending on the strain of PHV-1 and host cell system used. The mucous secreted is a complex of mucopolysaccharides and polyelectrolytes that help to alter these physiochemical microenvironmental adsorption factors[121] and provide for visco-elastic flow properties inherent with this product. It has been shown in vitro that various concentrations of agar added to susceptible monolayer tissue culture systems prevented the adsorption of PHV-1.[123] It should be pointed out that adsorption of the virus does not imply successful penetration of the virus into the cell. It is at this point that an intact mucocutaneous and mucous epithelial boundary is maintained.

Following adsorption of the virus to susceptible epithelial cell, penetration must occur to insure virus entry and replicative character in the host cell. Penetration, as opposed to adsorption, is a temperature-dependent phenomenon[122] and probably suggests an active enzymatic process. Evidence for the requirement of energy was demonstrated by the inhibition of penetration by the presence of cyanide, which was shown not to prevent adsorption in those studies.[124] During this early stage of virus introduction into the cell it appears that at least two more nonimmune mediated mechanisms may exist to limit or prevent infection. Interferon, a small polypeptide produced by viral infected cells of many types, has been shown to prevent adjacent cell penetration by virus. Also, interferon has been shown to enhance IgE-mediated histamine release of mast cells or basophils,[125] thereby activating local immunological mechanisms through an IgE-mediated inflammatory response. Included in this response are inflammatory mediators affecting vascular endothelium permeability, thus allowing for immunoglobulin transudation into the area, thus blocking adsorption of free virus. It has also been suggested that changes in cyclic nucleotides induced by herpesviruses could influence immune-cell reactivity, as well as interferon.[126] Lodmell and Notkins[127] have proposed that the local control of virus cell-to-cell spread involved specific immune recognition and nonspecific execution by various inflammatory mediators, which could affect virus infected as well as neighboring noninfected cells. Also, the effect of this local inflammatory response results in local hyperthemia which has been demonstrated to increase the survival of herpes simplex virus (HSV)-infected mice or newborn pups infected with canine herpesvirus.[128] As mentioned previously, this hyperthermia may affect the temperature-sensitive phase of penetration needed by PHV-1 and/or boosting of immune responses, e.g., blastogenic hyperresponsiveness.

Following penetration of PHV-1 into either fibroblast or epithelial cells the replicative cycle begins and may take approximately 9 to 15 hr. Once inside the cytoplasmic mileu, the envelope and capsid are disassembled, following which the viral DNA moves into the cell nucleus. It is here that transcription of viral DNA occurs, followed by synthesis of viral proteins and DNA, which ultimately leads to assembly of the virion. For a more detailed morphometric, morphologic, and biochemical scheme of the infective and viral replicative cycles the reader is referred to Ben-Porat/Kaplan[129] and Watson.[130] By 24 hr, infected epithelial cells will express herpesvirus-specific membrane antigens, which, in HSV infections, is the development of an Ig-Fc receptor,[131] although its effect in immune reactivity in this disease is not known. Also, herpesvirus-specific antigens have been shown to complex with histocompatability antigens,[132] thereby influencing the ability of the host cells to recognize a viral-infected cell. It appears that a humoral response in most herpes infections requires cooperation of B-cells, macrophages, and T-cells. Neutralizing antibody (IgM and IgG) can be detected in the serum by day 7,[93] and appear to correlate well with antibodies directed to membrane antigens on herpes-infected cells. These immunoglobulins can lyse herpesvirus-infected cells in combination with either complement, K-cells, and macrophages of polymorphonuclear leukocytes obtained from either immune or nonimmune individuals.

Pigs, however, have different serum immunoglobulin percentages than other species. For example, IgG makes up 80%, IgA about 15%, and IgM about 5%. Failure of the host to respond at this point in the infection may result in virus to absorb and replicate within the olfactory and oronasal epithelium and allow the generation of great amounts of virus to further assure adsorption (within 5 hr)[36] and invasion of the bipolar olfactory cells, transmission in the cytoplasmic extensions to the tufted glomeruli, and movement on to the mitral cells in the olfactory bulb for further dissemination. Also during this same time, nerve endings of the trigeminal, circumvallate papillae and glossopharyngeal nerve of the nasal and oral cavities provide entry points for virus into the gasserian ganglion (pons), medulla, and nucleus solitarius. It appears that PHV-1 thymidine kinase expression is essential for expression of the acute and latent trigeminal ganglion infection.[133] Thus, the virus spreads to the cerebral interpretive centers from three portals of entry. Virus can be found at the level of the pons and medulla by 10 to 18 hr of the experimental inoculation. It is believed that centripidal axoplasmic flow[134] of these nerves is responsible for the virus movement in these cells. Virus dissemination within nervous tissue is rapid, and, by the 7th day after infection sites such as atrioventricular node, celiac, ganglion, stellate ganglion, lumbar and sacral segments of the spinal cord have been found to contain virus.[93] Virus is usually replicating and sheds from oronasal areas for up to 10 days and intermittently for another 7 days.[93] During this time, local lymphatic drainage picks up free virus and transports it to regional lymph nodes of the nose and pharynx.[91] Virus has also been demonstrated within phagocytes and peripheral leukocytes.[135] Dissemination into general circulation[91] via thoracic duct drainage has been documented (on occasion) and it appears to be of minor importance with regard to the neural pathogenesis and is also found to be strain dependent. Once the virus has been incorporated into the axonal processes the immune response will be responsible for the neural pathology seen at this stage. It is at this stage that the animal will either die or become a carrier.

Lastly, the immunobiology of the neonate born to a recovered or vaccinated sow will be considered. The epitheliochorial placentation of the sow interposes a large number of epithelial layers between maternal and fetal circulation, thereby ensuring there is virtually no transfer of immunoglobulin to the fetus. This feature is shared with equines and ruminants, and immune protection in these species is initially determined by antibody passively acquired from maternal colostrum by absorption from the small intestine. The ability to absorb high levels of colostral immunoglobulin is fundamental to survival. Intestinal permeability in the pig is of relatively short duration and most of the passive immunity is acquired in the first few hours of life.[136] The efficiency of absorption decreases very rapidly, having a half-life of about 3 hr.[137]

At present the mechanism responsible for cessation of absorption is unknown, though several factors are thought to be involved. Leece et al.[138] found that efficiency of absorption was inversely proportional to the amount of colostrum ingested, thus in starved pigs maintained by parenteral administration of nutrients, the ability to absorb protein was retained for at least 5 days. Leece[139] carried out further investigations extending the thesis that the cessation of intestinal absorption of large molecules (closure) was a function of the feeding regimen. Closure activity was associated with proteins but also glucose was surprisingly incriminated, and piglets that ate more than 300 meq showed accelerated closure in a period of 18 to 24 hr. Thus, it seems that the pinocytotic capacity of the neonatal gut can be nonspecifically stimulated with loss of availability of surface membrane for invagination, thus quickly exhausting the capacity for intestinal absorption.

Maternal antibody to PHV-1 virus can be transferred in the colostrum of immune sows.[140] The majority of the immunoglobulin type is IgG, however, IgA and IgM are also found. The titer of the antibody is 11 to 16 times higher than that of the dam's serum, but a rapid fall in the titer of the colostral antibody occurs over the first 36 hr[140] as described above. If suckling pigs absorb enough colostrum containing anti-PHV-1 immunoglobulin to obtain a virus neutralizing titer of 1:8 12 hr prior to expo-

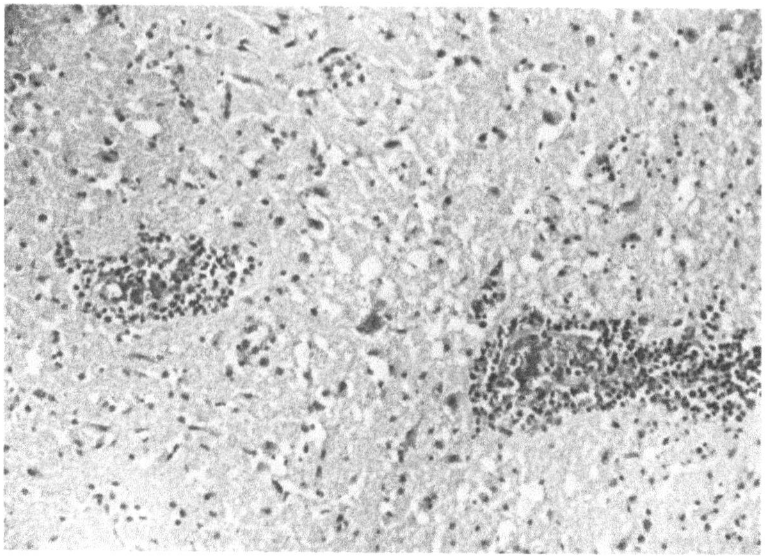

FIGURE 1. Typical histopathological section through the semilunar ganglion of a PHV-1 infected animal. Lymphoplasmacytic perivascular cuffing with multifocal to diffuse neuronal gliosis and necrosis are evident.

sure, they are generally considered resistant against viral challenge but not viral excretion. This kind of titer may last from 10 to 12 days depending on total amount obtained. Early in lactation IgG levels in the sow are somewhat related to increased serum levels and fall precipitously within the first few days, after which local synthesis in the mammary tissue takes over. It is at this time that IgA is the predominant immunoglobulin for the remainder of the lactation. Thus, it appears that the young pig can be very susceptible or very resistant, depending on the maternal immunological inheritance it receives. This fact is of prime importance when applied to production situations where a vaccination program will allow for priming of pregnant sows prior to lactation.

D. Gross Pathology

The most striking findings at post-mortem may be the occasional lack of gross changes in the affected pig. If animals are examined during the period when the CNS symptoms are becoming dominant, there is marked congestion of the meninges, especially over the cerebral hemispheres, accompanied by an excess of cerebrospinal fluid. If the animals are terminal, these features are less likely to be observed. Small hemorrhages and mild congestion in several or many widely separated lymph nodes are regular findings. Petechiation of renal papillae and cortex is often seen in severely affected swine and less so among those in which the disease may not be fatal. A common finding in severely affected swine is congestion of the nasal mucosa and pharynx. Pulmonary edema may be expected in such cases along with necrotic tonsillitis, pharyngitis, tracheitis, and sometimes esophagitis. In cases complicated by bacterial problems, one may anticipate findings related to their presence such as hemorrhagic enteritis or other gastrointestinal changes that are not to be expected in PrV infections. Occasionally, on opening the body cavities, the viscera may be slightly congested in general.

E. Histopathology

The histopathologic discussion of this disease is predominantly centered around the CNS pathology. It consists of a diffuse, nonsuppurative meningoencephalomyelitis and ganglioneuritis. There is marked perivascular cuffing and diffuse and focal gliosis associated with extensive neuronal and glial necrosis (Figure 1).

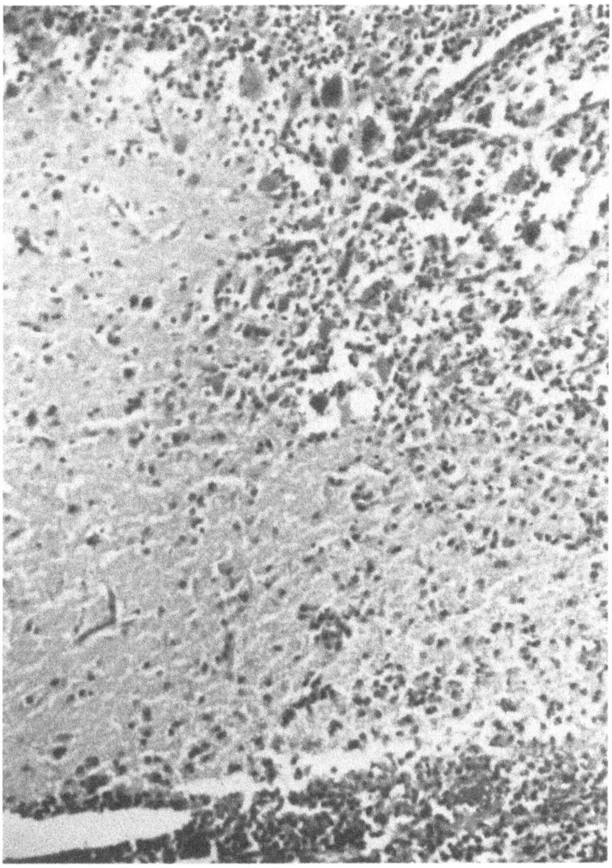

FIGURE 2. Prominent perivascular cuffs with evidence of neuronophagia, perineuronal gliosis, and astrocytosis in the dorsal root ganglion of an animal infected naturally with PHV-1.

Examination of CNS tissue obtained from various naturally occurring cases suggests that correlation between severity of signs and extent of lesions is difficult. The cerebrum is involved in all cases, with gray and white matter being essentially equally affected. Lesions are most pronounced in the frontal and temporal sections. Although the white matter is involved, lesions in the brain stem are most prominent in the areas of the various ganglia. Generally there is less extensive involvement of the medulla and cerebellum than of the anterior regions of the cerebellum.

Lesions are usually present in the spinal cord and spinal ganglia only when there is marked reaction in anterior sections. White and gray columns are approximately equally involved, with no discernible pattern of distribution. Semilunar ganglioneuritis may be seen in some individuals with extensive encephalomyelitis.

Throughout the CNS, meningitis is to be found adjacent to parenchymal lesions. In the cerebellum the meningitis may be the most striking feature. When spinal meninges are involved, cellular accumulations are most noticeable about the roots of spinal nerves.

In general, the severity of changes at the cellular level correlates with the extent of distribution of lesions. Neuronal necrosis is always present and usually extensive. It is accompanied by neuronophagia, perineuronal gliosis, and perivascular cuffing (Figure 2). The neuronal changes and gliosis are usually focal and widely disseminated, although in some there may be diffuse neuronal necrosis, perineuronal and perivascular edema and glial proliferation. In these cases there is prominent glial cell degeneration and necrosis as well.

Perivascular cuffs are of variable thickness; some may be composed of as many as eight layers of cells. The cuffs are made up mostly of small mononuclear cells of the lymphoid series. Most appear to be infiltrative, yet occasional mitotic figures may be found. In all cuffs, pleomorphism is due to the presence of a few neutrophils and occasionally eosinophils and macrophages. In severely damaged tissues the presence of granulocytes and macrophages is increased. However, there are sufficient degenerative and necrotic lymphoid and glial elements to create an appearance of even greater pleocellularity through karyorrhexis, resulting in the presence of nuclear debris. In most cases the cuffed vessels are patent, and although endothelial cells may be swollen or rather prominent, there does not appear to be proliferation or necrosis. In less cellular perivascular cuffs there may be some edema and slight extravasation of erythrocytes.

Cellular infiltration of the meninges is composed of cells comparable to those in perivascular cuffs. Most cells are lymphocytes having vesicular nuclei, the remainder are neutrophils, eosinophils, and macrophages. In some instances the infiltrating cells may extend directly from the cortical parenchyma into the meninges in cerebral or cerebellar sections. More commonly, the cells of perivascular cuffs extend directly into the meningeal areas.

Reaction in the gasserian and spinal ganglia is similar to that seen in the brain and cord. Cellular infiltration centers about capillaries next to necrotic ganglion cells and extends along epineural planes throughout the ganglia and into nerve bundles.

Meningitis may be observed to extend along nerve to the sclera. Usually, the intraocular changes are slight with mild lymphoreticular proliferation in the adventitia of retinal veins and mild gliosis associated with neuronal degeneration in the ganglion cell layer.

Nuclear inclusion bodies in glial cells are not often found readily in animals submitted from the field. Most are found between 4 and 11 days after infection and are associated with neurones, astrocytes, and oligodendrogia cells of the cerebral cortex and subcortical white matter.

In the lymph nodes there is commonly hyperplasia of lymphoid elements and hemorrhage of varying degree, especially in the peripheral cell-poor zones and sinusoids. Occasionally hemorrhage is associated with coagulation necrosis and neutrophilic infiltrations. Also, necrotic arteritis, phlebitis, and lymphangitis accompanied by thrombosis, are present in the capsule and adventitia. Intranuclear inclusions can be found in the large reticular cells adjacent to hyperplastic germinal centers or in the sinusoidal areas of the node. The inclusions are large, irregular, slightly eosinophilic masses of protein separated from the nuclear membranes by a clear or vesicular halo.

Severe lesions of the mucous membranes lining the nasal cavity and pharynx have been found in pigs of various ages that died of PHV-1 infection. Superficial or deep necrosis of epithelial cells in small foci or large areas can be observed. There is syncytial formation and the presence of large numbers of cells with intranuclear inclusions. Also, a marked lymphocytic infiltration accompanied by neutrophils and macrophages are found in these necrotic areas.

The lungs are regularly congested in this disease and characterized by alveolar filling with proteinaceous edema fluids; proliferation of reticuloendothelial cells; and rarely, focal necrosis of alveolar septums. Reactive changes in the lungs are, however, of uncertain origin in some cases due to the frequency of porcine viral pheumonia[141] and bacterial bronchopneumonia present in this species. Also, necrosis of the pulmonary intramural ganglion cells of the parabronchiolar innervation as well as necrosis and karyorrhexis of nerve cells of the stellate ganglion, celiac plexus, and mesenteric plexus of the autonomic nervous system have been reported.[142] Occasionally, necrotic foci may be found with attendant inflammatory cells in the parenchyma of liver, cortex of the spleen, and renal papillae.

IV. CLINICAL ASPECTS OF THE DISEASE

A. Clinical Signs

The clinical response to PHV-1 can be, for the most part, a predictable series of events, however, variations do exist as with any biological system. The disease in mature swine has a low rate of mortality (2%) associated with it and, in most instances, losses are not experienced. Sneezing and coughing are usually the earliest signs but are usually passed off by the producer as a mild and transient "cold" in the pigs. Soon after this (30 to 48 hr) a fever (103 to 107°F) will be present along with partial to complete anorexia. This is usually followed by decreased daily activities by the 3rd day following exposure to the virus. Also at this time, stool characteristics will change from a pasty consistency to a firmer and dryer character, with overt constipation characterizing this stage of the disease. Excessive salivation and vomiting ensue by the 4th to 5th day. Lastly, neurologic involvement will be characterized by more pronounced tremors with periods of incoordination, especially in the pelvic limbs. Tonoclonic spasms of the pelvic and thoracic limbs will make the affected pig seem to bounce. At this same time, convulsions will be characterized by raising of the head with the nares being retracted, head trembles, and staring eyes, the back is arched, and pile erection of the hair over the entire back and tail area ensue. As the animal loses its balance, it usually pivots in a tight circle and falls, whereupon stiffness of the limbs in tonic spasm becomes more apparent. This is followed by paddling movements of the limbs. During the convulsion there is excessive salivation from the half-opened mouth and may last approximately 45 sec. The animal will then usually recover early in the neuropathy and walk off. As many as three such convulsions may occur in a 10 min interval. By the 6th day the neuropathy deepens, whereupon the affected animals are prostrate and the limbs are stiffened in the extended position. The animals cannot stand and become moribund and die within the subsequent 12 hr from the onset of CNS signs. Occasionally, but rarely, some animals may become prostrate, cachetic, and convulsive but yet recover after a 5- to 6-week period, however, they never regain their prior health status. The total length of the infection will vary again depending on many factors but may average 11 days in most cases.

Pregnant swine that become infected will show signs as described previously, however, the involvement of the unborn in the uterus will be considered briefly. PHV-1 can and will cross the placental barrier and infect the developing embryos. If this occurs before or around 30 days of gestation, embryos are likely to be resorbed and the pregnancy terminated. At approximately 40 days of gestation the embryo becomes a fetus with respect to completed development of its skeletal system and infection here will result in death of all or some fetuses. This will be characterized by premature expulsion of the fetus if infection occurs between the 40th to 80th day of gestation. Sows infected in their last trimester usually go to term and will deliver macerated, stillborn, or weak piglets. Occasionally prolonged farrowing dates will be associated with the delivery of similar fetus 2 to 3 weeks later. If sows are infected very close to farrowing the pigs may be farrowed normally but become exposed to virus present in the sow's milk. The most economically important aspect of infections of the sow during gestation is a 20% infertility index at the next breeding period.

Suckling pigs born to infected susceptible sows will exhibit the greatest mortality (40 to 60%) of any age group. Baby pigs in this situation will show some clinical signs by 36 hr. Initially the piglets will vomit, show dyspnea and diarrhea, followed by depression, anorexia, trembling, incoordination, spasms of opisthotonus, and prostration. Death occurs within an additional 36 hr. Some pigs can only move backwards, or circle, and some in recumbency keep moving their legs in movements simulating a trotting gait. Temperatures rarely exceed 105.5°F during the maximum reaction phase of the syndrome and terminally decline to less than 100°F. Total leukocyte numbers remain in the normal range in nearly every case.

B. Diagnosis

Porcine herpesvirus 1 outbreaks are usually suspected from the clinical signs displayed by the affected swine as previously mentioned. Nervous signs of muscular tremors, incoordination, and epileptiform convulsions are almost invariable present, though these may not be seen unless the pigs are forced to move. General depression and high pyrexia are constant features, and sneezing and dyspnea are outstanding signs with some strains of virus. Outbreaks of abortion or a high incidence of stillbirths in a herd, especially if associated with illness in younger animals simultaneously, should always be regarded as being suspicious of PHV-1 until proven otherwise. Reinforcement may be gained from the history of the episode and the reaction of cats and/or dogs on the premises, that being a uniform fatality in these species within 3 days.[143] It should also be remembered that PHV-1 can and does occur concurrently with other conditions which renders diagnosis difficult; such an example is influenza A infection,[141] and other secondary bacterial infections of the upper respiratory tract.

Diseases which should be considered in the differential diagnosis are swine fever, foot and mouth disease, hemagglutinating encephalitis virus infection,[142] salt poisoning, pasteurellosis, Talfan/Teschen disease,[144] bacterial meningitis, and coliform septicemia. Since the aforementioned diseases may present a complex clinical picture, the application of laboratory techniques is essential for a conclusive etiologic diagnosis.

As mentioned previously, histologic examination of the brain and spinal cord is one of the most important and sometimes practical methods of diagnosing PHV-1. The presence of intranuclear inclusion bodies enable it to be differentiated from other diseases of the CNS or the pig. Also, the presence of the characteristic necrotizing lesions with inclusion body formation in the upper respiratory tract and lungs is specific for PHV-1.

A remarkable array of procedures is available to demonstrate the presence of PHV-1 or antibodies in the tissues or serum of infected swine. Those of lesser specificity and sensitivity are often used to screen large numbers of samples or individuals so that the expenses of application of more sophisticated techniques can be spared.

1. Virus Isolation

Diagnosis based on virus isolation or demonstration of presence of the virus in tissues requires, under optimum conditions, that fresh tissues be handled with care and precision. The tissue of choice for these purposes are tonsils, brain stem, and cerebrum, in that order. Tissues to be used for virus isolation may be frozen prior use. Tissues suspected of containing PHV-1 may be shown to contain virus with a variety of techniques in vitro. Extracts of tissues are placed in primary or subcultured cell cultures and incubated at 34 to 37°C for 4 to 7 days. Although a wide range of sources has been used, the most frequently employed cells are from swine kidneys. Explant techniques may be employed by themselves, with the cytopathic indicator effect occurring in the outgrowth, or by placing the explant on an indicator cell sheet. The tissues may be minced and co-cultivated with an indicator cell monolayer. In the latter two cases the cytopathic effect (CPE) is observed in the indicator cells. Fluids harvested from cultures may be identified through the virus neutralizing test, although the CPE of PHV-1 is typical enough for presumptive identification.

2. Virus Neutralization

A procedure has been approved for in vitro virus neutralization (VN) tests.[145] VN is used in official tests by both state and federal agencies in the U.S. It is very sensitive, but not the most sensitive of those available. Controversy is often aroused over threshold values. The procedure requires serum from the animal at test to be sent to a laboratory. Care must be taken in this step to assure maximum results and thus the following reminder is included: blood samples from swine should be handled with much care; the erythrocytes lyse readily, especially in cold weather. Serum samples used for VN tests are often toxic for the indicator cells. Toxicity is the result of unknown factors, coated tubes, ion-releasing stoppers, residues from syringes rinsed with antiseptics and detergents, and bacteria. Blood samples should be allowed to clot at room temperature long enough for the serum to begin to separate from the clot before refrigerating to prevent bacterial proliferation and to assist the clot to retract.

The time for this procedure usually involves 1 week. It is estimated that 0.5 to 1% of serum samples (higher in summer) are toxic for the indicator cells, making it necessary for the sample to be obtained again. The expense of time and effort in this test make it less attractive for routine diagnosis.

3. *Immunofluorescence*

The science of fluorescent demonstration of viral antigen in tissue or cell monolayer has been reviewed.[146] A procedure for fluorescent antibody (FA) detection of PHV-1 antigens in swine tissues has been recommended for use in diagnostic efforts.[142] The indirect test may be used to resolve indefinite reactions and to specify a swine herpesvirus in species other than swine.

Samples for FA studies should be chilled to prevent autolysis, but freezing disrupts cellular integrity and renders preparations unsatisfactory. The direct FA test has the advantage of being rapid and simple, as it can be completed in about 1 hr. It lacks the specificity of the VN test, but it is at least as sensitive as virus isolation tests. However, the procedure does not lend itself to large-scale sample interpretation as well as the VN test.

4. *Enzyme-Linked Immunosorbent Assay (ELISA)*

This test is one of the most sensitive, economical, and practical of the methods described. In today's technology ELISA tests are being put into long shelflife kits that can be done right in the office setting. The application of this test has been reported by others.[146-148] The test relies on an enzyme as an immunoglobulin marker. Antigen (PHV-1 in this case) is adhered to a well in a microtiter plate and serum is then added. If there is antibody in the serum, it will react with the coated antigen. After rinsing off excess antigen, antiswine globulin labeled with alkaline phosphatase or perovidase is added. The amount of antiimmunoglobulin bound is determined by adding substrate that will be converted by the enzyme-labeled antibody to a visible color. The amount of color is read and interpreted. This is a simple, reliable, and reproducible test for anti-PHV-1 antibody (to the nanogram level), therefore making it more sensitive than VN tests. Results can be obtained in about 6 hr and the test requires only small amounts of serum (<0.5 mℓ).

5. *Agar Gel Immunodiffusion*

The development and evaluation of an immunodiffusion test for PHV-1 antibody has been reported.[149] The test is simple, results can be obtained in 24 to 48 hr, and it is economical to perform. The test can assay cytotoxic, hemolyzed, or contaminated serums and requires modest expertise to conduct. The serious handicap is that it has been found to be less sensitive than the VN or ELISA tests.

6. *Complement Fixation*

The complement-fixation (CF) test has been compared to a modified VN test, in which serum and PHV-1 are preincubated for 24 hr, and to the conventional VN test.[150] The CF test was consistent with VN tests, although about 5% of the CF tests could not be read because of CF with control antigen. Tests for antibody in heat-inactivated swine serum by CF requires supplementation of the serum with unheated normal serum or porcine complement component C18. In lower dilutions most porcine immune sera must be treated with mercaptoethanol. The CF test requires about 24 hr and considerable expertise. The reagents used require special handling and monitoring.

7. Skin Test

The skin test based on cell-mediated immunity to PVH-1 was investigated and the technique presented by Scherba et al.[151] It is not nearly as sensitive as the VN or ELISA tests. However, it does have value and some attractive uses as a herd-screening test. It is reliable if interpreted as positive only when the reaction is clear-cut. It can be conducted on swine in the field with results available beginning at about 16 hr, with maximum reactivity at around 24 hr. The test is inexpensive, rapid, reproducible, and not difficult to conduct. Its use reduces the number of serum samples needed for diagnosis, for only those negative to the skin test, according to the author, need further testing to determine their immune status.

8. Indirect Hemagglutination

The technique for indirect hemagglutination (IHA) has been reported by Palmer et al.[152] The IHA test is at least as sensitive as the VN test and it is accurate. It will detect early antibody formation in contrast to the VN test. Cytotoxic serum samples can be resolved in the IHA test; the results are available in 2 hr and it is inexpensive to perform.

9. Biologic Tests

If facilities are not available for laboratory tests on swine, laboratory animals may be used with reasonable expectation for sensitivity. Ten percent suspensions of tissue are prepared in saline, and 1- or 2-m*l* amounts are injected subcutaneously into rabbits.[9] Rabbits may be expected to die within 48 to 72 hr, but it sometimes requires 4 to 5 days. One to four-week-old mice[153] may be employed as an aid in diagnosis also. Inoculated subcutaneously with 1.0 m*l* of the supernatant fluid from a 10% suspension of tissue specimen, the mice may be expected to die in 2 to 10 days, but most die between 3 and 5 days. The mice make grooming movements about their mouths with their forepaws, or they remain relatively motionless, interrupted occasionally by clonic spasms that appear to involve all voluntary muscles. Also, PHV-1 can be cultivated in embryonating chicken eggs[97] and seems to provide a convincing means of effecting a diagnosis from swine tissue. Plaques appear on the chorioallantoic membrane about 4 days after exposure. This occurs prior to invasion of the CNS of the embryo by the virus.

C. Treatment

Once an infection has been initiated, little can be done to alter the course of the incipient syndrome. However, some modalities of treatment have been reported.[154] Anti-PHV-1 serum given intraperitoneally as long as 12 hr following exposure was clearly beneficial in weanlings receiving virulent virus. Thirty milliliters of antiserum having 5×10^2 neutralizing capacity against 300 plaque-forming units of PrV in pigs up to 50 lb is an effective dose under similar conditions. Field studies by Crandell et al.[155] have confirmed its usefulness. It seems that the protective halflife of homologous serum is greater than that produced in a heterologous host.

A final treatment modality, of speculation, may be the use of antiviral compounds.

It appears that nucleoside analogs appear to hold the greatest promise for efficacy in ameliorating PHV-1 infections in baby pigs. There is significant activity in research on the value of such chemicals in herpesvirus infections. Some examples of antiherpes drugs include idoxuridine, cytosine arabinoside, and trifluorothymidine, however, only one of these drugs has been investigated with respect to PHV-1. Kolb et al.[156] investigated the use of idoxuridine on PHV-1 in rabbits. Following intracorneal administration of PHV-1 virus, the application of idoxuridine increased the incubation period of the disease, but did not influence the morbidity or mortality. One final antiviral treatment protocol involved the use of interferon in PHV-1 disease. Wawrzkiewcz[157] found that the virus showed little sensitivity to interferon and concluded that it was unlikely to be of value in the control of the disease. Preobranzhenskaya et al.[158] found that virus propagated in chick embryo fibroblasts were susceptible to interferon produced by Newcastle disease virus in chick embryo allantoic fluid. There is hope that these compounds and others to be developed, while not preventing infection, may hopefully serve to reduce the severity of disease, thus avoiding some of the vast economic losses incurred by the swine industry during outbreaks of this disease.

D. Prevention and Control

The epidemiology of Aujeszky's disease, as mentioned earlier, depends upon several factors, one of the most important being the number of pigs on the farm. In areas where small farms predominate the disease appears suddenly and involves the majority of susceptible pigs, usually causing the highest mortality in the piglets under 6 weeks of age. In these circumstances the disease also causes abortion and stillbirths in pregnant sows. Akkermans[159] suggested that duration of disease on a single farm is approximately 1 month in duration. Once an outbreak has passed there will not be a subsequent outbreak of the disease on the premises for many years. This is predominantly due to two reasons. One is that resident farrowing sows acquire immunity from active infections and passively transfer it to the next replacement generations. The other factor being the variation in the viral strains encountered by the pigs and their level of susceptibility. Under these conditions the use of vaccines is not entirely justified. However, the trend today is toward the larger production facilities, such as is exemplified by the poultry industry. This environment now permits the influx of greater numbers of susceptible pigs into the units, thereby setting up the production facility for disastrous losses should PHV-1 be introduced. On these types of farms the disease becomes endemic and, if economical pig raising is to be maintained, vaccination is essential.

Management schemes should include isolation of all recent additions to the herd until freedom from infection is ascertained through serologic testing prior to the addition and/or a test made 2 weeks after arrival on the premises. Eliminate contact access to the herd by human and fomite traffic that has not been thoughtfully evaluated. As much as is economically feasible, service each group of animals with utensils private to that group. Prevent attraction of scavenger feral animals to the premises as much as possible and avoid direct contact between them and the swine.

Antigens applied to swine for the development of resistance to infection have been produced by a variety of techniques. Vaccination in endemic areas of Europe has been practiced for more than 15 years, and the necessity for it continues. Losses have been reduced by vaccination, but the virus is thereby maintained in the environment. A vaccine has not been produced that will prevent infection. The vaccinated animal that is exposed to virulent virus survives the infection and then sheds virulent virus for awhile. Then, more likely than not, it harbors a latent infection that may exacerbate into shedding again.

Live attenuated and inactivated vaccines are available commercially in the U.S. These have been approved as safe and efficacious by the appropriate agency of the

USDA. Directions for their use are presented in the package insert and should be followed.

REFERENCES

1. Hanson, R. P., The history of pseudorabies in the United States, *J. Am. Vet. Med. Assoc.*, 124, 259, 1954.
2. Aujeszky, A., Ueber eine neue Infektions-Krankheit bei Haustieren, *Zentralbl. Bakteriol.*, 32, 353, 1902.
3. Marek, J., Klinische Mittheilungen, *Z. Tiermed.*, 8, 389, 1904.
4. Schmiedhoffer, J., Beitrage zur Pathologie der Infektiosen Bulbarparalyse (Aujeszksche Krankheit), *Z. Infektionskr. Haust.*, 8, 385, 1910.
5. Shope, R. E., An experimental study of "mad itch" with special reference to its relationship to pseudorabies, *J. Exp. Med.*, 54, 233, 1931.
6. Bang, D., Pseudowat (Akute infektiose Bulbarparalyse) beim Rinde in Danemark, *Acta Pathol. Microbiol. Scand.*, Suppl. 11, 180, 1932.
7. Burggraaf, A. and Lourens, L. F. D. E., Infectieuse bulbair-paralyse, (Zeikte van Aujeszky), *Tijdschr. Diergeneeskd.*, 59, 981, 1932.
8. Steiner, A. and Lopez, C., La maladie d'Aujeszky en Espangne, *Rev. Gen. Med. Vet.*, 44, 257, 1935.
9. Koves, J. and Hirt, G., Uber die Aujeszkysche Krankheit der Schweine, *Arch. Wiss. Prakt. Tierheilk*, 68, 1, 1934.
10. Masic, M., Untersuchungen uber die Aujeszkysche krankheit in Jugoslawien, *Zentrabl. Bakteriol. Parasitenkde. I*, 179, 383, 1961.
11. Bartosz, B., Aujeszky's disease in the Dobiegniewo district of Poland during 1958-60, *Med. Wet.*, 18, 393, 1961.
12. Bendinger, G., Zur Oathologie der Ferkelencephalitiden, *Arch. Wiss. Prakt. Tierheilk*, 70, 427, 1936.
13. Braga, A. and Faria, A., Paralysia bulbar infectuosa, *Rev.-Zootech. Vet.*, 18, 149, 1932.
14. Hipolito, O., Lamas da Silva, J. M., Batista, J. A., and Nascimento, L. S., Aujeszky's disease in pigs, in Brazil, *Args Esc. Vet. Minas Gerais*, 13, 61, plate 1, 1962.
15. Lamont, H. G. and Shanks, P. L., An outbreak of Aujeszky's disease amongst pigs, *Vet. Rec.*, 51, 1407, 1939.
16. Gordon, W. A. M. and Luke, D., Aukeszky's disease in Northern Ireland, *Vet. Rec.*, 64, 81, 1952.
17. McFerran, J. B. and Dow, C., The excretion of Aujeszky's disease virus by experimentally infected pigs, *Res. Vet. Sci.*, 5, 405, 1964.
18. McErlean, B. A., An outbreak of Aujeszky's disease in piglets, *Ir. Vet. J.*, 14, 160, 1960.
19. Johnston, J. B., Wittrick, D. A., Roberts, H. E., and Done, J. T., An outbreak of Aujeszky's disease, *Vet. Rec.*, 73, 818, 1961.
20. Mackey, R. R., Done, J. T., and Burrows, R., An outbreak of Aujeszky's disease in pigs in Lincolnshire, *Vet. Rec.*, 74, 669, 1962.
21. Shahan, M. J., Knudson, R. L., Seibold, H. R., and Dale, C. N., Aujeszky's disease (pseudorabies). A review, with notes on two strains of the virus, *N. Am. Vet.*, 28, 511, 1947.
22. Howarth, J. A. and DePaoli, A., An enzootic of pseudorabies in swine in California, *J. Am. Vet. Med. Assoc.*, 152, 1114, 1968.
23. Burgess, G. W., Stevenson, B. J., and Buddle, J. R., Demonstration of a herpesvirus from piglets with lesions of Aujeszky's disease in New Zealand, *N. Z. Vet. J.*, 24, 214, 1976.
24. Durham, P. J. K. and Ohara, P. J., A survey of antibodies to Aujeszky's disease virus in pigs, *N. Z. Vet. J.*, 28, 179, 1980.
25. Lin, S. C., Tung, M. C., Liu, C. I., Chang, C. F., Huang, W. C., and Cheng, C. W., An outbreak of pseudorabies in swine in Pintung, *Chin. J. Microbiol.*, 5, 56, 1972.
26. Lukashev, I. I. and Rotov, V. I., Meterialy Kepizootologii bolezni Aujeszky, *Sov. Vet.*, 7, 51, 1939.
27. Ljubashenko, S. J., Tjulpanova, A. F., and Grishin, V. M., Aujeszky's disease in minks, polar foxes and silver foxes, *Veterinariya*, 35, 37, 1960.
28. Von Hartung, J. and Fritsch, W., Augeszkysche Krankheit bei Nerz und Fuchs, *Monatsh. Veterinaermed.*, 19, 422, 1964.
29. Bitsch, V. and Munch, B., On pseudorabies in carnivores in Denmark. I. The red fox (*Vulpes vulpes*), *Acta Vet. Scand.*, 12, 274, 1971.
30. Nitolitsch, M., Aujeszky's disease in roe deer in Yugoslavia, *Wein. Tieraerztl. Monatsschr.*, 41, 603, 1954.

31. Grunert, Z. and Skoda, R., Epidemiology of Aujeszky's disease in Czechoslovakia. I. History and geographical distribution, *Vet. Med.*, 9, 351, 1964.
32. Bitsch, V. and Knox, B., On pseudorabies in carnivores in Denmark. II. The blue fox (*Alopex Lagopus*), *Acta Vet. Scand.*, 12, 274, 1971.
33. Geurden, L. M. G., DeVos, A., Viaene, N., and Staelens, M., Aujeszky's disease in mink, *Vlaams Diergeneesk Tijdschr.*, 32, 36, 1963.
34. Terpsta, J. I., Akkermans, J. P. W. M., and Duwerkerk, H., Notes on post-mortem examinations done at the Central Veterinary Institute, Rotterdam during 1961, *Tijdschr. Diergeneesk*, 87, 1246, 1962.
35. Christodoulou, T., Tsiroyiannis, E., Papadopoulos, O., and Tsaargaris, T., An outbreak of Aujeszky's disease in minks, *Cornell Vet.*, 60, 65, 1970.
36. Gustafson, D. P., *Pseudorabies in Diseases of Swine*, 5th ed., Dunne, H. W. and Leman, A. E., Eds., Iowa State University Press, Ames, Iowa, 1975, 391.
37. Wright, S. C. and Thawely, D. G., Role of the raccoon in the transmission of pseudorabies: a field and laboratory investigation, *Am. J. Vet. Res.*, 41, 581, 1980.
38. Tuncman, Z. M., La Maladie d'Aujeszky observee chez Dhomme, *Ann. Inst. Pasteur Lille*, 60, 95, 1938.
39. Plummer, G., Serological comparison of the herpes virus, *Br. J. Exp. Pathol.*, 45, 135, 1967; comparative virology of the herpes group, *Prog. Med. Virol.*, 9, 302, 1967.
40. Bachenheimer, S. L., Kieff, E. O., Lee, L. F., and Roizman, R., in *Oncogenesis and Herpesviruses*, Biggs, P. M., De The', G., and Payne, L. N., Eds., Intl. Agency Res. Cancer, Lyon, 1972, 74.
41. Ludwig, H., Untersuchungen am genetischen Material von Herpesviren. II. Genetische Verwentschaft verschiedner herpesviren, *Med. Microbiol. Immunol.*, 157, 212, 1972.
42. Hutyra, F., Beitraq zur Atiologie der infektiosen Bulbarparalyse, *Berl. Muench. Tieraerztl. Wochensehr.*, 26, 149, 1910.
43. Lamont, H. G., Observations on Aujheszky's disease in Northern Ireland, *Vet. Rec.*, 51, 1, 1947.
44. McFerran, J. B. and Dow, C., Experimental Aujeszky's disease, *Br. Vet. J.*, 126, 173, 1970.
45. Beddard, F. E., *Mammalia*, Macmillan, New York, 1909, chaps. 11 and 15.
46. Subak-Sharpe, J. H., Buck, R. R., Crawford, L. V., Morrison, J. M., Hay, J., and Keir, H. M., *Cold Spring Harbor Symp. Quant. Biol.*, 31, 737, 1966.
47. Reissig, M. and Kaplan, A. J., The morphology of non-infective pseudorabies virus produced by cells treated with 5-fluorouracil, *Virology*, 16, 1, 1962.
48. Rubenstein, A. S. and Kaplan, A. S., Electron microscopic studies of the DNA of defective and standard pseudorabies virions, *Virology*, 16, 1, 1962.
49. Ben-Porat, T. and Kaplan, A. S., The chemical composition of herpes simplex and pseudorabies viruses, *Virology*, 16, 261, 1962.
50. Stevely, W. S., Inverted repetition in the chromosomes of pseudorabies virus, *J. Virol.*, 22, 232, 1977.
51. Ben-Porat, T., Rixon, F. J., and Blankenship, M., Analysis of the structure of the genome of pseudorabies virus, *Virology*, 95, 282, 1979.
52. Ben-Porat, T. and Kaplan, A. S., Studies on the biogenesis of herpesvirus envelope, *Nature (London)*, 235, 165, 1972.
53. Ben-Porat, T., Veach, R. A., and Hartmut, H., Functions of the major nonstructural DNA binding protein of a herpes virus (Pseudorabies), *Virology*, 124, 411, 1983.
54. Kaplan, A. S. and Vatter, A. E., A comparison of herpes simplex and pseudorabies viruses, *Virology*, 7, 734, 1959.
55. Ivanicova, S., Inactivation of Aujeszky's disease (pseudorabies) virus by fluorocarbon, *Acta Virol.*, 5, 328, 1961.
56. Bartha, A., Experimental reduction in virulence of Aujeszky's disease virus, *Magy. Allatorv. Lapja*, 10, 42, 1961.
57. Bartha, A., Belak, S., and Benred, J., Trypsin and heat-resistance of some stains of the herpes virus group, *Acta Vet. Hung.*, 19, 97, 1969.
58. Sun, I. L., Gustafson, D. P., and Scherba, G., Comparison of pseudorabies virus inactivated by bromo-ethylene-imine, ^{60}Co irradiation and acridine dye in immune assay systems, *J. Clin. Microbiol.*, 8, 604, 1978.
59. Ivanicova, S., Skoda, R., Mayer, V., and Sokol, F., Inactivation of Aujeszky's disease (pseudorabies) virus by nitrous acid, *Acta Virol.*, 7, 7, 1963.
60. Reissig, M. and Kaplan, A. S., The morphology of non-infective pseudorabies virus produced by cells treated with 5-fluorouracil, *Virology*, 16, 1, 1962.
61. Gainer, J. H., Long, J., Jr., Hill, P., and Capps, W. I., Inactivation of the pseudorabies virus by dithiothreitol, *Virology*, 45, 91, 1971.
62. Wallis, C. and Melnick, J. L., Thermostabilization and thermosensitization of herpesviruses, *J. Bacteriol.*, 90, 1632, 1965.
63. Huang, Shou-sen and Cheng, Yu-chuan, Aujeszky's disease in cattle. I. Isolation of the virus and some of its biological properties, *Proc. Soc. Exp. Biol. Med.*, 116, 863, 1964.

64. Wallis, C., Trulock, S., and Melnick, J. L., Inherent photosensitivity of herpes virus and other enveloped viruses, *J. Gen. Virol.*, 5, 53, 1969.
65. Pfefferkorn, E. R., Rutsen, C., and Burge, B. W., Photoreactivation of pseudorabies virus, *Virology*, 27, 457, 1965.
66. Sun, I. L., Gustafson, D. P., and Scherba, G., Comparison of pseudorabies virus inactivated by bromo-ethylene-imine, ^{60}Co irradiation and acridine dye in immune assay systems, *J. Clin. Microbiol.*, 8, 604, 1978.
67. Gentry, G. A. and Randall, C. C., *The Herpesviruses*, Kaplan, A. S., Ed., Academic Press, New York, 1973, 45.
68. Yates, R. A. and Brunert, R. R., Activity of cyanate and labile carbamyl compounds against the pseudorabies virus: mode of action studies, *J. Infect. Dis.*, 116, 353, 1966.
69. Traub, E., Cultivation of pseudorabies virus, *J. Exp. Med.*, 58, 663, 1933.
70. Ivanovies, G., Beladi, I., and Slollosy, E., Interference between variants of pseudorabies virus demonstrable in tissue culture, *Nature (London)*, 176, 972, 1955.
71. Flir, K., Zur Pathologie des Morbus Aujeszky deim Hund, *Arch. Exp. Vet. Med.*, 9, 949, 1955.
72. Szent-Fuany, I., Laboratory diagnosis of a typical Teschen disease and Aujeszky's disease by virus isolation on kidney cell monolayers, *Acta Microbiol. Hung.*, 7, 177, 1960.
73. Zuffa, A. and Skoda, R., Growth of Aujeszky's virus in tissue culture, *Vet. Cas.*, 9, 65, 1960.
74. Ceccarelli, A. and Del Mazza, I., Coltura del virus d'Augeszky su cellule renati di agnello, *Zooprofilassi*, 13, 159, 1958.
75. Gagliardi, G., Borghi, G., and Girotto, V., Isolamento e coltirazione del virus die Aujeszky su embrioni di pollo e monostrati cellulari di differente origine, *Attl. Soc. Ital. Sci. Vet.*, 14, 703, 1960.
76. McFerran, J. B. and Dow, C., Growth of Aujeszky's disease virus in rabbits and tissue culture, *Br. Vet. J.*, 118, 386, 1962.
77. Torlone, V., Colture del virus della pseudo-rabbia su cellite renali di cane "in vitro", *Arch. Vet. Ital.*, 9, 501, 1958.
78. Horvath, Z. and Papp, L., Clinical manifestations of Aujeszky's disease in the cat, *Acta Vet. Hung.*, 17, 49, 1967.
79. Kerekjarto, B. and Rohde, B., Uber die vermehrung des Aujeszky-virus auf Affenieren-Epithelkulturen, *Z. Naturforsch.*, 12, 292, 1957.
80. Tokulmaru, T., Pseudorabies virus in tissue culture: differentiation of two distinct strains of virus by cytopathogenic pattern induced, *Proc. Exp. Biol. Med.*, 96, 55, 1957.
81. Deinhardt, F. and Henle, G., Studies on the viral spectra of tissue culture lines of human cells, *J. Immunol.*, 79, 60, 1957.
82. Petrescu, A., Popescu, A., and Constantinescu, S. P., Revue roum, *Microbiology*, 6, 301, 1969.
83. Brauner, S. and Skoda, R., Welche Moglichkeiten bietet die Gewebkulter fur das Studium der Epizootologie der Aujeszkyschem Krankeit, *Arch. Exp. Vet. Med.*, 15, 385, 1961.
84. Burrows, R., Aujeszky's disease virus, *Vet. Rec.*, 78, 769, 1966.
85. Scherer, W. F. and Syverton, S. T., The viral range in vitro of a malignant human epithelial (strain HeLa, Gey). I. Multiplication of herpes simplex, pseudorabies, and vaccinia viruses, *Am. J. Pathol.*, 30, 1057, 1954.
86. Stoker, M. and MacPhearson, I., Syrian hamster fibroblasts cell line BHK21 and its derivatives, *Nature (London)*, 203, 1355, 1964.
87. McFerran, J. B., Clarke, J. K., Knox, E. R., and Connor, T. J., A study of the cell lines required to detect a variety of veterinary viruses in routine diagnostic conditions, *Br. Vet. J.*, 128, 627, 1972.
88. Kaplan, A. S. and Valter, A. E., A comparison of herpes simplex and pseudorabies viruses, *Virology*, 7, 394, 1959.
89. Beladi, I., Study on the plaque formation and some properties of the Aujeszky disease virus on chicken embryo cells, *Acta Vet. Hung.*, 12, 417, 1962.
90. Kanitz, C. L. and Gustafson, P. P., Persistent pseudorabies virus infection of a cell line derived from kidney of a pig affected with myoclonia congenita, *Fed. Proc.*, 28, 697, 1969.
91. McFerran, J. B. and Dow, C., The distribution of the virus of Aujeszky's disease (pseudorabies virus) in experimentally infected swine, *Am. J. Vet. Res.*, 26, 631, 1965.
92. Baskerville, A., Ultrastructural changes in the pulmonary airways of pigs infected with a strain of Aujeszky's disease virus, *Res. Vet. Sci.*, 13, 127, 1972.
93. Baskerville, A., McFerran, J. B., and Dow, C., Aujeszky disease in pigs, *Vet. Bull.*, 43, 465, 1973.
94. Skoda, R., Brauner, I., Sadecky, E., and Mayer, V., Immunization against Aujeszky's disease with live vaccine. I. Attenuation of virus and some properties of attenuated strains, *Acta Virol.*, 8, 1, 1964.
95. Williams, P. P., Pirtle, E. D., and Coria, M. F., Isoelectric focusing of infectious particles of porcine pseudorabies virus strains in granulated dextran gels, *Can. J. Comp. Med.*, 46, 65, 1982.
96. Rixon, F. J., Benporat, T., and Feldman, L. T., Expression of the genome of defective interfering pseudorabies virions in the presence or absence of helper functions provided by standard virus, *J. Gen. Virol.*, 46, 119, 1980.

97. Ladin, B. F., Ihara, S., Hampfl, H., and Benparat, T., Pathway of assembly of herpesvirus capsids — an analysis using DNA + temperature sensitive mutants of pseudorabies virus, *Virology*, 116, 544, 1982.
98. Gustafson, D. P., Factors involved in the spread of pseudorabies among swine, *Proc. 71st Ann. Meet. U.S. Livestock Sanit. Assoc.*, 1967, 349.
99. Shope, R. E., Experiments on the epidemiology of pseudorabies. II. Prevalence of the disease among middle-western states and the possible role of rats in herd to herd infection, *J. Exp. Med.*, 62, 101, 1935.
100. Nikitin, M. G., Isolation of Aujeszky's disease virus from wild *Rattus norvegicus*, *Zool. Zh.*, 39, 282, 1960.
101. Fraser, G. and Ramachandran, S. P., Studies on the virus of Aujeszky's disease. I. Pathogenicity for rats and mice, *J. Comp. Pathol.*, 79, 435, 1969.
102. Kanitz, C. L., Hand, R. B., and McCrocklin, S. M., Pseudorabies in Indiana: current status, laboratory confirmation, and epizootiologic considerations, *Proc. 78th Ann. Meet. U.S. Anim. Hlth. Assoc.*, 1974, 346.
103. Kirkpatrick, C. M., Kanitz, C. L., and McCrocklin, S. M., Possible role of wild mammals in transmission of pseudorabies to swine, *J. Wildl. Dis.*, 16, 601, 1980.
104. Crandell, R. A., Pseudorabies (Aujeszky's Disease), *Vet. Clin. N. Am.: Large Anim. Practice*, 4 (2), 321, 1982.
105. Hagermoser, W. A., Hill, H. T., and Moss, E. H., Nonfatal pseudorabies in cattle, *J. Am. Vet. Med. Assoc.*, 173, 205, 1978.
106. Toma, B. and Gilet, J., Etude d'un foyer de maladie d'Aujeszky chez les bovins avec cas de querison spontanee, *Rec. Med. Vet.*, 154, 425, 1978.
107. Crandell, R. A., Mesfin, G. M., and Mock, R. E., Horizontal transmission of pseudorabies virus in cattle, *Am. J. Vet. Res.*, 43, 326, 1982.
108. Boucher, J. D. and Beran, G., Pseudorabies in the dog and cat, *Iowa State Univ. Vet.*, 1, 22, 1977.
109. Harris, A. L., Aujeszky's disease in a dog, *J. Am. Vet. Med. Assoc.*, 152, 54, 1968.
110. Hugoson, G. and Rockborn, G., On the occurrence of pseudorabies in Sweden. II. An outbreak in dogs caused by feeding a battoir offal, *Zentrabl. Veterinaermed. Reihe B*, 19, 641, 1972.
111. Ramachandran, S. P. and Fraser, G., Studies on the virus of Aujeszky's disease. II. Pathogenicity for chicks, *J. Comp. Pathol.*, 81, 55, 1971.
112. Medveczky, Z. and Szabo, I., Isolation of Aujeszky's disease virus from boar semen, *Acta Vet. Acad. Sci. Hung.*, 29, 29, 1981.
113. Hill, H. T., Pseudorabies: clinical signs and incidence in swine, *J. Am. Vet. Med. Assoc.*, 171, 1091, 1977.
114. Pirtle, E. C., Pseudorabies virus antibodies in swine slaughtered in Iowa, *Can. J. Comp. Med.*, 46, 128, 1982.
115. Sabo, A. and Grunert, Z., Persistence of virulent pseudorabies virus in herds of vaccinated and nonvaccinated pigs, *Acta Virol.*, 15, 87, 1971.
116. Sabo, A. and Rajcami, J., Latent pseudorabies virus infection in pigs, *Acta Virol.*, 20, 208, 1976.
117. Gutekunst, D. E., Latent pseudorabies virus infection in swine detected by RNA-DNA hybridization, *Am. J. Vet. Res.*, 40, 1568, 1979.
118. Mock, R. E., Crandell, R. A., and Mesfrin, G. M., Induced latency in pseudorabies vaccinated pigs, *Can. J. Comp. Med.*, 45, 56, 1981.
119. Kojnok, J., Mother's milk and the spread of Aujeszky's disease in suckling pigs, *Acta Vet. Hung.*, 7, 273, 1957.
120. Bolin, S. R., Runnels, L. J., Sawyer, C. A., and Gustafson, D. P., Experimental transmission of pseudorabies virus in swine by embryo transfer, *Am. J. Vet. Res.*, 43, 278, 1982.
121. Farnham, A. E. and Newton, A. A., The effect of some environmental factors on herpes virus grown in HeLa cells, *Virology*, 7, 449, 1959.
122. Farnham, A. E., The formation of microscopic plaques by herpes simplex virus in HeLa cells, *Virology*, 6, 317, 1958.
123. Youngner, J. S., Virus adsorption and plaque formation in monolayer cultures of trypsin-dispersed monkey kidney, *J. Immunol.*, 76, 288, 1956.
124. Yates, R. A. and Grunert, R. R., Activity of cyanate and labile carbamyl compounds against the pseudorabies virus: mode of action studies, *J. Infect. Dis.*, 116, 353, 1966.
125. Ida, S., Hooks, J. J., Siraganian, R. P., and Notkins, A. L., Enhancement of IgE-mediated histamine release from human basophils by viruses: role of interferon, *J. Exp. Med.*, 145, 892, 1977.
126. Stanwick, T., Anderson, R., and Nahmias, A., Interactions between herpes simplex virus and cyclic nucleotides: productive infection, *Infect. Immun.*, 18, 342, 1977.
127. Lodmell, D. L. and Notkins, A. L., Cellular immunity to herpes simplex virus mediated by interferon, *J. Exp. Med.*, 140, 764, 1974.
128. Ashman, R. B. and Nahmias, A. J., Enhancement of human lymphocyte responses to phytomitogens in vitro by incubation at elevated temperatures, *Clin. Exp. Immunol.*, 29, 464, 1977.

129. Ben-Porat, T. and Kaplan, A. S., *The Herpesvirus,* Kaplan, A. S., Ed., Academic Press, New York, 1973, 163.
130. Watson, D. H., *The Herpesvirus,* Kaplan, A. S., Ed., Academic Press, New York, 1973, 27.
131. Kerbal, R. S., Herpesvirus induction of Fc receptors, *Nature (London),* 263, 192, 1976.
132. Russell, A. S. and Schlant, J., HLA transplantation antigens in subjects susceptible to recrudescent herpes labialis, *Tissue Antigens,* 6, 257, 1976.
133. Tenser, R. B., Ressel, S. J., Fralish, F. A., and Jones, J. C., The role of Pseudorabies virus thymadine kinase expression in trigeminal ganglion infection, *J. Gen. Virol.,* 64, 1369, 1983.
134. McCracken, P. M., McFerran, Z. B., and Dow, C., The neural spread of pseudorabies virus in calves, *J. Gen. Virol.,* 20, 17, 1973.
135. Baskerville, A., A Study of the Pulmonary Tissue of the Pig and its Reaction to the Virus of Aujeszky's Disease, Ph.D. thesis, Queen's University, Belfast, 1, 1971.
136. Nordbring, F. and Olsson, B., Electrophoretic and immunological studies on sera of young pigs, *Acta Soc. Med. Ups.,* 63, 25, 1958.
137. Speer, V. C., Brown, H., Quinn, L., and Catron, D. V., The cessation of antibody absorption in the young pig, *Immunology,* 83, 632, 1959.
138. Leece, J. G., Matrone, G., and Morgan, D. D., Porcine neonatal nutrition: absorption of unaltered porcine proteins and polyvinyl pyrrolidone from the gut of piglets, *J. Nutr.,* 73, 158, 1961.
139. Leece, J. B., Morgan, D. O., and Matrone, G., Immunoelectrophoretic serum protein changes from birth to maturity in piglets fed different diets, *J. Nutr.,* 77, 349, 1962.
140. Kojnok, J. and Surjan, J., Investigations concerning the colostral immunity of pigs in the case of Aujeszky's disease, *Acta Vet. Hung.,* 13, 111, 1963.
141. Popovici, I., Wynohradnyk, U., Berbinschi, C., Feteanu, A., and Schuler, G., A severe outbreak of Aujeszky's disease associated with influenza virus in pigs, *Probl. Epiz. Vet. Bueuresti.,* 9, 37, 1960.
142. Mitchell, D., Encephalomyelitis of swine caused by a hemagglutinating virus. I. Case histories, *Res. Vet. Sci.,* 4, 506, 1963.
143. Shell, L. G., Ely, R. W., and Crandell, R. A., Pseudorabies in a dog, *J. Am. Vet. Med. Assoc.,* 178 (11), 1159, 1981.
144. Harding, J. D. J., Done, J. T., and Kershaw, G. F., A transmissable polioencephalomyelitis of pigs (Talfan disease), *Vet. Rec.,* 69, 441, 1957.
145. Hill, H. T., Crandell, R. A., Kanitz, C. L., McAdaragh, J. P., Seawright, G. L., Solorzano, R. F., and Stewart, W. C., Recommended minimum standards for diagnostic tests employed in the diagnosis of pseudorabies (Aujeszky's disease), *Proc. 20th Ann. Meet. Am. Assoc. Vet. Lab. Diagn.,* 375, 1977.
146. Holobrow, E. J., *Standardization in Immunofluorescence,* Blackwell Scientific, Oxford, 1970, 1.
147. Snyder, M. L. and Stewart, W. C., Applications of an enzyme-labeled antibody test in porcine pseudorabies, *Proc. Ann. Meet. Am. Assoc. Vet. Lab. Diagn.,* 17, 1977.
148. Briaire, J., Bartelin, S. S., and Meloen, R. H., Enzyme-linked immunosorbent assay for the detection of antibody against Aujeszky's disease virus in pig sera, *Zentralb. Veterinaermed. Reihe B,* 26, 76, 1979.
149. Gutekunst, D. E., Pirtle, E. C., and Mengeling, W. L., Development of a microimmunodiffusion test for detection of antibodies to pseudorabies virus in swine serum, *Am. J. Vet. Res.,* 39, 207, 1978.
150. Bitsch, V. and Eskildsen, M., A comparative examination of swine sera for antibody to Aujeszky's virus with the conventional and a modified virus serum neutralization test and a modified direct complement fixation test, *Acta Vet. Scand.,* 17, 142, 1976.
151. Scherba, G., Gustafson, D. P., Kanitz, C. L., and Sun, I. L., Delayed hypersensitivity reaction to pseudorabies virus as a field diagnostic test in swine, *J. Am. Vet. Med. Assoc.,* 173, 1490, 1978.
152. Palmer, D. F., Cavalaro, J. J., Herrman, K., Stewart, J. A., and Wells, K. W., Eds., *A Procedural Guide to the Serodiagnosis of Toxoplasmosis, Rubella, Cytomegalic Inclusion Disease, Herpes Simplex,* Immunol. Ser. No. 5, Center for Disease Control, Atlanta, 1979, 23.
153. Fraser, G. and Ramachandran, S. P., Studies on the virus of Aujeszky's disease. I. Pathogenicity for rats and mice, *J. Comp. Pathol.,* 79, 435, 1969.
154. Popescu, A., Aujeszky's disease virus. I. Culture in trypsinized cells. II. Preparation of hyperimmune serum by using culture virus, *Lucr. Inst. Cercet. Vet. Bioprep. Pasteur,* 2, 143, 1965; *Abstr. Vet. Bull.,* 37, 25, 1965.
155. Crandell, R. A., Doby, P. B., Hill, R. O., Hoefling, D. C., Jelly, G. G., Norton, H. W., Spencer, P. L., Starkey, A. L., and Wu, C. H., Use of pseudorabies hyperimmune serum in naturally occurring pseudorabies in Illinois swine herds, *J. Am. Vet. Med. Assoc.,* 17, 59, 1977.
156. Kolb, K. E., Bower, R. K., and Duffy, C. E., Effect of 5-iodo-2-deoxy-urine on pseudorabies infection in rabbits, *Proc. Soc. Exp. Biol. Med.,* 113, 476, 1963.
157. Wawrzkiewicz, J., Interferon formation by strains of Aujeszky's disease virus, *Med. Weter.,* 22, 657, 1966.

158. Preobrazhenskaya, E. A., Gannushkin, M. S., Furer, N. M., and Ermoleva, Z. V., Preparation of interferon and its action in tissue culture on the virus of avian infectious laryngotracheitis and on the virus of Aujeszky's disease, *Tr. Mosk. Vet. Acad.*, 51, 49, 1967.
159. Akkermans, J. P. W. M., *Ziekte van Aujeszky bij het varken in Nederland,* Centraal Diergeneeskundig Institut, Rotterdam, 1963, 1.

Chapter 7*

MALIGNANT CATARRHAL FEVER

Werner P. Heuschele and Anthony E. Castro

TABLE OF CONTENTS

* Portions of this chapter have been included in "Foreign Animal Diseases — Their Prevention, Diagnosis and Control" (revised 1984), published by the Committee on Foreign Animal Diseases of the United States Animal Health Association, Richmond, Va., which has given permission for publication in this volume.

I. INTRODUCTION

Malignant catarrhal fever (MCF, malignant head catarrh, malignant catarrh, snotsiekte) is a generalized viral disease of domestic cattle and buffaloes and many species of wild ruminants characterized by high fever, profuse nasal discharge, corneal opacity, ophthalmia, generalized lymphadenopathy, leukopenia, and severe inflammation of the conjunctival, oral, and nasal mucosa with necrosis in the oral and nasal cavities, sometimes extending into the esophagus and trachea. Occasionally CNS signs, diarrhea, skin lesions, and nonsuppurative arthritis are observed.[1-6]

II. ETIOLOGY

The etiologic agent of MCF in Africa is a highly cell-associated lymphotropic herpesvirus of the subfamily Gammaherpesvirinae. Two viral strains have recently been designated: alcelaphine herpesvirus-1 and alcelaphine herpesvirus-2,[7] although some continue to designate this agent as bovid herpesvirus-3.[8] The complete enveloped virion is 140 to 220 nm, with a DNA core enclosed in a capsid approximately 100 nm in size.[9,10] The morphology of MCF virus is like that of other herpesviruses when examined by electron microscopy (Figure 1). This agent is carried as a latent infection by African antelope of the family Bovidae, subfamily Alcelaphinae which includes wildebeest (*Connochaetes sp.*), hartebeest (*Alcelaphus sp.*), and topi (*Damaliscus sp.*).[3] The African antelope herpesvirus of MCF was first isolated by Plowright et al.[6] from a blue wildebeest (*Connochaetes taurinus taurinus*) in 1960.

Epidemiologic evidence from other parts of the world and Africa suggests that domestic sheep may also be reservoirs of a virus causing MCF. Recent serologic evidence suggests that this putative virus may be related but not identical to the alcelaphine herpesvirus-1; but a sheep-associated MCF virus has not yet been isolated.[11,12]

Viruses identical or closely related to alcelaphine herpesvirus-1 have recently been isolated from several captive wild ruminants in two U.S. zoos located in Oklahoma City and San Diego. Animals involved were white-tailed gnu, white-bearded gnu, gaur, greater kudu, Formosan sika deer, axis deer, nilgai, and topi.[13-17] Alcelaphine MCF virus has been successfully cultivated in cell cultures of fetal bovine kidney, spleen, thyroid, lung and adrenals, in fetal aoudad (a wild ovine) kidney, deer thyroid and kidney cells, and African green monkey kidney (Vero) cell line cultures.[9,13-16]

MCF virus in cell cultures causes typical herpesviral cytopathic effects (CPE) characterized by formation of large syncytia which gradually contract and round-up becoming thereby very refractile. These syncytia detach and resulting cell-free areas in the monolayer may fill in with normal cells. Gradually the syncytia become more abundant and are accompanied by rounded cell clusters at their periphery. These ultimately detach destroying the monolayer. As the virus becomes more adapted to cell culture, the CPE changes from syncytial formations to foci of rounded cells which spread rapidly and destroy the monolayer within 7 to 10 days postinoculation (DPI).[9,13,14,17]

Cell-free MCF virus, once developed appears to replicate equally well at 33°C or 37°C, producing peak extracellular viral titers at 4 DPI. Titers achieved are usually between 10^5 and $10^{6.8}$ median infectious doses per 1.0 mℓ of culture fluid.[17] Titers for cell-associated (usually the early passages) remain relatively low at $< 10^2$ $TCID_{50}/m\ell$.

Purification of MCF virus has been reported by Schloer and Breese[18] and was accomplished by a method employing separation of virus in a two-phase aqueous polymer system of 10% polyethylene glycol and 8% dextran, followed by centrifugation into a 20% Ficoll cushion with subsequent further density gradient purification in continuous 15 to 36% CsCl or 30 to 60% sucrose. Herring et al.[19] reported on the purification of MCF virus by gradient centrifugation in potassium tartrate and metrizamide.

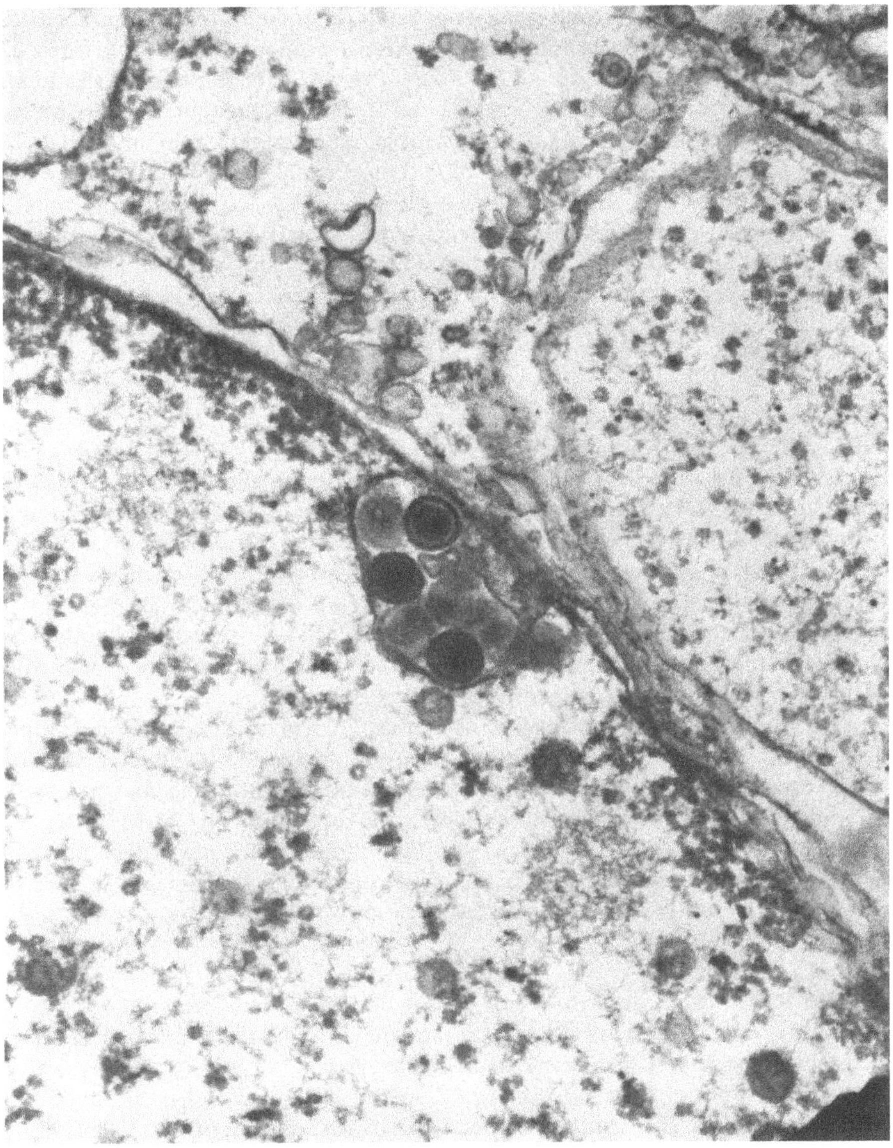

FIGURE 1. Mature malignant catarrhal fever virions within a nuclear vesicle budding from the nucleus, with immature virus particles nearby within the nucleus. Magnification × 64,000.

In this study, preliminary characterization of the viral genome was done on DNA from purified virions, which yielded two components in CsCl density gradients; a major peak with a density of 1.710 and a minor with a density of 1.730. These peaks were found to resemble the M and H DNA components found in *Herpesvirus saimiri* and *H. ateles*, also lymphotropic gammaherpesviruses.

Restriction endonuclease cleavage of these DNAs showed that the major component has a complexity similar to that of other herpesviruses.

MCF virus may be considered an oncogenic herpesvirus similar to other gammaherpesviruses such as *Herpesvirus saimiri* of squirrel monkeys, *Herpesvirus ateles* of spider monkeys, Mareks disease virus, *Herpesvirus sylvilagi* and Epstein-Barr virus of humans.[20,21]

Justification for MCF virus oncogenicity is based upon the following characteristics: (1) highly cell-associated, (2) produces typical herpetic cytopathic effect in vitro, i.e., in cell cultures including syncytia and Cowdry Type A inclusion bodies, (3) does not produce inclusions or syncytia in vivo, i.e., in infected animals, (4) the disease is not contagious between cattle or rabbits, (5) the incubation period is highly variable, (6) the virus is associated with lymphocytes, (7) the principal pathological feature of the infection is lymphocytic proliferation and infiltration.[21]

Two reports of in vitro cultivation and characterization of lymphocytes derived from tissues of rabbits experimentally infected with MCF virus indicate that the majority of proliferating lymphoid cells obtained were T lymphocytes or T lymphoblasts.[22,23] Interestingly, one of these involved alcelaphine MCF and the other sheep-associated MCF.

A. History

Sheep-associated MCF was described in Europe as early as 1798. A number of European workers described the clinical signs and lesions in cattle during the late 19th century. Successful transmission of MCF by inoculation of blood from sheep-associated bovine cases was demonstrated in 1929 in Europe.[3] In Africa, MCF has been known to Maasai tribesmen for centuries. These herdsmen have recognized that MCF outbreaks in their cattle were associated with grazing of these cattle in areas where wildebeest (gnu) also grazed during the wildebeest calving season. In South Africa, settlers and hunters encountered MCF frequently in the mid-19th century and called it "snotsiekte", which in Afrikaans means snotting sickness.[3]

The first isolation of a wildebeest-associated MCF virus, i.e., alcelaphine herpesvirus-1 in the U.S. was made in 1979 from a gaur and greater kudu with clinical MCF at the Oklahoma City Zoo.[13] These isolates were readily transmitted and passaged by inoculation into cattle and white-tailed deer, and produced typical MCF signs and lesions.[14,24] MCF virus was reisolated from these animals in fetal bovine kidney cell cultures. A herpesvirus similar to alcelaphine herpesvirus-1 was isolated in 1977 from the blood of dairy cattle involved in an outbreak of MCF in a Minnesota dairy.[25]

B. Clinical Signs[1-3,5]

Clinical MCF in cattle has arbitrarily been divided into four forms:

1. **Peracute form** — Severe inflammation of the oral and nasal mucosa and hemorrhagic gastroenteritis with a course of 1 to 3 days.
2. **Intestinal form** — Pyrexia, diarrhea, hyperemia of oral and nasal mucosa with accompanying discharges, and lymphadenopathy with a course of 4 to 9 days.
3. **Head and eye form** — This is the typical syndrome of MCF with pyrexia, nasal and ocular discharges progressing from serous to mucopurulent and purulent. Encrustation of the muzzle and nares occurs in later stages, causing obstruction to the nostrils and dyspnea, open-mouthed breathing, and drooling. There is intense hyperemia and multifocal or diffuse necrosis of the oral mucosa, usually on the lips, gums, hard and soft palate, and buccal mucosa. Sloughing of the tips of buccal papillae leaving them reddened and blunted is often encountered.

Ocular signs referrable to opthalmia include lacrimation progressing to purulent exudation, photophobia, hyperemia and edema of the palpebral conjunctiva, and injection of scleral vessels. Corneal opacity, starting peripherally and progressing centripetally results in partial to complete blindness. Hypopyon may also be seen. Corneal opacity is usually bilateral, but occasionally is unilateral. Pyrexia is common and usually high (104 to 107°F) until the animal becomes moribund, at which time it is hypothermic. Increased thirst accompanies the pyrexia and anorexia is seen in late stages. Constipation is common in this form of MCF, but terminal diarrhea is sometimes observed.

Nervous signs are not frequently seen but may be manifested by trembling or shivering, uncoordinated gait and terminal nystagmus.

Necrotic skin lesions occasionally are seen and horn and hoof coverings may be loosened or sloughed in some cases. The course of the head and eye form, which is invariably fatal, is usually 7 to 18 days.

4. Mild forms — These are syndromes caused by experimental infection of cattle with attenuated viruses, and are usually nonfatal.

There is considerable variation and overlap among these artificial categories and their use, and in the opinion of the authors has little value. The pathologic description of MCF is related to symptomatology by the degree of pathology which occurs in the principal organ systems affected by MCF virus.

While the manifestations of the "head and eye" form of MCF are considered the typical syndrome of MCF in cattle, clinical signs in exotic ruminants are less dramatic or specifically diagnostic, except in members of the subfamily Bovinae, i.e., wild cattle.[17] MCF in deer and antelope species tends to be more subtle clinically and usually is manifested by conjunctivitis, photophobia, moderate corneal clouding (often unilateral), fever, depression, variable lymphadenopathy, occasional diarrhea, and usually a mild serous nasal discharge. Death may be sudden, following a brief course of hemorrhagic diarrhea. Inflammation of the oral and nasal cavity is usually less severe than in cattle and only occasionally progresses to mucosal erosions.[16]

There is some suggestion from studying cases of MCF among exotic ruminants that there is a host-dependent modification with respect to the clinical and pathologic manifestations of MCF virus infection.[15,26]

C. Incubation Period

The incubation period in natural cases is not known but epidemiologic evidence indicates it may be as long as 200 days. Experimentally, the incubation period has varied from 9 to 77 days.[2-5,27-30]

III. PATHOLOGY

A. Post-Mortem Lesions

Gross lesions vary considerably depending on the form or severity and course of the disease.[1,3,30,31] Animals that die of the peracute disease may have few lesions other than a hemorrhagic enterocolitis.

In the more protracted acute to subacute disease (intestinal and head and eye forms) the carcass may be normal, dehydrated, or emaciated. The muzzle is often encrusted and raw. Cutaneous lesions sometimes occur as a generalized exanthema with exudation of lymph with crusting and matting of the hair. Where skin is unpigmented, hyperemia is apparent. The lesions are frequently seen in the ventral thorax and abdomen, inguinal region, perineum and loins, and sometimes on the head.

Lesions in the respiratory system range from mild to severe. When the clinical course is short there is slight serous nasal discharge and hyperemia of the nasal mucosa. Later the discharge becomes more copious, mucopurulent to purulent accompanied by intense nasal mucosal hyperemia, edema, and focal small erosions. Occasionally a croupous pseudomembrane formation is seen. Lesions in the nasal passages and turbinates may extend to the frontal sinuses. The pharyngeal and laryngeal mucosa are hyperemic and edematous and later develop multiple erosions, often covered with gray-yellow pseudomembranes. Inflammation and sometimes petechiation and ulceration are seen in the tracheobronchial mucosa. The lungs are often edematous and sometimes emphysematous, but may appear normal. A bronchopneumonia may complicate chronic cases.

The alimentary tract mucosa may have no gross lesions in peracute cases. When the course is longer, alimentary lesions are commensurately more severe and include mild to severe mucosal inflammation (hyperemia and edema), erosions and ulcerations especially on the dental pad and gingival surfaces, the palate, tongue and buccal papillae. Mucosal inflammation, hemorrhage, and erosions may also be found in the rest of the digestive tract including the esophagus, abomasum, small intestines, colon, and rectum. Petechiation may be seen. Feces are usually scant, dry, pasty, or blood-stained.

Urinary tract lesions include hyperemia and sometimes marked distention and prominence of bladder mucosal vessels and mucosal edema, sometimes with petechial to severe hemorrhage and occasionally epithelial erosion and ulceration. Kidneys may appear normal or mottled with patches of beige discolored raised areas. Petechiae or ecchymoses may occur in the renal pelvis and ureters.

The liver is usually slightly enlarged, and, upon close examination, has a prominent reticular pattern. There may be hemorrhages and erosions in the gall bladder mucosa.

Enlarged lymph nodes are characteristic findings in MCF. All nodes may be involved, but those in the head and neck and periphery are the most consistently prominent. Affected nodes are grossly enlarged, edematous and sometimes have patchy reddened or beige-brown areas on cut surface. Hemolymph nodes are also enlarged and prominent. The spleen is slightly enlarged and Malpighian corpuscles are prominent. Pale areas of mononuclear cell infiltration may be seen in the heart muscle.

In most cases, small arterioles are very prominent, tortuous, and have thickened walls. This is usually seen in subcutaneous vessels and those in the thorax, abdomen, and CNS.

Fibrinous polyarthritis with characteristic mononuclear cells in synovial fluid, is seen in many cases of MCF.

B. Microscopic Pathology[1,3,30-33]

Histopathologic lesions provide a consistent basis for the diagnosis of MCF. The changes are found in the epithelium of the gastrointestinal and urinary tracts, the lymphoid tissues, and in the adventitia and walls of small blood vessels in any organ. These lesions are seen in MCF-infected cattle, wild ruminants, and experimentally infected rabbits.

Vascular lesions consist of a fibrinoid necrotizing vasculitis with chiefly mononuclear cell infiltration of the adventitia. If the intima is involved, there is often endothelial swelling and mononuclear cell infiltration. There is a necrosis, karyorrhexis of fixed cells of the vessel wall, and infiltrating mononuclear cells. These vascular lesions are present in all cases and forms of MCF. The best organs to examine for vascular lesions and lymphoid cell infiltration are brain, leptomeninges, carotid rete, kidney, liver, adrenal gland, and areas of skin and gross lesions.

All lymphoid tissues show destruction and loss of small mature lymphocytes with concomitant increase in macrophages ingesting cellular debris, and lymphoblast proliferation. Proliferation and infiltration of lymphoid cells is sometimes very profuse in many organs. It is usually associated with blood vessels, i.e., perivascular and may become so dense that it resembles neoplasia (lymphoma). Indeed, cases have recently been seen in wild ruminants in which frank lymphomatous tumors arising from Peyer's patches in the small intestine were seen in association with MCF, with metastatic or multicentric tumor foci in lymph nodes, kidneys, liver, lungs, thyroid, and myocardium. In many cases, it is difficult to ascertain whether the extensive proliferation and infiltration of lymphoreticular cells is a hyperplastic or neoplastic process. These observations further support the speculation that MCF virus is an oncogenic herpesvirus. Indeed MCF infection in rabbits provides a very suitable model for the study of herpesvirus-induced vasculitis, lymphoproliferation, and oncogenesis.

Microscopic lesions in the nervous system are largely a reflection of vasculitis with perivascular extension of mostly mononuclear cells, and sometimes associated necrosis in meninges and plasma exudation in Virchow-Robin spaces.

Oral epithelium, esophagus, and intestinal mucosa have foci of dense basilar infiltration and mononuclear cells and accompanying necrosis. Degeneration and desquamation of epithelium, edema, and hemorrhage in the walls of the alimentary tract organs are common. Similar lesions are usually present in the urinary bladder.

Ophthalmitis is fairly consistent, usually involving mononuclear cell infiltrations in all parts of the globe especially the ciliary processes. The cornea is edematous and may be ulcerated and vascularized. There is retinal vasculitis, sometimes with hemorrhage and focal retinal detachment.

IV. DIAGNOSIS

A. In the Field

A history indicating contact with sheep or alcelaphine antelope, especially around the period of parturition, associated with typical clinical features of MCF provides grounds for a tentative diagnosis of MCF.

Gross necropsy lesions revealing inflammation and erosions in nasal passages, alimentary tract mucosa, and urinary bladder, enlarged lymph nodes, corneal opacity, and prominent tortuous small arteries in the subcutis, thorax, and abdomen provide further evidence for a presumptive diagnosis of MCF.

B. Laboratory

Microscopic lesions of extensive vasculitis, perivasculitis, and lymphoreticular proliferation in lymphoid organs with mononuclear infiltrations in kidney, liver, adrenals, CNS, etc., are pathognomonic for a diagnosis of MCF.

Virologic and serologic examinations provide the means for confirming the histopathologic diagnosis. Methods used consist of:

1. Virus isolation in fetal bovine or ovine thyroid, kidney, or spleen cells, or by intraperitoneal or intravenous inoculation of domestic rabbits or cattle.[3,4,9,13,16,34]
   a. Identification of virus isolates by typical herpesvirus cytopathic effects,[16] by immunofluorescence,[35] immunoperoxidase,[36] and by electron microscopy.[3,9,10]
2. Demonstration of MCF antibodies by indirect immunofluorescence,[36] CF,[37] ELISA,[36] and virus neutralization.[3,5,9,38,39]

C. Differential Diagnosis

Clinical MCF must be distinguished from other diseases which produce inflammation, erosions, and ulcerations of the nasal and alimentary tract mucosa: BVD-mucosal disease, bluetongue, rinderpest, vesicular diseases (FMD, VS), ingested caustics, and some poisonous plants and mycotoxins.

D. Collection of Specimens for Laboratory Confirmation

1. For animal transmission/inoculation and virus isolation at least 300 to 500 m*l* blood in EDTA (1 mg/m*l* blood), heparin, or ACD solution should be collected and carried or shipped iced, *not frozen.*
2. Tissues for virus isolation, FA, or immunoperoxidase examination should also be refrigerated (iced) but *not frozen*; and should include pieces of spleen, lung, lymph nodes, adrenals and thyroids, as well as unclotted blood. These should be collected as soon after death as possible as the virus becomes inactivated rapidly in a carcass dead more than 1 hr.

3. Tissues for histopathology fixed as thin pieces in 10% neutral buffered formalin: lung, kidney, liver, adrenals, lymph nodes, eyes, oral epithelium, esophagus, Peyer's patches, urinary bladder, brain, carotid rete, thyroid, and heart muscle and skin if skin lesions are present.
4. Serum for serology should consist of paired samples taken 3 to 4 weeks apart, i.e., the first during the acute phase of disease and the second during convalescence or at death.

Buffy coat leukocytes from unclotted blood provide the best source for virus isolation in cell cultures. Fusion of leukocytes with suspended tissue culture cells using polyethylene glycol (1000 or 4000) enhances the sensitivity of MCFV isolation.[17,34]

Characteristic herpesvirus CPE in inoculated cell cultures may require several subculture passages of cells. These changes consist of early formation of syncytia followed by contraction and rounding of syncytia with detachment leaving bare cell-free areas which are filled in by normal-appearing cells. As the MCF virus isolate becomes adapted to cell culture there will be formation of focal clusters of shrunken rounded cells which detach, and as the virus becomes cell-free, fewer syncytia are formed. The development of cell-free virus is enhanced by incubation of cultures at 33 to 34°C.[40] Cowdry type A intranuclear inclusion bodies are formed in infected cell cultures, but have not been seen in tissues of MCF virus-infected animals.[3,9,13,14]

V. PROGNOSIS

The prognosis in MCF is poor. The case fatality rate is usually greater than 95% (90 to 100%).[2]

VI. EPIZOOTIOLOGY

A. Geographical Distribution

Sheep-associated MCF occurs worldwide. The alcelaphine antelope-associated form in cattle occurs chiefly in Africa, in the natural habitat of wildebeest, hartebeest and topi. This form of MCF has, however, occurred in zoos and wild animal parks which also kept wildebeest.[15] Their increasing popularity in North America and other areas of the world of wild game animal ranches, often in association with domestic cattle raising, increases the possibility of MCF becoming a more prevalent disease in cattle and ranched exotic ruminants.

B. Transmission

MCF virus in wildebeest,[3,6] hartebeest,[42] and topi[17] is largely cell-associated in adult animals and hence rarely transmissible.[42] Transmission to cattle or other susceptible species may occur by inhalation of cell-free virus in infectious aerosol droplets, ingestion of feed or water contaminated with infectious secretions or feces, or possibly mechanically by arthropods. Maasai herdsmen believed cattle acquired MCF by contact with wildebeest placentas or birth hair of neonates. Recent studies have failed to demonstrate infective MCF virus in fetal fluids or placentas of wildebeest.[47] The mode of transmission of sheep-associated MCF remains unknown, although relatively close contact between cattle and sheep, especially lambing ewes, is believed necessary. MCF-affected cattle appear to shed only cell-associated virus and thus cattle-to-cattle transmission is thought to be rare or nonexistent.[45]

C. Hosts

All species of wildebeest,[3,16] hartebeest,[41] and topi[17] are considered carriers of alcelaphine MCF virus. There is serologic evidence that several other African wild ruminants such as various species of oryx and addax may also be reservoir hosts, although MCF virus has not been isolated from these species.[16]

Domestic sheep are considered reservoir hosts for MCF virus, but the associated virus has not yet been isolated.[3,46] Recent serologic and epidemiologic evidence suggests that some wild sheep and goat species may also be reservoirs of a MCF virus.[17] Domestic sheep have been experimentally infected with alcelaphine MCF virus, as have rabbits, guinea pigs, and hamsters.[38]

Many exotic ruminant species in zoos have been reported affected with MCF including several wild bovines such as bison, water buffalo, gaur and banteng, and several deer and antelope species.[16] Interestingly, no cases of MCF have been reported from antelope species which normally cohabitate wildebeest grazing areas in Africa.

In cattle and susceptible wild ruminants, MCF affects all ages, breeds and sexes.

VII. CONTROL AND ERADICATION

A. Preventative Measures

Cattle should be kept separated from potential reservoir hosts such as sheep and wildebeest, especially during lambing or calving seasons, respectively.

The stocking of cattle ranches with alcelaphine antelope, wild sheep or goats should be discouraged, or should require a negative MCF serologic test, preferably by the serum-virus neutralization method, for any wild ruminants destined for such a facility. Similar serologic testing of such wild ruminants before being placed in or transferred between zoos is also recommended as a means to prevent the introduction of potential carriers of MCF virus.

B. Natural Immunity

Cattle and experimentally infected rabbits recovered from MCF have a solid immunity against all known strains of MCF virus.[3]

C. Induced Immunity

An effective vaccine is not available for MCF. Some viral strains have undergone some attenuation after serial passage in cell cultures and offer hope for a future modified live virus vaccine. Experimental killed virus vaccines have been inconsistent in inducing protection against virulent virus challenge, although some have induced significant titers of serum virus neutralizing antibodies.[3,39]

VIII. PUBLIC HEALTH ASPECTS

There is no evidence that MCF is infectious for humans.

REFERENCES

1. Jubb, K. V. F. and Kennedy, P. C., *Pathology of Domestic Animals*, Vol. 2, Academic Press, New York, 1970, 27.
2. Pierson, R. E., Hamdy, F. M., Dardiri, A. H., Ferris, D. H., and Schloer, G. M., Comparison of African and American forms of malignant catarrhal fever: transmission and clinical signs, *Am. J. Vet. Res.*, 40, 1091, 1979.
3. Plowright, W., Herpesviruses of wild ungulates, including malignant catarrhal fever virus, in *Infectious Diseases of Wild Mammals*, 2nd ed., Davis, J. W., Karstad, L. H., and Trainer, D. O., Eds., Iowa State University Press, Ames, 1981, 126.

4. Mare, C. J., Malignant catarrhal fever, an emerging disease of cattle in the USA, *Proc. USAHA 81st Ann. Mtg.*, Minneapolis, 81, 151, 1977.
5. Kalunda, M., Dardiri, A. H., and Lee, K. M., Malignant catarrhal fever. I. Response of American cattle to malignant catarrhal virus isolated in Kenya, *Can. J. Comp. Med.*, 45, 70, 1981.
6. Plowright, W., Ferris, R. D., and Scott, G. R., Blue wildebeest and the aetiological agent of bovine malignant catarrhal fever, *Nature (London)*, 188, 1167, 1960.
7. Roizman, B., *The Herpesviruses*, Vol. 1, Plenum Press, New York, 1982, 7.
8. Ludwig, H., Bovine herpesviruses, in *The Herpesviruses*, Vol. 2, Roizman, B., Ed., Plenum Press, New York, 1983, 135.
9. Plowright, W., McCadam, R. F., and Armstrong, J. A., Growth and characterization of the virus of bovine malignant catarrhal fever in East Africa, *J. Gen. Microbiol.*, 39, 253, 1963.
10. Castro, A. E. and Daley, G. G., Electron microscopic study of the African strain of malignant catarrhal fever virus in bovine cell cultures, *Am. J. Vet. Res.*, 43, 576, 1982.
11. Rossiter, P. B., Antibodies to malignant catarrhal fever virus in sheep sera, *J. Comp. Pathol.*, 91, 303, 1981.
12. Rossiter, P. B., Antibodies to malignant catarrhal fever virus in cattle with non-wildebeest-associated malignant catarrhal fever, *J. Comp. Pathol.*, 93, 93, 1983.
13. Castro, A. E., Whitenack, D. L., and Goodwin, D. E., Isolation and identification of the herpesvirus of malignant catarrhal fever from exotic ruminant species in a zoological park in North America, *Proc. 24th Ann. Mtg. Amer. Assn. Vet. Lab. Diagnosticians*, St. Louis, 24, 67, 1981.
14. Castro, A. E., Daley, G. G., Zimmer, M. A., Whitenack, D. L., and Jensen, J., Malignant catarrhal fever in an Indian gaur and greater kudu: experimental transmission, isolation, and identification of a herpesvirus, *Am. J. Vet. Res.*, 43, 5, 1982.
15. Heuschele, W. P., Malignant catarrhal fever in wild ruminants — review and current status report, *Proc. 86th Ann. Mtg. USAHA*, Nashville, 86, 552, 1982.
16. Heuschele, W. P., Diagnosis of malignant catarrhal fever due to alcelaphine herpesvirus-1, *Proc. III Intl. Symp. Vet. Lab. Diag.*, Ames, Iowa, 707, 1983.
17. Heuschele, W. P., unpublished data, 1984.
18. Schloer, G. M. and Breese, S. S., Jr., Purification of malignant catarrhal fever virus using a two-phase aqueous polymer system, *J. Gen. Virol.*, 59, 101, 1982.
19. Herring, A. J., Berrie, E., Reid, H. W., and Pow, I., Malignant catarrhal fever virus: purification and preliminary characterisation of the genome, *Abstr. 8th Int. Herpesvirus Workshop*, Oxford, July 31, 1983, 53.
20. Edington, N., Patel, J., Russell, P. H., and Plowright, W., The nature of the acute lymphoid proliferation in rabbits infected with the herpesvirus of bovine malignant catarrhal fever, *Comp. J. Cancer*, 15, 1515, 1979.
21. Hunt, R. D. and Billups, L. H., Wildebeest-associated malignant catarrhal fever in Africa: neoplastic disease of cattle caused by an oncogenic herpesvirus?, *Comp. Immun. Microbiol. Infect. Dis.*, 2, 275, 1979.
22. Rossiter, P. B., Proliferation of T lymphoblasts in rabbits fatally infected with the herpesvirus of malignant catarrhal fever, *Clin. Exp. Immunol.*, 54, 547, 1983.
23. Reid, H. W., Buxton, D., Pow, I., Finlayson, J., and Berrie, E. L., A cytotoxic T lymphocyte line propagated from a rabbit infected with sheep-associated malignant catarrhal fever, *Res. Vet. Sci.*, 34, 109, 1983.
24. Whitenack, D. L., Castro, A. E., and Kocran, A. A., Experimental malignant catarrhal fever (African form) in white-tailed deer, *J. Wildlife Dis.*, 17, 443, 1981.
25. Hamdy, F. M., Dardiri, A. H., Mebus, C., Pierson, R. E., and Johnson, D., Etiology of malignant catarrhal fever outbreak in Minnesota, *Proc. USAHA 82nd Ann. Mtg.*, Buffalo, N.Y., 82, 248, 1978.
26. Hatkin, J., Endemic malignant catarrhal fever at the San Diego Wild Animal Park, *J. Wildlife Dis.*, 16, 439, 1980.
27. Piercy, S. E., Studies in bovine malignant catarrh. I—II. Experimental infection in cattle, *Br. Vet. J.*, 108, 35, 1952.
28. Plowright, W., Malignant catarrhal fever, *J. Am. Vet. Med. Assoc.*, 152, 795, 1968.
29. Mushi, E. Z. and Rurangirwa, F. R., Epidemiology of bovine malignant catarrhal fevers, a review, *Vet. Res. Commun.*, 5, 127, 1981.
30. Liggitt, H. D., De Martini, J. C., McChesney, A. E., Pierson, R. E., and Storz, J., Experimental transmission of malignant catarrhal fever in cattle: gross and histopathologic changes, *Am. J. Vet. Res.*, 39, 1249, 1978.
31. Zimmer, M. A., McCoy, C. P., and Jensen, J. M., Comparative pathology of the African form of malignant catarrhal fever in captive Indian gaur and domestic cattle, *J. Am. Vet. Med. Assoc.*, 179, 1130, 1981.
32. Liggitt, H. D. and De Martini, J. C., The pathomorphology of malignant catarrhal fever. I. Generalized lymphoid vasculitis, *Vet. Pathol.*, 17, 59, 1980.

33. Liggitt, H. D. and De Martini, J. C., The pathomorphology of malignant catarrhal fever. II. Multisystemic epithelial lesions, *Vet. Pathol.*, 17, 74, 1980.
34. Castro, A. E., Schramke, M. L., Ramsay, E. C., Whitenack, D. L., and Dotson, J. F., A diagnostic approach in the identification and isolation of malignant catarrhal fever virus from in apparent carriers in a wildebeest herd, *Proc. III Intl. Symp. Vet. Lab. Diag.*, Ames, Iowa, 715, 1983.
35. Ferris, D. H., Hamdy, F. M., and Dardiri, A. H., Detection of African malignant catarrhal fever virus antigens in cell cultures by immunofluorescence, *Vet. Microbiol.*, 1, 437, 1976.
36. Rossiter, P. B., Immunofluorescence and immunoperoxidase techniques for detecting antibodies to malignant catarrhal fever in infected cattle, *Trop. Anim. Health Prod.*, 13, 189, 1981.
37. Hamdy, F. M., Dardiri, A. H., and Ferris, D. H., complement fixation test for diagnosis of malignant catarrhal fever, *Proc. 84th Ann. Mtg. USAHA*, Louisville, Ky., 84, 329, 1980.
38. Kalunda, M., Ferris, D. H., Dardiri, A. H., and Lee, K. M., Malignant catarrhal fever. III. Experimental infection of sheep, domestic rabbits and laboratory animals with malignant catarrhal fever virus, *Can. J. Comp. Med.*, 45, 310, 1981.
39. Rossiter, P. B., Mushi, E. Z., and Plowright, W., The antibody response in cattle and rabbits to early antigens of malignant catarrhal fever virus in cultured cells, *Res. Vet. Sci.*, 25, 207, 1978.
40. Harkness, J. W. and Jessett, D. M., Influence of temperature on the growth in cell culture of malignant catarrhal fever virus, *Res. Vet. Sci.*, 31, 164, 1981.
41. Reid, H. W. and Rowe, L., The attenuation of a herpesvirus (malignant catarrhal fever virus) isolated from hartebeest (*Alcelaphus buselaphus cokei*), *Res. Vet. Sci.*, 15, 144, 1973.
42. Rweyemamu, M. M., Karstad, L., Mushi, E. Z., Otema, J. C., Jessett, D. M., Rowe, L., Drevemo, S., and Grootenhuis, J. G., Malignant catarrhal fever virus in nasal secretions of wildebeest: a probable mechanism for virus transmission, *J. Wildlife Dis.*, 10, 478, 1974.
43. Mushi, E. Z., Rossiter, P. B., Karstad, L., and Jessett, D. M., The demonstration of cell-free malignant catarrhal fever herpesvirus in wildebeest nasal secretions, *J. Hyg. Camb.*, 85, 175, 1980.
44. Mushi, E. Z., Rurangirwa, F. R., and Karstad, L., Shedding of malignant catarrhal fever virus by wildebeest calves, *Vet. Microbiol.*, 6, 281, 1981.
45. Mushi, E. Z. and Rurangirwa, F. R., Malignant catarrhal fever virus shedding by infected cattle, *Bull. Anim. Health. Prod. Afr.*, 29, 111, 1981.
46. Piercy, S. E., Studies in bovine malignant catarrh. V. The role of sheep in the transmission of the disease, *Br. Vet. J.*, 110, 508, 1954.
47. Rossiter, P. B., Role of wildebeest fetal membranes and fluids in the transmission of malignant catarrhal fever, *Vet. Rec.*, 150, 1983.

Chapter 8

EQUINE HERPESVIRUS TYPE 1

James R. Blakeslee, Jr. and David Jasko

TABLE OF CONTENTS

I. INTRODUCTION

A. Historical Perspective

In 1936, Dimock and Edwards[14] described outbreaks of abortions in Kentucky horses. Diagnosis was based on similar characteristics found in each abortion case: (1) no bacterial agent was isolated from the aborted fetuses, (2) mares generally aborted between the 8th and 9th month of gestation, (3) the placenta was passed immediately after abortion, and (4) no other signs of disease were noted. Characteristic lesions were described in the fetuses which included the presence of small white, degenerated areas on an enlarged congested liver and large amounts of yellow fluid in the thoracic cavity. Bacteria-free filtrates prepared from the livers of the aborted fetuses induced abortion in susceptible mares, and the fetuses displayed the characteristic lesions found under natural conditions. Based on this evidence, Dimock and Edwards concluded that the epizootic abortion they reported in mares was due to an infectious process caused by a viral agent.

In 1949, Manninger[33] further demonstrated that this agent that caused abortion produced a mild form of "influenza" in exposed horses. The diagnosis of influenza was based on clinical symptoms and any horse with fever and upper respiratory infection was diagnosed as having "influenza". Manninger further showed that horses recovered from "influenza" were also resistant to respiratory infections when injected with liver suspensions prepared from aborted fetuses while normal control horses were susceptible. Horses which aborted during outbreaks of these epizootic abortions were found to be resistant to respiratory disease when inoculated with defibrinated blood from horses with "influenza". The results of these studies led Manninger to conclude that "influenza" and abortion were different manifestations of the same agent.

In a continuation of these studies, Doll et al.[19,20] induced abortions in mares with defibrinated blood from a horse with "influenza". The association between virus abortion and respiratory infection was further studied by Doll et al.[18] These investigators showed that suckling hamsters infected with either liver suspensions from aborted fetuses or defibrinated blood from upper respiratory diseases, both led to a fatal hepatitis in the susceptible hamster. Immunological evidence provided by experiments in which infected hamster livers were used as a source of antigen and equine antiserum collected from horses that either aborted or experienced upper respiratory infection, showed reciprocal complement fixation indicating that the agents inducing upper respiratory infection and the abortion were similar or identical. It was then shown by Doll et al.,[21,22] that the viral agents of both human and swine influenza were not related to the virus found from the equine virus abortion disease in horses. Based on these results, Doll suggested that the term "influenza" no longer be used to describe the respiratory disease in the horse. Rather, he proposed the term "rhinopneumonitis" be used to describe the disease induced by equine abortion virus because of the characteristic signs of rhinitis in young horses and pneumonitis in the aborted fetuses. Further, it wasn't until the discovery of the equine influenza virus in Czechoslovakia in 1956,[43] that the term "equine influenza" was related to a specific virus and not used to describe the symptoms of disease. Electron microscopic analysis of the equine rhinopneumonitis virus revealed the morphology of a herpesvirus type 1 (EHV-1).[4] Two other species (types) of the herpesvirus group are indigenous to the Equidae family and are designated EHV-2 and EHV-3.[36]

In addition to upper respiratory infections, a paralytic syndrome is occasionally seen following EHV-1 infection. In 1949, Manninger[33] described a case of myelitis in an experimentally infected horse. The horse exhibited paralysis of the hind quarters and displayed urinary incontinence, and in 1966 Saaxegaard[40] reported a naturally occurring outbreak of posterior paralysis in mares and stallions on a Norwegian breeding farm. Virus isolated from the liver, brain, spinal cord, and spleen of one stallion was neutralized by a reference antiserum to EHV-1. Naturally occurring outbreaks have been reported by Thorsen and Little[46] in Canada, and by Greenwood and Simpson[28] in the United Kingdom.

Experimentally, Jackson and Kendrick[30] induced a neurologic disease in pregnant mares following inoculation with an EHV-1 strain.

In addition to abortion or respiratory disease in adult horses, a neonatal foal disease associated with EHV-1 infection has been described. The disease is manifested in still-births, weak foals which die within 24 hr of birth, or those which die 2 to 3 days after developing respiratory distress soon after birth.[16] Bryans et al. described another syndrome in which respiratory distress and diarrhea resulted in death of the foal at about 2 weeks of age.

B. Host Range

The natural host range for EHV-1 is the Equidae family although virus replication has been shown in kittens, rabbit cornea, baby mice, and on the chorioallantoic membrane of embryonated eggs.[36] Additionally, the virus was adapted to many different primary cultures, cell strains, and cell lines.[36]

II. CHARACTERISTICS OF EHV-1

The EHV-1 isolates are members of the Herpetoviridae family.[35] Members of this family have cubic symmetry and are enveloped. Capsid assembly takes place in the nucleus and envelopment occurs at the nuclear membrane. These viruses are sensitive to lipid solvents, capsids are composed of 162 capsomers, the diameter of the virion is approximately 100 nm, and the molecular weight of the DNA in the virion is 93×10^6 daltons. EHV-1 is further subclassified in the alpha-herpesvirinae subfamily. Alpha-herpesvirinae are rapidly growing, highly cytolytic viruses with short reproductive cycles. In cell culture, a rapid spread of infection occurs resulting in mass destruction of susceptible cells. Latent infections occur frequently, but not exclusively in ganglia.[35]

III. PATHOLOGY

A. Respiratory Form

The lesions found in the respiratory form of the disease include edema and fibrinous infiltration of the interstitial lobular tissue in the lung, consolidated areas of the apical lung lobes, and necrosis of bronchial epithelium. Also found are edema, congestion, and petechial hemorrhage of the nasal mucosa and lymph nodes of the head and mediastinum. Diagnosis is often based on the combination of clinical signs, rising serum titers to virus, and virus isolation. This virus is most readily isolated from nasal swabs during the first 3 to 4 days of the illness.[37]

B. Abortion

It has been suggested that the leading cause of equine abortions in the U.S. is due to EHV-1 infections. For example, McKercher[34] reported that anywhere from 10 to 90% of pregnant mares may abort during and outbreak of EHV-1. Gross lesions of aborted fetuses have been described.[12,15,44,48] In the aborted fetus, the most characteristic lesions appear in the liver, lung, and lymph nodes. The liver is usually congested and fatty, containing multiple grayish-white necrotic foci throughout the parenchyma. The thoracic cavity is generally filled with the serous fluid; the lungs are pale, heavy, and very edematous and congested. Fibrin casts are found in the trachea and bronchi. Most lymph nodes are hyperemic and necrotic. Other lesions may be icterous with congested spleen and intestines, and petechial hemorrhages in the heart and conjunctivitis. The placenta may be thick and edematous. The lesions are more pronounced in fetuses aborted after the 6th month. It has been suggested that the pronounced lesions in aborted fetuses after 6 months of age may be due to the immune system of the fetus being developed enough to mount an inflammatory and immune response.[44] The pathologic lesions found in aborted fetuses are also very characteristic of EHV-1 infections and have been described by numerous investigators.[12,22,44,48,49]

In an aborted fetus the lungs are edematous and congested with fibrin present in the alveoli and bronchi. The bronchial epithelium is necrotic and characteristic eosinophilic intranuclear inclusion bodies can be found in the alveolar and bronchiolar epithelial cells. Inclusion bodies involve the entire nucleus. Microscopic liver lesions are found in about 50% of the cases. Livers are fatty, there is an increased amount of intralobular connective tissue which contains areas of ventrilobular focal necrosis. The inclusion bodies found in the lung epithelium are also found in liver hepatocytes, the interlobular bile ducts, epithelium, and endothelial cells. The lymph nodes are hyperemic and inclusion bodies are found in the supporting reticular endothelial cells. Inclusion bodies have also been reported in the reticular cells of the spleen and thymus.

C. Neurological Form

The microscopic lesions have been described again by a number of investigators.[11,30,-32,38] Gross lesions include congestion of central nervous system blood vessels, occasional meningeal hemorrhage, and focal areas of malacia in the brain and spinal cord. The microscopic lesions are more characteristic and are best described as a nonsuppurative, disseminated, necrotizing, meningeal encephalomyelitis composed of two discrete entities. The first is a necrotizing vasculitis and the vessels most often involved are those along the meninges and their penetrating branches entering into the gray matter of the brain stem and white matter of the spinal cord. The small arteries and arterioles are most severely affected, and exhibit endothelial proliferation and necrosis resulting in thrombosis. In areas of vessel wall necrosis, a paravascular infiltrate of lymphocytes and macrophages is usually found extending through the tunica intima and into the tunica media and adventitia. The vasculitis is similar to that seen in the endometria of inoculated pregnant mares examined prior to abortion, but appears to be much more severe than the abortigenic form. It has been proposed by Jackson and Kendrick[30] that this vasculitis leads to a regional hypoxia or an infarction, causing focal areas of malacia in the brain and spinal cord. Such focal areas of necrosis are seen and represent the second characteristic microscopic lesion. The areas of focal necrosis exhibit swollen axons, demyelination, and necrosis of neurons. The parenchyma is partially replaced with neutrophils and glial cells but there is little evidence of which would be found in a hypoxic state.

D. Perinatal Disease

Both stillborn and foals born alive but surviving up to 72 hr were examined pathologically.[6,29] As stated by Hartley and Dixon:[29] "Most foals had a mild to severe pulmonary edema, histiocytic alveolitis and/or a necrotizing bronchitis/bronchiolitis with inclusions. Massive atelectasis was a feature of all neonatal deaths. Focal adrenocortical necrosis and depletion and/or degeneration of thymic and splenic lymphocytes were present in roughly the same proportions in aborted or perinatal deaths. Intranuclear inclusions were a constant finding in the lungs, were frequently seen in the thymus, rarely seen in the liver and not identified in spleen or adrenal."

In foals that survived longer than 6 hr after birth, hyaline membrane formation was observed in a high percentage of animals.

Bryans et al.[6] reported interstitial pneumonitis with lymphocyte depletion in the spleen and thymus with thymic depletion permitting potentially lethal bacterial infections.

IV. CLINICAL ASPECTS

A. Clinical Signs

1. Respiratory Disease

The respiratory disease syndrome following infection has been described by Doll et al.,[19,20,22] Manninger,[33] and McKercher.[34] Following infection, a biphasic fever 102 to 103°F which peaks on the second and third to fifth day postinoculation or postinfection, and generally lasts 8 to 10 days has been a consistent finding. Leukopenia consisting mostly of a neutropenia develops over this same time period. Absolute lymphocyte numbers generally remain normal over this period of time, however a leukocytosis may follow infection. A mild serous or seromucous discharge and cough are often reported to occur during the first 7 days. The nasal mucosa is often hyperemic and congested. Occasionally there is an associated conjunctivitis and a few horses become anorectic; however it is generally a mild illness but stressed or debilitated animals may develop secondary streptococcal, staphylococcal infections that can lead to a mucopurulent rhinitis, pleuritis, bronchopneumonia, or an enteritis.[22,38] Often these horses have enlarged submaxillary lymph nodes and it is believed that inapparent infections may be common in older horses, although to date no carrier state has been proved.

2. Abortion

The aborted fetus does not appear to have been dead very long *in utero* before being expelled, and it is usually expelled within the fetal membranes. In view of this, it has been proposed that the allantochorion separates from the endometrium prior to the expulsion of the fetus.[5,44] This has been supported histologically by examining uteri of experimentally infected mares prior to abortion. The intramicrovillar spaces between the epithelium of the trophoblast and endometrium are abnormally separated by large amounts of edematous fluid, and there is widespread necrosis of the chorion villi. This supports the theory that the allantochorion membrane becomes separated from the endometrium prior to the fetus being aborted.

The mononuclear perivascular cuffing around the subendometrial blood vessels and hyperplasia of regional lymph nodes is thought to arise from the reaction of fetal viral antigens and lymphocytes in the endometrium leading to histamine release and damaged intravillus capillaries. This, in turn, is thought to cause the chorion epithelial necrosis, thus, the abortion may be in part caused by an immune-mediated mechanism and not solely as a result of fetal viral infection.

If the infected fetuses are born alive, a different set of events takes place. In this neonatal disease resulting from EHV-1 infection, the foals are believed to be infected *in utero* during the last 2 months of gestation.[6,13,33] At the time of birth these foals can appear either weak or normal, but within one week will develop clinical signs. They stop nursing, becoming lethargic and weak. Many develop diarrhea and may show vestibular signs. There are signs of lymphopenia and leukocytosis. The foals do not respond to treatment and may develop multiple secondary bacterial infections, including bronchopneumonia, polyarthritis, enteritis, and colitis; death usually occurs within 2 weeks. It is difficult to isolate virus or demonstrate inclusion bodies from these foals, unlike the aborted fetuses in which one can demonstrate inclusion bodies. However, in this case the mares that foaled often show a fourfold rise in serum antibody titer to EHV-1 virus.

3. Neurologic Disease

The neurologic form of the disease has been reported in the U.S. and Europe.[11,13,28,30,38,40,46] The disease has been reported to occur in mares, stallions, and foals. Historically, the affected horses often originate from farms in which either respiratory disease or abortion had occurred recently. The horses present with varying degrees of neurological involvement. Early signs often start with atony of the bladder and slight ataxia, but commonly progress to a posterior paresis and even tetraparesis. Affected animals may become recumbent in as short a time as 24 hr but usually do not deteriorate any further after 24 to 48 hr. Recovery is generally slow and may not be complete. Cerebrospinal fluid analysis may show an increase in protein.

4. Perinatal Disease

Premature and term foals are affected. Foals born alive are weak and depressed and unable to stand without aid. Heart and respiratory rates are elevated as are some temperatures.[29] In foals that survive 48 to 72 hr, a severe inspiratory and expiratory dyspnea is observed with marked cyanotic mucous membranes.[16]

Inclusion bodies are rare and limited to endothelial cells when found, and there is no evidence of a direct neurotropism of the virus. It has been proposed[30,31] that an immune-mediated mechanism in which virus bound to antibody might be responsible for the lesions, however Thorsen and Little[46] failed to demonstrate viral antigens or equine immunoglobulin in sections of brain and spinal cord using labeled antibodies. However, isolation of virus has been made from brain and spinal cord by a number of investigators.[28,30,46] Unlike aborted mares, horses affected with a neurologic condition often show a rise in antibody titer during the time that clinical signs are manifested.[30,34,37]

V. EPIDEMIOLOGY

Doll and co-workers[19,23-26] investigated multiple equine herpesvirus outbreaks on breeding farms in Kentucky. It was shown that by December of each year, 80 to 100% of the weanlings and yearlings had measurable antibody titers to equine herpesvirus 1. On one breeding farm in August, 14% of the weanling horses had antibody titers; in November the incidence was 45%, and by December, 78% were showing an antibody response to EHV-1. In fall and early winter, an epidemic of upper respiratory desease spread through the weanlings and yearlings on those farms. The infection spread rapidly by direct contact, aerosol transmission, or by ingestion of contaminated food or water; and it was believed that the yearly infections of the weanlings and yearlings would also be a source of infection for the mares on the same farm. Sherman et al.[41,42] reported outbreaks of respiratory disease in 2 and 3-year-olds at Canadian race tracks for a 3-year period in which EHV-1 was isolated, predominantly from nasal swabs taken from horses during the annual winter epidemics. Thirty percent of the clinically ill horses showed a rise in virus neutralizing titer to EHV-1, and the winter outbreaks that were found at the tracks were usually mild and unlike those seen during the spring, because the spring outbreaks are usually attributed to true influenza epidemics. Thus, it appears that mild outbreaks of respiratory disease caused by EHV-1 infection are usually found during the early winter in the majority of weanlings and yearlings on breeding farms, and in a large percentage of 2- and 3-year-olds at race tracks. Spring epidemics of upper respiratory infection, therefore, most likely are caused by influenza virus. It is believed that epidemics of EHV-1 abortions occur in mares that were exposed during the late fall and early winter at the time weanlings and yearlings are showing signs of respiratory disease.[19,25] Doll et al.[25] showed rising serum titers in mares during these respiratory outbreaks. Mares rarely show any signs of disease at this time, nor do they show any illness at the time they abort. The abortions occur with no premonitory signs. Fetal membranes are expelled with the fetus or quickly after the abortion, and the mares do not suffer any uterine infections, and will breed back as quickly as if they had given birth to a full term live foal.

It has been reported that EHV-1 induced abortions have occurred in every month of the year except July and August.[23,24,26] Five percent of the abortions occur during the 8th to 11th month, 65% occur during the 9th and 10th month of gestation. Eight percent of the epidemics occurred during December through April, while 63% occurred from January to March. Most abortions occurred between January and March at the time when most mares were in their 9th to 11th month of gestation. In epidemics, 36% lasted less than 30 days, another 38% lasted less than 30 days, while another 38% lasted between 31 and 60 days. It has been estimated that the natural immunity against the abortigenic form may last 2 to 3 years.[3] Support for this is given by Doll and Bryans[23,24,26] who reported that the epidemic of abortion seems to cycle about every 3 years. Epidemiological analysis of EHV-1 infections is complex because of the different disease syndromes associated with infections. Initial experiments were designed on the basis of quantitive differences in cross-neutralization titers between virus strains. The results of these studies indicated two specific serotypes: subtype-1 was associated with fetal isolate and subtype-2 was associated with respiratory isolates.[1,8,9] Additionally, strain differences were also associated with viral growth rates, host range, and viremia.[9] Recent studies by a number of investigators where strain differences were analyzed by restriction endonuclease maps showed distinct patterns for fetal and respiratory isolates. That is, all isolates typed as fetal isolates had similar restriction patterns as did the respiratory isolates.[36,39,40,45,47,48-50]

VI. PATHOGENESIS OF INFECTION

Inoculation with the wild type strain of EHV-1 leads to clinical respiratory disease with the appearance of serum-neutralizing antibodies at 7 to 10 days and maximum peaking in 3 to 5 weeks. The titers decline gradually and persist for only 4 to 5 months. Inoculation virus can be isolated from the nasal pharynx. A viremia develops following inoculation but the virus can only be isolated from the buffy coat fraction of a blood sample.[3]

It has been proposed that the virus remains intracellular in macrophages during this time. A virus can be isolated from this buffy coat beginning 4 to 6 days, and for as long as 20 days, postinfection. Horses that showed neither a rise in virus neutralizing antibody titers nor viremia, were considered to be protected.[3,34] McKercher[34] and Bryans[3] showed horses that had a neutralization index (NI) of greater than 2 were rarely able to be infected after intranasal inoculation of a virulent strain of virus; however those with a neutralization index of less than 2 were not protected. The criteria that was used was either an animal became viremic or had an anamnestic response. However, when an intramuscular inoculation of challenge virus was used, even those horses with a neutralization index of greater that 2 were not protected and became viremic. This led to the proposal that a local immunity — that is, that provided by secretory IgA in the nasopharyngeal mucosa is very important for protection against natural EHV-1 infection.[5]

It was believed that high serum virus netralizing antibody titers indirectly reflected an inadequate amount of secretory IgA antibodies. This was shown by Bryans[3] in which he produced infection in the presence of high VN antibody titers. Low VN antibody titers are not protective as it appears that once the virus has passed the nasal pharynx it requires both a cellular and a humoral response to eliminate the infectious agent. It appears that protection from natural infection lasts only about 3 to 6 months. Doll and Bryans[25] showed that EHV-1-exposed mares on breeding farms seroconverted during the late fall and early winter epidemics of respiratory disease in weanlings and yearlings, and were confirmed by Studdert.[44] At this period of time, mares are usually at about their 8th month of gestation; however most mares abort during late winter or early spring and show no evidence of respiratory disease or a rise in virus neutralizing antibody titer at that time. The incubation period for abortion to occur following experimental infection by intranasal, intramuscular, intravenous, or subcutaneous route has been reported.[25,26,44] The incubation periods ranged from 14 to 120 days but most appear to abort between 20 to 90 days postinfection. These experimental results are consistent with the natural incubation time if one considers that mares are infected in the late fall or early winter, as proposed by Doll and Bryans.[25]

Direct inoculation of the fetus, however, will lead to abortion, usually within 3 to 5 days. In a like manner as seen with respiratory disease, abortion can be produced in mares with high serum virus neutralizing antibody titers, again indicating that the protection is not dependent necessarily entirely on humoral immunity. The natural immunity is believed to last anywhere from 2 to 3 years.

VII. VACCINES

Serum therapy was initially used in attempts to prevent abortion epidemics in which convalescent serum from the initial cases was used in an attempt to control the epidemic. However, due to the time of initial exposure and collection of serum, the convalescent serum probably had very low levels of antibody titer and would not be efficacious in preventing the initial rounds of infection. Doll and Bryans[23] derived killed virus vaccine from aborted fetuses, and these vaccines failed to protect against the disease, and occasionally resulted in a neonatal isoerythrolysis in foals of the vaccinated dams. They then attempted to produce a killed vaccine from hamsters. This vaccine also was not protective, and Bryans[7] reported that such vaccine was useful if mixed with an adjuvant and used at 60-day intervals. An attenuated vaccine has been produced in hamsters by Doll and Bryan.[24] The attenuated vaccine was produced in Syrian hamsters and has been used by intranasal inoculation in the mares, and the strain that was used would still produce abortion if inoculated directly into the fetus, but would confer protection and had a reduced virulence when used intranasally. The immunity produced from this attenuated vaccine lasted approximately 6 months. The virus then was used in a planned infection vaccination program. All weanlings, yearlings, mares, and stallions were vaccinated in July and October of each year. The reason for this was to reduce the virus shedding and disease in young horses during the winter and thus prevent the mares from being exposed to native indigenous virus strains. This approach is recommended only for endemic areas because the vaccine strain can produce abortions, however in a very low percentage of vaccinated mares (about 0.3%).[24] Its benefit was in preventing explosive epidemics and reducing the incidence of naturally occurring abortion to about 0.6% The vaccine virus was shed for as long as 3 weeks and during this time it was suggested that one confine the vaccinated horse for 3 weeks following its use.

Attenuated (modified-live) vaccines have been produced in cell cultures. Dutta and Shipley[27] evaluated such a vaccine and found it to be of questionable value in preventing abortion when used according to the manufacturer's recommendation. Few attempts have been made to produce a vaccine specifically for the prevention of the neurological form of EHV-1 infection. A highly effective and safe vaccine is still to be developed.

REFERENCES

1. Bartha, A., in *Proc. 2nd Int. Conf. Equine Infections Diseases,* Bryans, J. and Gerber, H., Eds., S. Karger, Basel, 1970, 18.
2. Bracken, E. C. and Norris, J., *Proc. Soc. Exp. Biol. Med.,* 98, 747, 1958.
3. Bryans, J., *J. Am. Vet. Med Assoc.,* 155, 294, 1969.

4. Bryans, J. T., *Proc. Am. Assoc. Equine Pract.*, 1 (968), 119, 1969.
5. Bryans, J. T. and Prickett, M., in *Proc. 2nd Int. Conf. Equine Infections Diseases*, S. Karger, Basel, 1970, 34.
6. Bryans, J., Swerczek, T., Darlington, R., and Crowe, M., *J. Equine Med. Surg.*, 1, 20, 1977.
7. Bryans, J., in *Proc. 4th Int. Conf. Equine Infections Diseases*, Bryans, J. and Gerber H., Eds., Veterinary Publ., Princeton, N.J., 1978, 83.
8. Burrows, R., in *Proc. 2nd Int. Conf. Equine Diseases*, Bryans, J. and Gerber, H., Eds., S. Karger, Basel, 1970, 1.
9. Burrows, R. and Goodridge, D., in *Proc. 3rd Int. Conf. Equine Infectious Diseases*, Bryans, J. and Gerber, H., Eds., S. Karger, Basel, 1973, 306.
10. Campbell, T. and Studdert, M., *Vet. Bull.*, 53, 135, 1983.
11. Charlton, K., Mitchell, D., and Corner, A., *Vet. Pathol.*, 13, 59, 1976.
12. Corner, A., Mitchell, D., and Meads, E., *Cornell Vet.*, 53, 78, 1963.
13. DeLahunta, A., *Cornell Vet. Suppl.*, 7, 122, 1978.
14. Dimock, W. W. and Edwards, P. R., *Cornell Vet.*, 26, 231, 1936.
15. Dimock, W. W., *J. Am. Vet. Med. Assoc.*, 96, 656, 1940.
16. Dixon, R., Hartley, W., Hutchins, D., Lepherd, E., Feilen, C., Jones, R., Love, D., Sabine, M., and Wells, A., *Aust. Vet. J.*, 54, 103, 1978.
17. Doll, E., *Cornell Vet.*, 42, 505, 1952.
18. Doll, E., Richards, M., and Wallace, E., *Cornell Vet.*, 44, 133, 1954.
19. Doll, E., Wallace, M., and Richards, M., *Cornell Vet.*, 44, 181, 1954.
20. Doll, E. and Kintner, J., *Cornell Vet.*, 44, 355, 1954.
21. Doll, E., McCollum, W., Bryans, J., and Crowe, E., *Am. J. Vet. Res.*, 17, 262, 1956.
22. Doll, E., Bryans, J., McCollum, W., and Crowe, E., *Cornell Vet.*, 47, 3, 1957.
23. Doll, E. and Bryans, J., *Cornell Vet.*, 53, 24, 1963.
24. Doll, E. and Bryans, J., *Cornell Vet.*, 53, 249, 1963.
25. Doll, E. and Bryans, J., *J. Am. Vet. Med. Assoc.*, 141, 351, 1962.
26. Doll, E. and Bryans, J., *J. Am. Vet. Med. Assoc.*, 142, 31, 1963.
27. Dutta, S. and Shipley, W., *Am. J. Vet. Res.*, 36, 445, 1975.
28. Greenwood, R. and Simpson, A., *Educ. Vet. J.*, 12, 113, 1980.
29. Hartley, W. and Dixon, R., *Equine Vet. J.*, 11, 215, 1979.
30. Jackson, T. and Kendrick, J., *J. Am. Vet. Med. Assoc.*, 158, 1351, 1971.
31. Jackson, T., Osborn, B., Cordy, D., and Kendrick, J., *Am. J. Vet. Res.*, 38, 709, 1977.
32. Little, P. and Thorsen, J., *Vet. Pathol.*, 13, 161, 1976.
33. Manninger, R., *Acta Vet. Hung.*, 1, 62, 1949.
34. McKercher, D., in *The Herpesviruses*, Academic Press, New York, 1973, 527.
35. Melnick, J. L., Taxonomy and nomenclature of viruses, *Prog. Med. Virol.*, 28, 208, 1982.
36. O'Callaghan, D., Allen, G., and Randall, C., in *Proc. 4th Int. Conf. Equine Infectious Diseases*, Bryans, J. and Gerber, H., Eds., Veterinary Publ., Princeton, N.J., 1978, 1.
37. Platt, H., *Vet. Rec.*, 91, 31, 1972.
38. Platt, H., Singh, H., and Whitewell, K., *Equine Vet. J.*, 12, 118, 1980.
39. Sabine, M., Robertson, G., and Whalley, J., *Aust. Vet. J.*, 57, 148, 1981.
40. Saaxegard, F., *Nord. Vet. Med.*, 18, 504, 1966.
41. Sherman, J., Thorsen, J., Barnum, D., Mitchell, W., and Ingram, D., *J. Clin. Microbiol.*, Mar. 2977, 285, 1977.
42. Sherman, J., Mitchell, W., Martin, S., Thorsen, J., and Ingram, D., *Can. J. Comp. Med.*, 43, 1, 1979.
43. Sovinova, O., Tumova, B., Pouska, F., and Nemec, J., *Acta Virol.*, 2, 52, 1958.
44. Studdert, M. J., *Cornell Vet.*, 64, 94, 1974.
45. Studdert, M., Simson, T., and Roizman, B., *Science*, 214, 562, 1981.
46. Thorsen, J. and Little, P., *Can. J. Comp. Med.*, 39, 358, 1975.
47. Turtinen, L., Allen, G., Darlington, R., and Bryans, J., *Am. J. Vet. Res.*, 42, 2099, 1981.
48. Westerfield, C. and Dimock, W., *J. Am. Vet. Med. Assoc.*, 109, 101, 1946.
49. Wildy, P., in *The Herpesviruses*, Academic Press, New York, 1973, 1.

Chapter 9

CANINE HERPESVIRUS-1

Steven Krakowka

TABLE OF CONTENTS

I. INTRODUCTION AND HISTORY

In 1964, during an investigation of an acute, fatal, hemorrhagic disease of neonatal and infant puppies, Carmichael and colleagues isolated the causative agent in tissue culture.[1] The agent was tentatively identified as belonging to the herpesvirus group. Other veterinary virologists soon reported similar isolations from similar situations.[2-5] These early studies were quick to associate this viral isolate and infection, designated as canine herpesvirus-1 (CHV-1) with a distinct clinicopathologic syndrome in affected pups. Salient clinical features within an affected litter were stereotypic and consisted of rapid onset, high morbidity, high mortality. Almost invariably, affected pups were less than 3 weeks of age. The incidence of the disease within the canine population was sporadic but could become epizootic within a kennel. Typically, necropsy findings of fatally affected dogs revealed widespread areas of necrosis and hemorrhage in many parenchymal tissues. Subsequent study with this agent within the dog population has revealed that the neonatal fatal puppy syndrome was an unusual manifestation of the infection. In fact, the predominant clinical manifestation was a selflimiting and subclinical upper respiratory infection in adult dogs.[6-9] Exceptionally, adult dogs developed a genital form,[3,10] and it was from this genital form that the neonatal disease developed. The epidemiologic and biological similarities between this viral disease in the dog and a similar viral disease in man, namely herpesvirus-2 are obvious.

II. CHARACTERIZATION OF CANINE HERPESVIRUS

Canine herpesvirus is a typical member of the herpesvirus group.[1-3] Herpesviruses are medium-sized, cubic viruses composed of 162 structural subunits, or capsomeres, and assembled in isometric symmetry to form a protein capsid core around coiled viral DNA. In its typical form the entire virus particle, or virion, is surrounded exterior to the capsid by a loose flexible membranous envelope.[11] Canine herpesvirus DNA has been the subject of several studies.[12,13] Its buoyant density in cesium gradients have been determined to be 1.69 to 1.75 g/cm^3. The virus replicates readily in primary canine kidney cell cultures,[1,3,14-17] and harvests from that have been used to characterize the virus. The virus is a nonhemagglutinating agent that is ether and heat-sensitive. In addition, infectivity is completely abolished by incubation of CHV-1 at pH 5 or below. The virus as indicated readily is infectious for primary canine kidney cell cultures, and also Maden Darby canine kidneys (MDCK) cells. The dog is the only known host for this agent and viral growth occurs exclusively in cells of canine origin. In canine kidney cultures, viral cytopathology consists of foci rounded up refractile cells within 3 days after inoculation. These refractile cells are easily detached from glass.[18] With time these foci enlarge to total destruction of the monolayer within 6 to 7 days after infection. The incubation period can be shortened to within less than 24 hr by serial passage in dog kidney cells. Coincident with the appearance of infected foci are the appearance of variable numbers of polykaryocytes, i.e., syncytial giant cells, and intranuclear type A Cowdry inclusion bodies. As is typical of the herpesvirus group, these inclusion bodies contained herpesvirus DNA.

Different plaque types have been identified in inocula of CHV-1. One strain of CHV-1, the F-205 or mP variant of CHV, produces small plaques, approximately one half the size of the wild type of CHV.[19] This small plaque size is consistent over temperature ranges as low as 30° and small plaque characteristics are retained over multiple serial in vitro passages. Animal inoculation studies done with the small plaque and standard plaque variants of CHV have revealed that the small plaque variant has significantly reduced virulence for puppies. Since no immunological or serological difference can be detected between the small plaque variant and standard CHV strains, the F-205 strain of CHV shows promise as a prototype vaccine strain virus.

In vitro, CHV shows a marked temperature sensitivity to viral replication.[12,20,21] Peak replication has been found to occur between 35 and 36°C. It turns out for in vitro work that the optimal in vitro temperature for CHV replication is below the normal core body temperature of adult dogs, but equivalent to the body temperatures of neonatal puppies in the immediate postnatal period, that is, the temperature is between 70 and 80°F or 35 to 36°C.

Other variables noted with in vitro propagation are the following: canine pulmonary macrophages are susceptible to CHV infection but the amount of infectious virus is less than 0.1% when compared to similar viral productions performed on primary dog kidney cells.[20] For replication of virus, cells grown as monolayers are superior to those maintained in suspension.[16] Highly enriched media enhance viral replication, presumably by enhancing the viability of the host cells. Young, rapidly growing cultures, when infected tend to yield more virus than do older or stationary cultures.

III. SEROLOGY

A number of studies have evaluated the immune response to these important viral pathogens within the dog population. Antibody activity to CHV-1 may be measured by a standard neutralization test.[1,22,23] When measured by the neutralization method, serum titers are generally low and can be missed. Immunodiffusion in gel has also been used to measure antibodies to this virus. It is, as one would predict, somewhat laborious to do this on any great or extensive scale. The test is greatly enhanced by the addition of complement of the reaction mixture[24] and the sensitivity can be further increased by the application of a plaque assay into the procedure. Using the complement-dependent virus neutralization assay, dog populations have been surveyed for an incidence of antibody to CHV-1. Anywhere from 0 to 67% of the tested dog populations contained antibody to CHV-1.[8,9,24-26] It is important to stress that many of these dogs surveyed were adult dogs living under natural conditions and did not necessarily exhibit clinical signs or lesions associated with CHV-1 infection.

IV. PATHOGENESIS OF THE VIRAL INFECTION WITHIN THE AFFECTED ANIMAL

The development of this virus infection within the host and the consequences of infection are dependent upon the age of the host. As such, it is convenient to divide the results of pathogenesis experiments into that sequence of events observed in neonatal or young animals, and that sequence of events observed in the weanling or adult animal.

A. Adult Form of CHV-1 Infection

With adult animals, or at least animals older than 2 weeks of age, transmission from shedder or carrier animals to susceptible animals is accomplished most likely by the oronasal route. Based upon a systematic study of exposure to CHV in beagle dogs, it is apparent that infection by this route results in an inapparent or mild clinical infection.[27] Infectious virus can be recovered from these animals from the oral cavity and nasal regions for at least 2 weeks and probably slightly longer following infection. There is no evidence of viral infection in other tissues distant to the site of inoculation. Virus could occasionally be isolated from the respiratory tract, presumably as a result of subsequent spread of the virus from local tonsillar infection. It's important to point out that in this experimental study, the virus could only be recovered from the proximate sites of inoculation. Significantly, however, virus has been recovered from many tissues of diseased[28] or clinically normal dogs, including primary kidney cultures.[29] This finding implies that the virus is capable of being carried as a latent or inapparent infection in the adult animal. This also implies that under these circumstances carrier animals do,[30] in fact, impose some risk for infection of the neonatal dog.

B. Neonatal Form of CHV-1 Infection

In contrast to this mild and selflimiting upper respiratory disease noted in the adult or weanling animal, the pathogenesis of CHV-1 infection in neonatal puppies is much different and much more dramatic.[27] As indicated earlier, the clinical course in these animals is rapid and progressive with a high morbidity and very high mortality. Following oronasal installation, viral antigen can first be demonstrated within the epithelium of the nasal mucosa within 24 hr after inoculation. Three to four days after infection, virus was detected within regional lymph nodes of the respiratory tract and also within lymphoid tissues of the body in general. It is likely that systemic dissemination of the virus probably occurs by hematogenous-origin monocytes as is known to occur with other members of the herpesvirus group. Viremia, if detected, was found to be exclusively cell-associated. Beyond 4 days of infection the lesions progress from small areas of involvement to progressively larger ones, including production of the characteristic lesions of necrosis and hemorrhage (Figures 1 and 2). Virtually all tissues contain infectious virus and/or antigen and the presence of lesions can be correlated with the presence of virus. In most cases, death results in anywhere from 5 to 15 days after inoculation.

Under natural conditions, the route of exposure or inoculation to neonatal dogs is thought to occur perinatally. By that, one means that puppies acquire virus during transit through the birth canal of a bitch experiencing genital or inapparent herpesvirus infection.[18] It is known, however, that the virus, under both natural[31] and also experimental conditions,[32] can be acquired *in utero*. Like other members of the herpesvirus group, this is not a unique finding. Specific evidence of CHV-1 transmission across placenta has been acquired in recent years. In these animals, the placenta was shown to be a primary site of viral replication. Multiple gray to white foci of necrosis and hemorrhage containing CHV virus have been noted in these cases.[31,32] Some evidence of mineralization was observed.

Fetal infection has been established in seronegative pregnant dogs inoculated between the 48th to 53rd day of gestation. As before, lesions were similar to those observed in postnatally infected animals. In addition to placental lesions, viral antigen could be demonstrated in the walls of maternal and fetal blood vessels within the trophoblastic cells of the junctional zone of the placenta. Regardless of whether the young dogs acquire the infection in utero or in the postnatal environment, it is likely that proximate cause of death is an acute necrotizing viral encephalomyelitis.[30] It is thought that virus is brought to the brain by virus-infected monocytes. Confluent zones of hemorrhage and malacia have been observed in moribund animals. These areas invariably contain viral antigen or viral inclusion body-like structures in affected neural or parenchymal cells.

V. HISTOPATHOLOGIC LESIONS OF CHV-1 INFECTION

The histopathologic lesions associated with CHV infection have been well characterized. Lesions, if present in the adult, are mild and focal or multifocal in nature. They consist of patchy interstitial pneumonia, mild serous conjunctivitis and/or rhinitis, and exceptionally the presence of vesicular-type lesions on the genital[10] or upper respiratory mucosa. In contrast, lesions in fatally affected puppies reflect the consequences of a widespread and disseminated viral disease. The stereotypic lesion is parenchymal cell necrosis with accompanying local vasculitis and hemorrhage.[1,18,30-34] Inclusion bodies typical of herpesvirus are of the Cowdry type A morphology and may be seen within

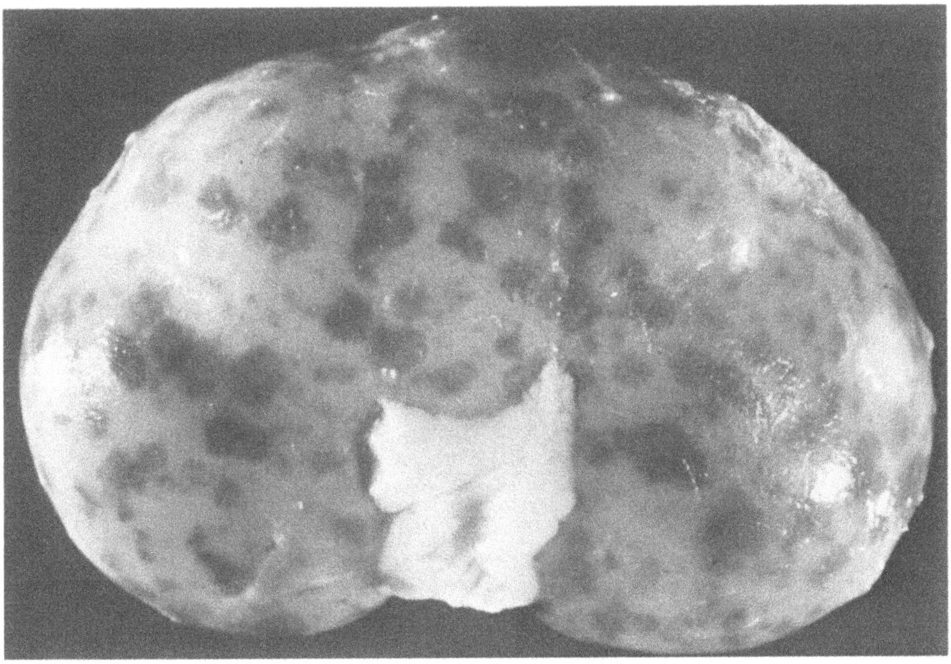

FIGURE 1. A characteristic lesion of CHV-1 infection in neonatal dogs. The lesion consists of bilateral necrosis and hemorrhages in the kidneys. (Photograph courtesy of Daniel Morton, D.V.M., The Ohio State University.)

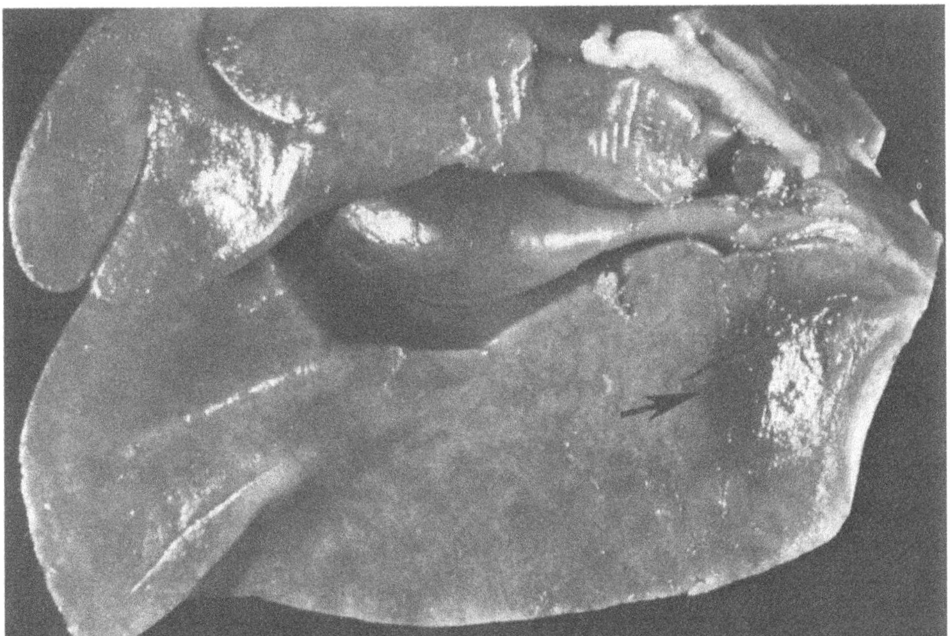

FIGURE 2. Hepatocellular hemorrhage and necrosis (arrow) in a pup fatally affected with CHV-1. (Photograph courtesy of Daniel Morton, D.V.M., The Ohio State University.)

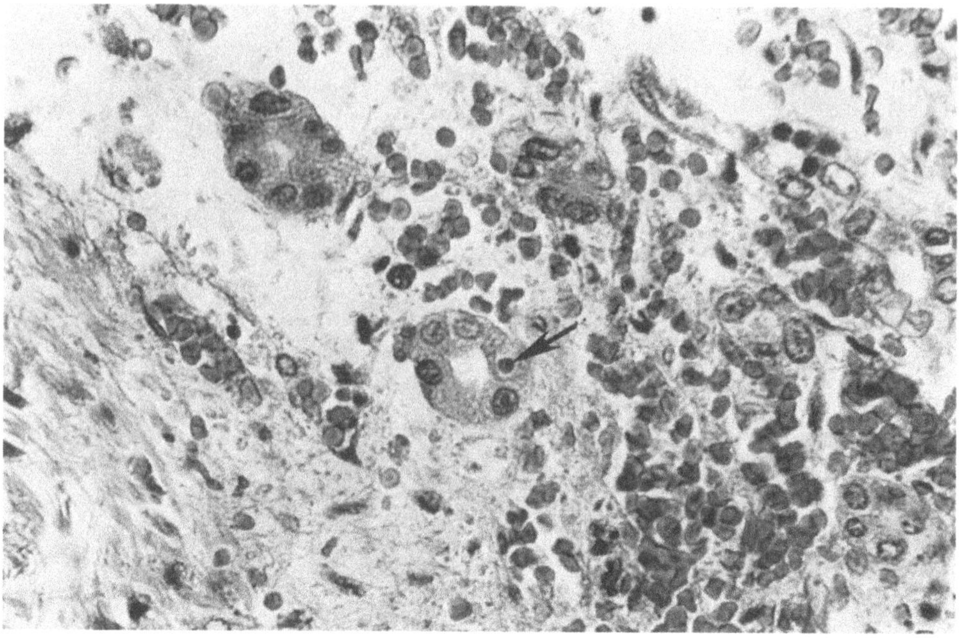

FIGURE 3. An intranuclear CHV-1 inclusion body (arrow) within a renal tublar epithelial cell in an area of hemorrhage.

and around these lesions (Figure 3). Within the brain, neuronal necrosis and degeneration are accompanied by microgliosis, moderate to extensive lymphoplasmacytic perivascular cuffing, occasional evidence of neutrophilic infiltration, some hemorrhage, and astrocytic intranuclear inclusions.[35] A similar lesion has been noted within the eye of affected dogs which shows a mononuclear infiltration into the uveal tract and the subsequent development of a panuveitis.[36] Inflammation, if present, is usually bilateral.

In spite of the high morbidity and mortality, not all animals neonatally infected with CHV-1 will die. Animals that recover either naturally or following artificial elevation of body temperature will show residual lesions that can be attributable to widespread CHV-1 infection.[37] These include multifocal granulomatous encephalomyelitis, a resolving interstitial pneumonia, and renal tubular necrosis and fibroplasia. In addition, segmental cerebellar hypodysplasia and retinal dysplasia may also occur. It is important to stress that even in these animals, CHV-1 antigen can be demonstrated within such tissues.

VI. TREATMENT AND PREVENTION

CHV infection within the dog population can be viewed as a sporadic infectious disease problem. As such, there is little pressure or economic incentive to develop vaccine products. By far and away the most important manifestation of the disease is a clinically inapparent, mild and selflimiting upper respiratory tract disease. As such, no treatment is called for and none is warranted. As indicated above, the chief biological significance involves CHV-1 infection in neonatal puppies. Carmichael and others have reported that the virus will be inhibited in its growth by elevating the ambient temperature during incubation.[12,16,20,21] This observation has been extended to the in vivo situation. Investigators report that some benefit can be achieved by artifically elevating the body temperature of suspect animals with heat lamps or other external heat sources.[21] The mortality rates will still be high, easily in excess of 50%. Further, any recovered animals will likely have residual lesions[37] and serve as either latent or inapparent carriers of the virus.[11,18,27,29,30] Bitches with a history of CHV-1 infected litters[30] may be difficult to manage. Probably the best thing to do is to follow the general recommendations for human herpesvirus-2 infection. Since the infection under most circumstances is acquired by transit through the genital tract, Cesarean section should prevent exposure during this critical period of maximal susceptibility. Obviously, hand-raising the litter away from the bitch will decrease the chance for exposure to CHV-1. Since the disease is innocuous in animals greater than 2 to 3 weeks of age, it is likely that animals at this age or older are not at risk for developing CHV-1-related disease.

REFERENCES

1. Carmichael, L. E., Strandberg, J. D., and Barnes, F. D., Identification of a cytopathogenic agent infectious for puppies as a canine herpesvirus, *Proc. Soc. Exp. Biol. Med.*, 120, 644, 1965.
2. Geldard, H., Geering, W. A., and Bagust, T. J., Isolation of a herpesvirus from neonatal dogs in Australia, *Aust. Vet. J.*, 47, 286, 1971.
3. Poste, G., Characterization of a new canine herpesvirus, *Arch. Ges. Virusforsch.*, 36, 147, 1972.
4. Watt, D. A., Spradbrow, P. B., and Lamberth, J. L., Neonatal mortality of puppies in Queensland caused by canine herpesvirus infection, *Aust. Vet. J.*, 50, 120, 1974.
5. Hahimoto, A., Hirai, K., Miyoshi, A., Shimakura, S., Yagami, K., Kato, N., Kunihiro, K., Fujiara, A., and Kitazawa, K., Naturally occurring canine herpesvirus infection in Japan, *Jp. J. Vet. Sci.*, 40, 157, 1978.
6. Wright, N. G., Cornwell, H. J. C., Thompson, H., and Stewart, M., Canine herpesvirus respiratory infection, *Vet. Rec.*, 25, 108, 1970.
7. Thompson, H., Wright, N. G., and Cornwell, H. J. C., Canine herpesvirus respiratory infection, *Res. Vet. Sci.*, 13, 123, 1972.
8. Wright, N. G., Thompson, H., Cornwell, H. J. C., and Taylor, D., Canine respiratory virus infections, *J. Small Anim. Pract.*, 15, 27, 1974.
9. Binn, L. N., Alford, J. P., Marchwicki, R. H., Keefe, T. J., Beattie, R. J., and Wall, H. G., Studies of respiratory disease in random-source laboratory dogs: viral infections in unconditioned dogs, *Lab. Anim. Sci.*, 29, 48, 1979.
10. Hill, H. and Mare, C. J., Genital disease in dogs caused by canine herpesvirus, *Am. J. Vet. Res.*, 35, 669, 1974.
11. Martin, W. B., The herpesviruses, *Vet. Rec.*, 99, 352, 1976.
12. Lust, G. and Carmichael, L. E., Suppressed synthesis of viral DNA, protein and mature virions during replication of canine herpesvirus at elevated temperature, *J. Inf. Dis.*, 124, 572, 1971.
13. Lust, G. and Carmichael, L. E., Properties of canine herpesvirus DNA, *Proc. Soc. Exp. Biol. Med.*, 146, 213, 1974.
14. Strandberg, J. D. and Carmichael, L. E., Electron microscopy of a canine herpesvirus, *J. Bacteriol.*, 90, 1790, 1965.
15. Strandberg, J. D. and Aurelian, L., Replication of canine herpesvirus. II. Virus development and release in infected dog kidney cells, *J. Virol.*, 4, 480, 1969.
16. Aurelian, L., Factors affecting the growth of canine herpesvirus in dog kidney cells, *Appl. Microbiol.*, 17, 179, 1969.
17. Poste, G., Lecatsas, G., and Apostolov, K., Electron microscopic study of the morphogenesis of a new canine herpesvirus in dog kidney cells, *Arch. Ges. Virusforsch.*, 39, 317, 1972.
18. Cornwell, H. J. C. and Wright, N. G., Neonatal canine herpesvirus infection: a review of present knowledge, *Vet. Rec.*, 84, 2, 1969.
19. Carmichael, L. E. and Medic, B. L. S., Small-plaque variant of canine herpesvirus with reduced pathogenicity for newborn pups, *Infect. Immun.*, 20, 108, 1978.
20. Carmichael, L. E. and Barnes, F.D., Effect of temperature on growth of canine herpesvirus in canine kidney cell and macrophage cultures, *J. Infect. Dis.*, 120, 664, 1969.

21. Carmichael, L. E., Barnes, F. D., and Percy, D. H., Temperature as a factor in resistance of young puppies to canine herpesvirus, *J. Inf. Dis.*, 120, 669, 1969.
22. Aurelian, L., Demonstration by plaque reduction technique of immunologic relationship between canine herpevirus and herpes simplex virus, *Proc. Soc. Exp. Biol. Med.*, 127, 485, 1968.
23. Binn, L. N., Koughan, W. P., and Lazar, B. A., A simple plaque procedure for comparing antigenic relationships of canine herpesvirus, *J. Am. Vet. Med. Assoc.*, 156, 1724, 1970.
24. Osterhaus, A., Berghuis-deVries, J., and Steur, K., Antiviral antibodies in dogs in the Netherlands, *Zbl. Vet. Med.*, 24, 123, 1977.
25. Binn, L. N., Marchwicki, R. H., Eckermann, E. H., and Fritz, T.E., Viral antibody studies of laboratory dogs with diarrheal disease, *Am. J. Vet. Res.*, 42, 1665, 1981.
26. Fulton, R. W., Ott, R. L., Duenwald, J. C., and Gorham, J. R., Serum antibodies against canine respiratory viruses: prevalence among dogs of eastern Washington, *Am. J. Vet. Res.*, 35, 853, 1974.
27. Carmichael, L. E., Herpesvirus canis. Aspects of pathogenesis and immune response, *J. Am. Vet. Med. Assoc.*, 156, 1714, 1970.
28. Kakuk, T. J., Conner, G. G., Langham, R. F., Moore, J. A., and Mitchell, J. R., Isolation of a canine herpesvirus from a dog with malignant lymphoma, *Am. J. Vet. Res.*, 30, 1951, 1969.
29. Smith, R.F., Yamashiroya, H. M., and Magis, J. M., Recovery of a canine herpesvirus from primary kidney cultures derived from a closed dog colony, *Appl. Microbiol.*, 20, 523, 1970.
30. Hunoll, D. L. and Hemelt, D. E., Clinical observations of canine herpesvirus, *J. Am. Vet. Med. Assoc.*, 156, 1706, 1970.
31. Hashimoto, A., Hirai, K., Okada, K., and Fujimoto, Y., Pathology of the placenta and newborn pups with suspected intrauterine infection of canine herpesvirus, *Am. J. Vet. Res.*, 40, 1236, 1979.
32. Hashimoto, A., Hirai, K., Yamaguchi, L., and Fujimoto, Y., Experimental transplacental infection of pregnant dogs with canine herpesvirus, *Am. J. Vet. Res.*, 43, 844, 1982.
33. Wright, N. G. and Cornwell, H. J. C., Experimental herpesvirus infection in young puppies, *Res. Vet. Sci.*, 9, 295, 1968.
34. Love, D. N. and Huxtable, C. R. R., Naturally-occurring neonatal canine herpesvirus infection, *Vet. Rec.*, 99, 501, 1976.
35. Percy, D. H., Olander, H. J., and Carmichael, L. E., Encephalitis in the newborn pup due to a canine herpesvirus, *Pathol. Vet.*, 5, 135, 1968.
36. Albert, D. M., Lahav, M., Carmichael, L. E., and Percy, D. H., Canine herpes-induced retinal dysplasia and associated ocular abnormalities, *Invest. Ophthalmol.*, 15, 267, 1976.
37. Percy, D. H., Carmichael, L. E., Albert, D. M., King, J. M., and Jonas, A. M., Lesions in puppies surviving infection with canine herpesvirus, *Vet. Pathol.*, 8, 37, 1971.

Chapter 10

CANINE PARVOVIRUS

Roy V. H. Pollock and Colin R. Parrish

TABLE OF CONTENTS

I. INTRODUCTION

A previously unknown disease of dogs, characterized by the acute onset of pyrexia, vomiting and diarrhea, first appeared in 1977 and spread rapidly.[1-18] A parvovirus was isolated from such cases almost simultaneously in North America,[1,3,4,10] Europe,[11,15] and Australia.[13] By 1981, the disease was panzootic (Table 1). A second novel syndrome, viral myocarditis in weanling pups, appeared simultaneously.[19-22,27,28,38,44,56,59,60,63,66-68,73]

The causal agent of both syndromes is a new viral species of the genus *Parvovirus.* Although the virus has not been formally classified by the International Committee for Virus Nomenclature, the designation will likely be canine parvovirus type-2 (CPV-2). The Minute Virus of Canines (MVC),[74] a serologically distinct parvovirus of dogs, will probably be designated CPV-1.

Retrospective serologic studies indicate that CPV-2 emerged only recently. Anti-CPV-2 antibody was not detected in sera collected prior to 1976 in Europe,[26,52] and 1978 in the U.S.,[4,75] Australia,[76] New Zealand,[76a] and Japan.[49,50] Although the number of samples in each study was small, the concurrence of results strongly supports the concept that CPV-2 is indeed a new canine pathogen.

Canine parvovirus-2 is closely related serologically to feline parvovirus (feline panleukopenia virus; FPV),[3,4,6,13,53,77-81] mink enteritis virus (MEV),[6,75,80,81] and a raccoon parvovirus,[82] but it is unrelated to CPV-1 (MVC).[75,83] Limited serologic cross-reactivity with porcine parvovirus also has been reported.[84,85] The antigenic similarities among CPV-2, FPV, and MEV were recognized early, prompting speculation about the origin of the canine virus[13] and leading to the practical use of FPV to immunize dogs.[3,86-91]

Early studies were concerned principally with the description of the natural disease and the practical problems of immunization and control. More recent studies have begun to elucidate viral structure and behavior. This chapter summarizes existing information on CPV-2 and the pathogenesis of canine parvoviral disease in dogs. Many aspects must be regarded as tentative, however.

II. BASIC AND MOLECULAR VIROLOGY

Canine parvovirus (CPV)-2 is an autonomous parvovirus, closely related both antigenically and in DNA sequence to the parvoviruses causing feline panleukopenia and mink enteritis.[3,13,77-81,92,93] It also is related in DNA sequence to the H-1 parvovirus[94] and probably to other autonomous parvoviruses.[94,99]

Studies of CPV-2 are reviewed where those exist. Studies of related parvoviruses are cited where studies of CPV-2 have not been reported, but where similarities are probable.

A. Origin

The disease caused by CPV-2 was unknown prior to 1977. Serologic evidence for infection of dogs by CPV-2 prior to that year has not been detected in studies conducted in the U.S., Australia, Japan, or Europe (Table 2). The CPV-2 is closely related to FPV and MEV, and it has been suggested that CPV-2 is a host range variant of one

Table 1
EARLY REPORTS ON THE OCCURRANCE OF CPV-2 ENTERITIS AND MYOCARDITIS IN VARIOUS COUNTRIES

Country	References
Australia	Huxtable et al.,[19] Johnson and Spradbrow,[13] Kelly and Atwell,[20] Robinson et al.,[21,22] Smith et al.,[23] Walker et al.[24]
Belgium	Burtonboy et al.,[5,25,] Schwers et al.[26]
Canada	Gagnon and Povey,[10] Hayes et al.,[27,28] Thomson and Gagnon[17]
Egypt	Bucci et al.[29]
Finland	Jalanka,[30] Neuvonen et al.[31]
France	Lescure et al.,[32] Moraillon et al.,[16] Touratier[18,33,34]
Germany, East	Becker and Becker,[35] Bergmann et al.[36]
Germany, West	Arens and Krauss,[37] Bohm,[38] Hoffmann et al.,[39] Kraft et al.,[40] Niemand et al.,[41] Theil[42]
Hungary	Boros and Bartha[43]
Ireland	Sheahan and Grimes,[44] McNulty et al.,[45] Neill et al.[46]
Israel	Perl et al.[47]
Italy	Cammarata et al.[48]
Japan	Azetaka et al.,[49] Mohri et al.[50]
Netherlands	Van den Ingh et al.,[51] Osterhaus et al.[52,53]
New Zealand	Gumbrell,[54] Horner and Chisholm,[55] Parrish et al.[56,57]
Norway	Krohn and Blakstad[58]
South Africa	Bastianello,[59] Van Rensburg et al.[60]
Sweden	Klingeborn and Moreno-Lopez,[61] Olson et al.[62]
Thailand	Tingpalapong[64]
United Kingdom	Else,[65] Hitchcock and Scarnell,[11] Jefferies and Blakemore,[66] McCandlish et al.,[15] Thompson et al.[67]
U.S.	Appel et al.,[1-3] Black et al.,[4] Carpenter et al.,[68] Eugster et al.,[8] Fritz,[9] Nelson et al.,[69] Pletcher et al.,[70] Pollock and Carmichael[71]
Zimbabwe	Blackburn and LeBlanc Smith[72]

Table 2
RESTROSPECTIVE SEROLOGIC SURVEYS FOR ANTIBODY TO CPV-2 IN DOGS

Country	No. of samples	First positive sample	Ref.
Australia	428	May, 1978	76
Belgium	156	October, 1976	26
Japan	796	July, 1978	49
	369	January, 1979	50
Netherlands	344	November, 1977	52
U.S.	757	June, 1978	4
	102	1978	75

of those viruses. Restriction enzyme mapping of the replicative form (RF) DNA of the CPV-2, FPV, and MEV revealed extensive sequence homology among the three viruses, but differences were noted in at least seven cleavage sites.[81,93]

Antigenic comparisons of CPV-2 and FPV or MEV isolates using conventional serology[78-80] or analyses with monoclonal antibodies to either the CPV-2 or FPV[80,92,95] revealed antigenic differences between the viruses (Figure 1). The monoclonal antibodies revealed at least one antigenic site on the CPV-2 isolates which was not present on the FPV or MEV isolates. An antigenic site present on the FPV and MEV isolates was not seen on the CPV-2 isolates. A further antigenic site not present on any of the CPV-2 isolates examined was detected on the FPV isolates and on some of the MEV isolates.

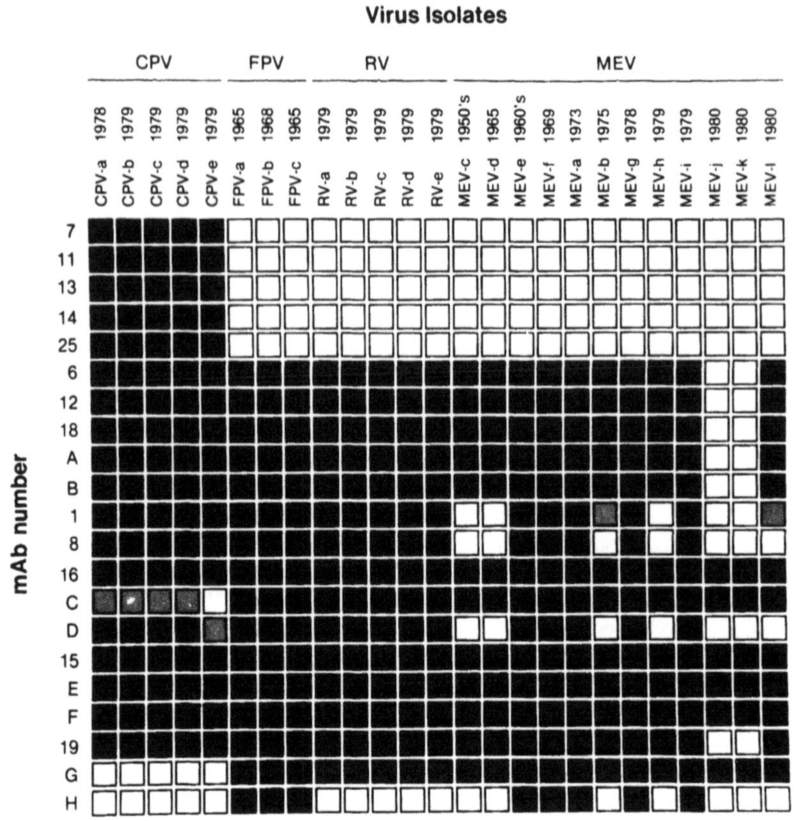

FIGURE 1. Reactions of isolates of CPV-2, FPV, RV, and MEV when titrated against a panel of monoclonal antibodies to these viruses using the HI test. Shaded: $^1/_4$ the titer with the homologous virus; crosshatched: $^1/_4$ but $^1/_{20}$ the titer with the homologous virus: open $^1/_{20}$ the titer with the homologous virus or no detectable reaction. (From Parrish, C. R. and Carmichael, L. E., *Virology* 129, 401, 1983. With permission.)

A parvovirus isolated from raccoons with enteric disease was serologically closely related to FPV.[82,92] Recently, monoclonal antibody studies have revealed slight antigenic variation among CPV-2 isolates.[99]

These data suggest that the CPV-2 arose or became widespread in the domestic dog population during or immediately prior to 1977, and that it may have arisen as a variant of either FPV or MEV.

B. Cultivation and Assay

1. In Vitro Cultivation

A number of different primary cells and cell lines are reported to support viral replication in tissue culture. These include both primary cells and cell lines of canine, feline, mink, raccoon, and bovine origin.[3,4,75,96] Cells that have been used most frequently to isolate and propagate CPV-2 are the Crandell feline kidney cell line (CRFK),[97] the Norden Laboratories Feline Kidney (NLFK) clonal derivitive of that cell line, primary and secondary canine kidney cells,[75,96] the A72 canine fibroma-derived cell line,[98] and an SV-40 transformed canine kidney cell line.[94] The CPV-2 appears to replicate poorly in the MDCK cell line.[75,99]

As with other autonomous parvoviruses, CPV-2 replicates in dividing cells. Efficient replication may be obtained when the virus is inoculated into thinly seeded (about 1 to

2×10^4 cells per cm^2) cell cultures or by passage of the infected cells after inoculation.[99] The cytopathic effect (CPE) observed in infected cultures is generally nondescript, particularly with primary or low-passaged isolates of the virus, frequently appearing as a general deterioration of the cell culture. The CPE may not be noticeable until the 2nd to 7th days after inoculation, depending on the virus titer in the inoculum and the mitotic state of the culture.[99]

Viral plaquing has been described using high-passage (170-190 passage) A72 cell line.[96] Cells were thinly seeded, inoculated, and overlayed with growth medium containing either 1% methylcellulose or 1% agarose. Plaques were observed after 5 to 7 days incubation. The SV-40 transformed dog kidney cell line has also been reported to plaque CPV-2 efficiently.[94]

2. *Hemagglutination by CPV-2*

Most CPV-2 isolates cause hemagglutination (HA) of erythrocytes of a number of species under certain conditions, although nonhemagglutinating strains may occur.[99] Erythrocytes agglutinated by most strains include those of the pig, rhesus macaque, cat, African green monkey, cyanomologous monkey, and crab-eating macaque.[3,5,13,49,75]

The HA of pig and rhesus monkey cells (and probably of others) occurs at pHs between 6.0 and 8.2.[75] The agglutination is more stable at a pH below pH 6.8. Hemagglutination also is favored by temperatures of 4°C or less; at higher temperatures the CPV may dissociate from the erythrocytes.[75]

3. *Identification of CPV-2*

Virus present in the feces or tissues of infected dogs may be identified by isolation in tissue culture or by direct detection of virus or viral antigen in the infected materials.

Viruses have been isolated in a variety of cell lines, or in primary or secondary cells of canine or feline origin.[3,4,75,96] The virus-containing material may be treated with chloroform or heated to 60°C for 30 min to reduce contamination by bacteria or other viruses. Up to three passages of the infected cells may be required before viral CPE or HA can be detected in cultures. The time required appears to be dependent on the titer of virus in the initial inoculum.[99]

Because of the limited CPE of most CPV-2 isolates in early passages, a variety of methods have been used to confirm the isolation of the virus. These include detection of viral HA and subsequent confirmation by hemagglutination-inhibition (HI) with specific antisera, fluorescent antibody (FA) staining of the cell cultures, or by neutralization of infectivity with specific antiserum (see below).

Virus in infected materials, if present in sufficient quantity, may be identified directly. Methods described have included examination of materials by electron microscopy after negative straining[1,5,7,8,10,37,65] or thin sectioning,[5,25,28,73] detection of HA and identification with HI,[5,53,75] and a competitive enzyme linked immunosorbant (ELISA) assay.[101]

4. *Viral Serology*

a. *Hemagglutination-Inhibition (HI)*

Generally, HI tests have been performed using four HA units of virus, and 0.5 or 1.0% pig or rhesus macaque erythrocytes.[3,49,75,76] Either fresh or formalin-fixed erythrocytes may be used.[103] Antisera should be treated by heat inactivation for 30 min at 56°C and then by absorption with a 1/10th volume of 50% packed erythrocytes from the species being used for HA. Hemagglutination-inhibition tests may be performed using buffers between pH 6.2 to 7.4 in phosphate buffered saline,[75] or in barbital buffered saline containing 2.5 mM Ca^{++} and 0.75 mM Mg^{++}.[80]

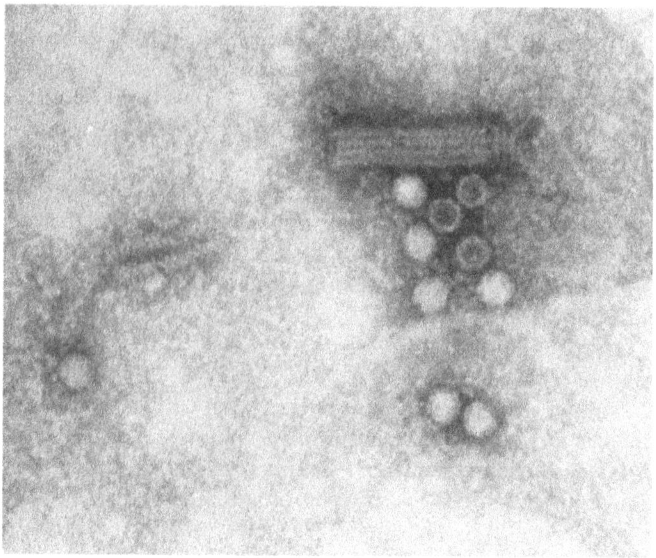

FIGURE 2. Canine parvovirus in the feces of a naturally infected dog. Note the presence of both full and empty particles.

b. Neutralization Assays

A variety of neutralization assays have been described. Some involve the neutralization of viral plaques,[100] with titers being expressed as the inverse of the dilution giving a 50% reduction in the number of plaque forming units. This assay is reported to yield 50 to 100-fold higher titers than the hemagglutination-inhibition (HI) test (see below). Other neutralization tests have assayed reduction in viral activity by endpoint dilution with a variety of methods being used to determine the endpoint of infectivity.[75,79,81,95]

c. Other Serologic Methods

A radioimmune assay (RIA) was used to titrate anti-CPV-2 or anti-FPV monoclonal antibodies (mAb) in studies of the antigenic relationships among CPV-2, FPV, and MEV isolates.[92] In that assay, viral antigen, partially purified by sucrose gradient centrifugation, was bound to polyvinyl chloride (PVC) microtiter trays. The binding of the mAb to the virus was detected using a ^{125}I-labeled goat anti-mouse or anti-rat IgG.

An ELISA test has been described for detecting antigen in the feces of infected animals.[101] Partially purified viral antigen was bound to PVC microtiter plates and the presence of virus in the specimens was revealed by decreased binding of a horseradish peroxidase-linked anti-CPV-2 mAb to the fixed antigen.

An immune-radial hemolysis assay has been used to detect antibodies to CPV-2 in dog sera[102] and has also proven useful for screening mAbs to CPV-2 or FPV in hybridoma supernatants.[80,92] In those assays, CPV-2 antigen was covalently coupled to sheep erythrocytes by 1-ethyl-3-(3-dimethylaminopropyl) carbodiimide HCl. The presence of antibody was detected by complement mediated lysis of the erythrocyte-virus-antibody complexes.

C. Virion Structure

The CPV-2 virion is spherical and nonenveloped (Figure 2). Diameters range from 18 to 26 nm in negatively stained preparations examined in the electron microscope.[1,5,7,15,37] It is probable, by comparisons with other parvoviruses, that the capsids are composed of either 12 or 32 capsomers.[104,106]

Empty capsids (which do not contain DNA) and DNA-containing infectious (full) virions are formed in various proportions in infected cells. The bouyant density of the empty virions is about 1.32 g/cm^3, [5,94] and that of the full capsids is about 1.44 g/cm^3 [5,94,105] in cesium chloride gradients. Two virus proteins (VP-1 and VP-2') with apparent molecular weights of 82.3 kD and 67.3 kD, respectively, are present in both the empty and full capsids.[92,94,105a] In the full capsids a proportion of the VP-2' may be cleaved off, apparently by cellular proteinases, to form a 63.5 kD product (VP-2).[94]

Studies of the capsid of the related parvovirus H-1 suggest that the arrangement of the different proteins in the capsid is not symmetrical. Crosslinking studies with polyethylene glycol and suberimidate revealed that the VP-1 molecules were extensively crosslinked to one another, and it appears that the VP-1 molecules may be clustered in the capsid or arranged in close contact.[106]

The VP-1, VP-2', and VP-2 of parvoviruses are all coded for by the same gene and overlap both in amino acid sequence and in antigenic structure.[107-109,114-116] The VP-1 and VP-2' are translated from different mRNA splice products and the VP-1 contains the amino acid sequence of both the VP-2' and VP-2, as well as a unique highly basic amino terminal peptide.[107,108] It is thought that the unique region of the VP-1 is involved in packaging the viral DNA.

The CPV-2 virion appears to be extremely stable, with little loss of infectious titer on heating to 60°C for 1 hr[13] or after long periods in the environment.[110]

D. DNA Structure

The genome of CPV-2 consists of about 5000 bases of DNA.[81,93,94] The infectious virions contain one copy of the negative strand of the viral genome.[94] In other parvoviruses, a small proportion (≤1%) of positive strands may be encapsidated under some circumstances.[111] Palindromic sequences at either end of the viral strand form foldback "hairpin" structures which are required for DNA replication. The 3' end terminal hairpin is utilized as a primer for synthesis of a double-stranded genome after the virus has infected a cell.[112,113]

E. Genome

Sequencing and genomic mapping studies of the parvoviruses MVM[114,115] and H-1,[116] and unpublished sequencing studies of the CPV-2,[99,117] have revealed that the parvovirus genomes encode two separate genes. One, represented by an open reading frame in the sequence between map unit (MU) 4 and 44, encodes a noncapsid protein of uncertain function, which is revealed in infected cells upon immunoprecipitation with sera from infected animals.[92,115,116] This protein may be involved in DNA replication. The other gene is encoded as two or more open reading frames in the DNA sequence between MU 38 and 95 with a possible use of sequences between 4 and 10 MU. It codes for two mRNAs, the products of different splicing reactions, which are translated to form the capsid proteins VP-1 and VP-2'.[115,116,118]

Repeated passage of parvoviruses in tissue culture often results in the formation of defective viruses. These have been characterized as having deleted genomes (MVM)[119] or genomes containing both deletions and duplications (H-1).[120] All of the replicating defective genomes retain both terminal palidromes, which appear to be required for replication. Most defective viruses require the coinfection of the same cell by a nondefective virus in order to supply essential replication functions.

F. Replication

The replication of autonomous parvoviruses is not yet fully understood, and little work has been done on the replication of CPV-2.

Studies of the minute virus of mice (MVM) revealed that the virions attach to specific cell receptors.[121,122] At 4°C the attachment was reversible, and up to 78% of the attached virions could be removed after 1 hr by washing the cells with solutions containing EDTA. At 37°C the association with the cells rapidly becomes irreversible, and aggregates of virions are seen in coated pits when the cells are examined by electron microscopy.[121]

After uptake into the cell, the infecting virions appear to be transported to the nucleus, although the mechanisms by which this occurs are unknown.

Parvoviruses replicate only in cells which are in S or early G2 phase of the mitotic cycle.[112,123,124] Beginning at the 3′ terminal palindrome, the ssDNA genome is filled in to form a double-stranded genome.[111,112,125,125a] This then replicates by a form of strand displacement replication,[111,112] possibly mediated by host DNA polymerases α and γ.[125-127] The DNA is nicked at a position adjacent to the 5′ end of the viral strand and then DNA synthesis occurs on both strands beginning from the nick site, without the formation of Okasaki-fragment intermediates.[125a,128] A protein is reported to be covalently attached to the 5′ end of each strand of the dsRF-DNA.[129] Neither the mediator of the specific nicking reaction or the origin of the 5′ terminal protein is known with certainty.

Messenger RNA transcripts are formed from the double stranded DNA, and are reported to encompass map units 4 to 95.[114-116] Three spliced, polyadenylated mRNAs are transported to the cytoplasm and translated to form the viral proteins.

It appears that the capsid proteins are transported to the nucleus and that viral assembly occurs there. Single-stranded DNA, mostly of the (−) orientation, is formed from the viral double-stranded genome. The mechanism by which this occurs is unknown, but may be due to interaction of viral proteins with the (−) strand as it is synthesized.[112] The viral capsids, both full and empty, remain in the nucleus and aggregates of virus may be observed as intranuclear inclusion bodies by light microscopy, as antigen by FA, or as virions by electron microscopy.

The dependence of parvovirus replication on the infection of dividing cells has been widely used to explain the pathogenesis of diseases caused by parvoviruses, including FPV[130-132] and CPV-2.[14,133-135] It appears, however, that other factors may determine which of the dividing cells will support virus replication. Studies of MVM in tissue culture revealed that dividing teratocarcinoma cells will support virus replication at some stages of differentiation but not at others.[136,137] In studies of H-1 virus infection of hamsters, it was found that only some of the populations of cells shown to be dividing by labeling with ^{3}H-thymidine could be shown to contain virus antigen when stained by FA methods.[138] Other studies of the Kilham Rat virus (KRV) and MVM suggest that strains of these viruses may replicate in some cell lines derived from their host animal but not in others.[139-141]

III. THE NATURAL DISEASE

A. Host Range

Canine parvovirus-2 infections have been documented in a number of canidae, including the domestic dog (*Canis familiaris*), bush dogs (*Speothus venaticus*),[142] coyotes (*Canis latrans*), [143] crab eating foxes (*Cerdocyon thous*),[142] maned wolves (*Chrysocyon brachyurus*),[144,145] and raccoon dogs (*Nyctereutes procyonoides*).[31] Serologic evidence of infection was produced by feeding infected feces to blue foxes (*Alopex lagopus*).[31]

Feline panleukopenia virus, MEV and CPV-2 cross-infect and cross-protect a number of carnivores and mustelids, but each shows a marked predeliction for its natural host. Domestic cats (*Felis domesticus*) can be infected with CPV-2 by parenteral inoculation.[53,99a] Viral titration and FA staining revealed limited viral replication in lymphoid organs.[99a] The amount and distribution of CPV-2 replication in cats, however, were much less than that of FPV in the cat or of CPV-2 in dogs. No clinical illness was observed in CPV-2-infected cats,[53,99a,146] and maximum serologic responses of cats to

CPV-2 were several-fold lower than those observed following FPV infection,[53,147] again suggesting less avid replication.

Mustelidae are variably susceptible to CPV-2. Oral exposure of nine ferrets (*Mustela furo*) to feces containing CPV-2 produced no serological or clinical evidence of infection.[31] Although Moraillon et al.[149] reported the experimental induction of disease in dogs with an agent originally isolated from mink and believed to be CPV-2, restriction endonuclease and antigenic mapping studies suggest that the virus in question is a strain of MEV rather than CPV-2.[81,92] Experimental inoculation of mink with known CPV-2 produced minimal viral replication and no disease.[99]

An early report which suggested that raccoons (Procyanidae: *Procyon lotor*) were susceptible to CPV-2 was based on indirect evidence.[148] More recent studies have suggested that parvoviral enteritis in raccoons is caused by a distinct raccoon parvovirus which shares antigenic cross-reactivity with CPV-2 and FPV.[82] Experimental oronasal exposure of six raccoons to CPV-2 failed to produce illness, viral shed, or seroconversion.

Man appears to be refractory to infection. No serologic evidence of infection was detected in heavily exposed kennel workers, laboratory personnel, or veterinary students.[6,83,146a]

B. Manifestation of Disease

The manifestations of parvoviral infection in dogs range from subclinical to acute and fulminating. Serologic evidence suggests that subclinical infections are common.[49,57,83,150-152a] Experimental infections, especially of specific pathogen-free (SPF) dogs,[3,110] have been characteristically mild, suggesting that secondary factors or pathogens influence the outcome of infection under natural conditions. Direct evidence for this hypothesis is lacking, however.

A similar range of clinical disease is observed in cats infected with FPV. The experimental disease is usually mild in SPF cats and subclinical in germ-free animals,[130,153,154] although it is often fulminating and rapidly fatal in conventional cats.[155-157] These differences have been attributed to differences in the rate of intestinal epithelial cell turnover.[130] While there is a correlation between the rate of epithelial cell regeneration and the severity of disease in cats, other hypotheses have not been excluded. For example, there are similar differences in the rate of lymphoid cell turnover between SPF and germ-free cats. The hypothesis that the severity of disease is determined by the rate of lymphocyte multiplication is also consistent with the above observations. Given the apparent tropism of FPV and CPV-2 for lymphocytes (see below) the latter hypothesis is attractive and ought to be explored.

Clinical CPV-2 disease in dogs usually takes one of two forms: enteritis or myocarditis. The enteric form of the disease occurs in dogs of all ages, but severe illness is encountered more often in pups.[6,64,71,158,159,162] Acute myocarditis, on the other hand, is believed to occur exclusively in puppies,[158,160,161] although older animals may develop cardiomyopathy and congestive heart failure as a sequella to neonatal infection.[158,161a,173,174] A third form of the disease, generalized necrotizing vasculitis has been described in young pups,[162a,162b] but it appears to be very rare.

1. Enteric Form

Prodromal signs of depression and anorexia are usually noted on the third or fourth day after oral infection.[162,163] Vomiting and diarrhea of variable severity commence on the 5th or 6th day postinfection (dpi). These are believed to result from the destruction of the germinal epithelium of the intestinal glands (see below) leading to loss of normal intestinal architecture and function.[163,164] Maldigestion, malabsorption and possibly endotoxin absorption result, which are manifested as vomiting and diarrhea. Bleeding into the gastrointestinal tract, leading to hematochezia or hematamesis, is reported in about one half of clinical cases.[163]

That only severely ill dogs are presented for veterinary care leads to overestimation of the frequency with which clinical findings are associated with infection. Thus, leukopenia is observed in up to 86% of clinical cases of parvoviral enteritis in dogs,[165] occasionally reaching profound levels, but it is an infrequent finding in experimental infections.[3,110,135,166,167] This differs from FPV infection in cats, in which panleukopenia, beginning 2 to 6 days after infection, is a constant finding even in germ-free animals.[153,154]

A reduction in the number of circulating lymphocytes is regularly observed in dogs on the 4th to 6th days after experimental CPV-2 infection.[96,110,135,167] Presumably, this reflects direct viral destruction of lymphocytes; extensive lymphoid necrosis is observed in the thymus, splenic nodules, and mesenteric or peripheral lymph nodes.[56,69,70,133,151,160,163,178a] In many cases, however, a relative lymphopenia or the presence of atypical "reactive" lymphocytes is the only hematologic abnormality observed.[91,110,168]

In severe cases, there may be a profound leukopenia characterized by lymphopenia, neutropenia, degenerative left shift, and toxic neutrophils.[163,165,169] The causes of these findings are not known. They are believed to result from direct viral destruction of stem cells in the bone marrow, together with depletion of mature forms by extensive tissue destruction and endotoxin absorption from the gut.[168,170]

The leukopenia resulting from CPV infection is transient. Dogs presented early in the course of the disease often did not manifest a drop in the white cell count until 12 to 24 hrs after hospitalization.[165] Peripheral white cell counts return to normal in 3 to 5 days and a leukocytosis often occurs during recovery.[71,162]

Other signs of CPV-2 infection have been described infrequently. Among these are abdominal pain, small oral ulcers, seizures, and necrosis at the site of therapeutic injections.[163] It is unclear whether these are primary CPV-2-induced lesions, or secondary and incidental findings.

2. Myocardial Form

Acute parvoviral myocarditis occurs most often in puppies 3 to 8 weeks of age.[22,28,51,57,59,60,68,158,163,171,172] As a result, the temporal incidence of the disease parallels the natality pattern of pups.[161]

Clinical signs appear suddenly and progress rapidly. Affected pups were often found dead without premonitory signs, or succumbed after short periods of dyspnea, crying, and retching.[22,28,51,57,60,66,68,171,172] Death apparently results from acute heart failure. Agonal electrocardiographs (EKGs) revealed marked arrythmias and paroxysmal premature ventricular contractions.[22,68,171] Parvoviral myocarditis usually occurs without concurrent enteritis, and the bitch is most often clinically unaffected.[28,57,158]

Mortality commonly exceeds 50% in an affected litter.[28,57,60,73,158,161] Surviving littermates may appear clinically normal, but have histologic lesions and EKG changes indicative of myocarditis.[22,135,173] These pups may develop congestive heart failure months or even years later,[158,161a] apparently as a result of secondary myocardial fibrosis. Cardiomyopathy, believed to be secondary to parvoviral myocarditis, has been reported in dogs up to $3^1/_2$ years of age.[158,161a,171,174]

Although the clinical signs of CPV-2-induced myocarditis appear suddenly, the heart lesions themselves develop slowly. Heart failure does not become manifest until several weeks after infection. Cases have been reported in pups 3 to 7 weeks after recovery from parvoviral enteritis.[68] In another study,[57] serological evidence of CPV-2 infection was obtained 3 to 5 weeks prior to the deaths of pups from acute heart failure.[57] Characteristic histopathologic lesions of CPV-2 myocarditis were observed. The bitch was

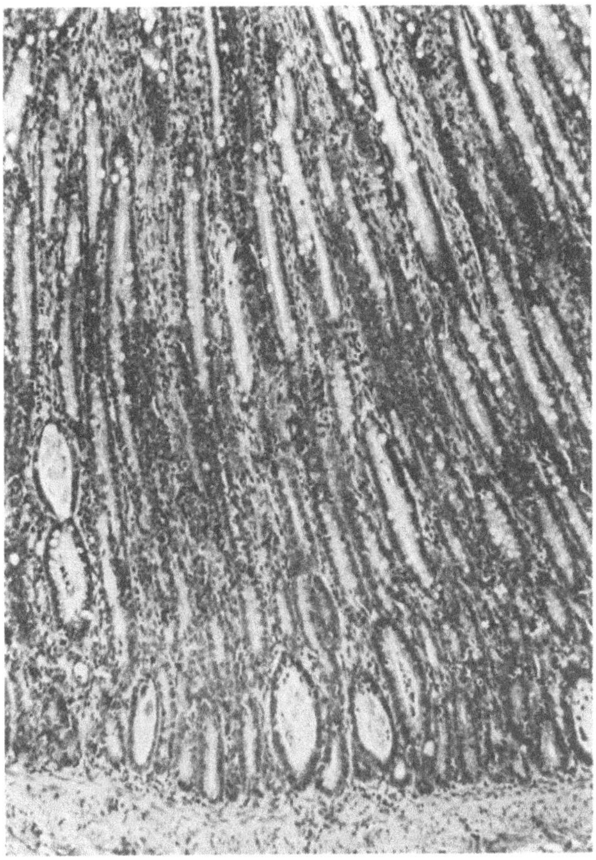

FIGURE 3. Section of jejunum from a natural case of CPV-2 enteritis. Scattered necrosis of the intestinal gland epithelium is evident accompanied by dilitation and cellular debris. (From Meunier, P.C. et al., *Cornell Vet.*, 71, 96, 1981. With permission.)

infected concurrently, but had no clinical signs. Heart failure from experimentally induced CPV-2 myocarditis first appeared 23 days after *in utero* infection of pups.[175]

C. Pathologic Lesions

1. Enteric Form

a. Macroscopic Lesions

The gross lesions of CPV-2 enteritis are variable and nonspecific.[135,151,163] In some cases, there are no gross lesions. When present, lesions of CPV-2 enteritis are segmentally distributed in the intestine. The ileum and jejunum are most often affected, while the stomach, duodenum, and colon are usually spared.[69,151,160] Affected segments may be flaccid with subserosal hemorrhage or congestion.[14,56,70,160,162] The bowel lumen is often empty or it may contain watery or hemorrhagic ingesta.

Mesenteric lymph nodes are often enlarged and edematous with multifocal petechial hemorrhages in the cortex.[14,24,56,70,151] Thymic cortical necrosis and gross thymic atrophy are common findings in young dogs.[69,151,160]

b. Microscopic Lesions

The microscopic lesions of CPV-2 infection are initially confined to areas of proliferating cell populations.[135] In the enteric form, the early lesions consist of necrosis of the intestinal gland (crypt) epithelium[14,39,56,132,160,162] (Figure 3). Gland lumena are often

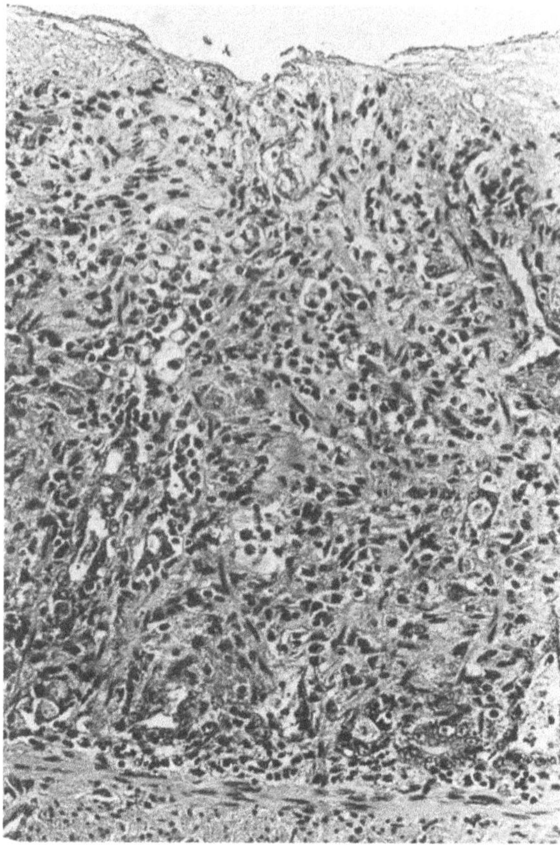

FIGURE 4. Jejunum from a natural, fatal case of parvoviral enteritis. There is severe necrosis of the intestinal epithelium with complete collapse of the villi and lamina propria. Nevertheless, inflammatory changes are minimal. (From Meunier, P. C. et al., *Cornell Vet.*, 71, 96, 1981. With permission.)

dilated and filled with necrotic debris.[56,60,132,151,162] The remaining epithelium is thinned and flattened; some areas of the basement membrane may be entirely denuded. Eosinophilic intranuclear inclusion bodies are occasionally observed in intact epithelial cells.[14,56,60,132,151] There are surprisingly few inflammatory changes associated with these early lesions.[70,132,162]

As the disease progresses, the villi become progressively blunted because the destruction of the gland epithelial cells interrupts the normal process of continuous intestinal epithelial replacement. Attenuated, immature epithelial cells cover the mucosal surface and there may be fusing of adjacent villi.[151,162] Complete collapse of the villi and lamina propria are observed in severe cases (Figure 4). These lesions may be focal, locally extensive, or diffusely distributed throughout the small intestine.[56,151] Rarely, similar lesions are seen in the large bowel.

Evidence of intestinal epithelial regeneration is often present, even in fatal cases. Remaining intestinal glands are elongated and lined by hyperplastic epithelium with a high mitotic index.[69,133,151]

There is widespread necrosis and depletion of lymphocytes in the gut-associated lymphoid tissue, the germinal centers of mesenteric lymph nodes, and the splenic nodules.[56,69,70,133,135,151,160,162] Diffuse cortical necrosis of the thymus occurs in young dogs,

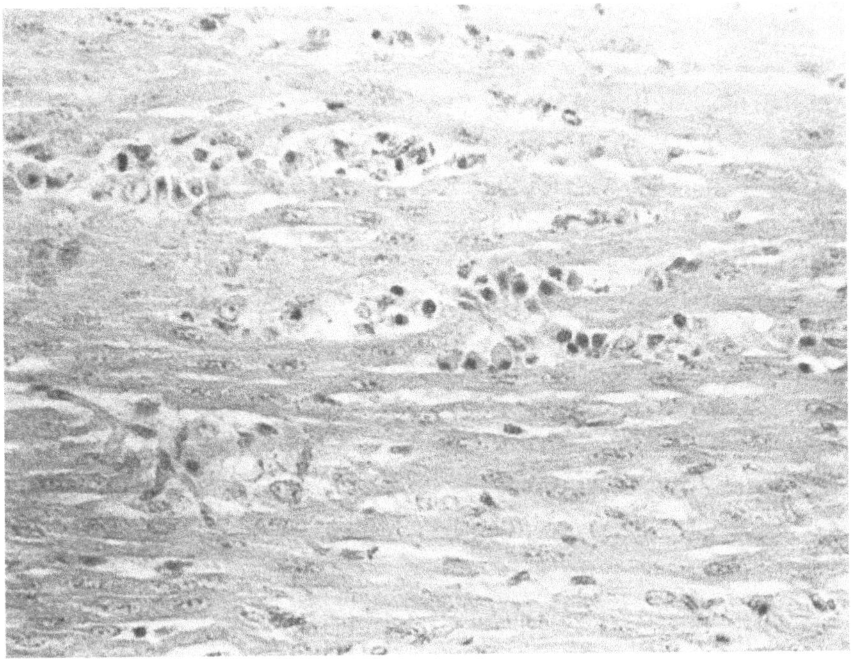

FIGURE 5. Section of the myocardium from a natural case of parvoviral myocarditis in a 6-week old coyote pup. Note myocardial necrosis and lymphocytic inflitrate. (Micrograph courtesy of C. W. Leathers, Washington State University.)

with an associated loss of thymic mass.[135,151,160] There is necrosis and depletion of stem and mature cells of both the myeloid and erythroid series in the bone marrow.[14,160,168,170] Regenerative hyperplasia of lymphoid, myeloid, and erythroid cells is seen during recovery.

2. *Myocardial Form*

a. *Macroscopic Lesions*

In cases of parvoviral myocarditis, the gross lesions are those of acute heart failure. The cardiac chambers are dilated and flaccid, and there is pulmonary edema and passive congestion of the liver.[28,56,73,175,176] Ascites, hydrothorax, and hydropericardium may be seen.[28,51,56,73] Pale white streaks associated with cellular infiltrates or fibrosis may be visible in the ventricular myocardium.

b. *Microscopic Lesions*

A nonsuppurative myocarditis is observed microscopically (Figure 5). Edema and myofiber loss are present, associated with a locally extensive lymphocytic infiltrate.[28,60,68,73,151,162] Lesions are most prominent in the left ventricle and interventricular septum. Occasional Feulgen-positive intranuclear inclusion bodies are seen, which stain intensely with anti-CPV-2 or anti-FPV fluorescent-antibody conjugates.[28,135] Healing occurs with intersitial fibrosis and scarring; extensive areas of fibrosis are observed in chronic cases.[73,135,175]

IV. PATHOGENESIS

There are few published studies of the pathogenesis of CPV-2 infection in dogs. The general sequence of events is known, but many specific questions are unresolved.

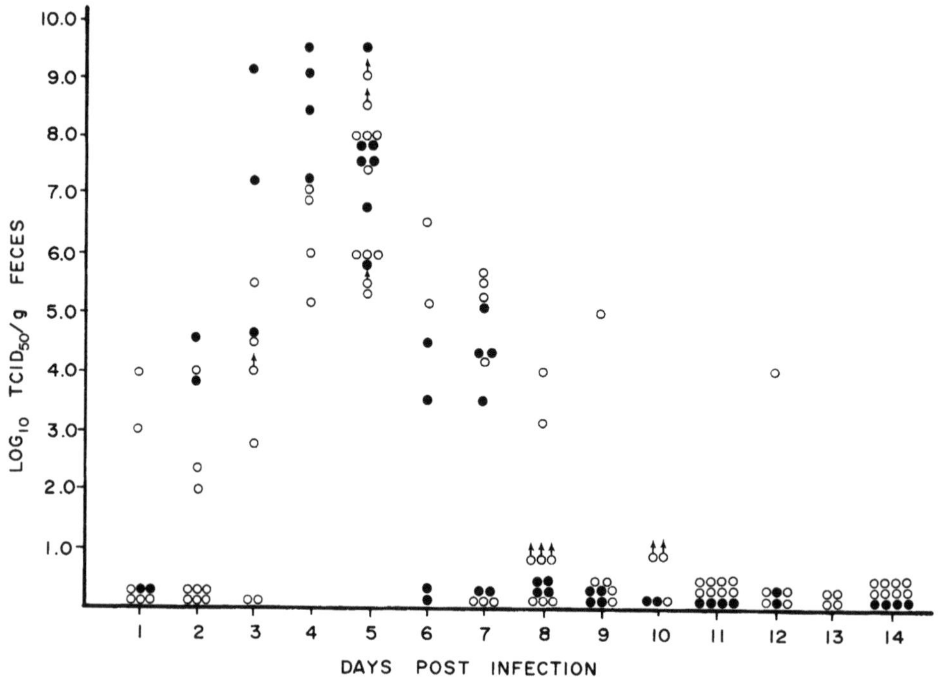

FIGURE 6. Fecal shed of CPV-2 after experimental infection by the oronasal (open circles) or parenteral (filled circles) routes. Arrows indicate viral titers equal to or greater than amounts indicated (end points not determined). (From Pollock, R. V. H., *Cornell Vet.*, 72, 103, 1982. With permission.)

A. Enteric Form

1. Route of Infection

The natural route of infection is presumed to be fecal-oral. As much as 10^9 $TCID_{50}$ of virus per gram of feces are shed during the period of acute illness.[96,110] (Figure 6) Infection is readily transmitted to susceptible animals by oral inoculation with infected feces or tissue culture fluids.[160,167,177,178,178a]

Whether there are other natural routes of transmission is not known, but the occurrence of infection in several closed dog colonies is indirect evidence of transmission on fomites.[9,152a,162,179] Virus may be present in all excretions and secretions during the viremic phase of infection. Arthropods could potentially carry virus after ingestion of infected blood or excrement, but appropriate studies remain to be done.

Experimentally, infection can be accomplished by a number of routes, including oral, nasal,[3] or oronasal exposure,[110,168] and intramuscular, intravenous,[3,11,160] or subcutaneous[166] inoculation. The minimum infectious dose is not known. At least one study would suggest that it is very small; immunization was accomplished by intramuscular inoculation of as few as 16 $TCID_{50}$ of an attenuated strain.[182]

Persistence of the virus in the environment is believed to be more important than chronic carriers in the epizootiology of the disease. Excreted virus is resistant to inactivation; infectious CPV-2 was demonstrated in feces held 6 months at room temperature.[110] The period of active virus excretion appears to be brief, however; in most instances, virus cannot be recovered from feces more than 12 days after infection.[96,110,167] The possibility still exists, however, that some dogs may excrete virus for long periods, as has been demonstrated to occur with FPV in cats[180] and certain rodent parvoviruses in laboratory animals.[181]

2. Viral Replication

Despite the association of CPV-2 with enteric disease, the virus appears to be primarily lymphocytotropic rather then enterotropic. The sites of primary replication are believed to be the lymphoid tissues of the oropharynx and mesenteric lymph nodes.[110,135,168,178a] Virus was isolated from the tonsils, retropharyngeal, bronchial, and mesenteric lymph nodes on days 1 and 2 after oral exposure. Virus was not recovered at this time from the other organs studied, which included the liver, spleen, heart, bone marrow, and intestinal tract, except in one instance in which a small amount (100 $TCID_{50}$) was detected in the ileum 24 hr after exposure.[135] The latter may represent residue of the oral inoculum, as there was no evidence of viral replication in the ileum at this time by FA staining.

In agreement with virus isolation studies, examination of tissues by FA during the first 2 days after oral exposure revealed infected cells only in the tonsils, and retropharyngeal and mesenteric lymph nodes. Importantly, viral antigen was not found in intestinal epithelial cells until late in the course of the disease (see below). Infection can apparently occur across the intestinal mucosa, since infection has resulted from direct instillation of virus by gastric intubation[168] or by use of enteric coated capsules.[147] Nevertheless, intestinal lymphoid tissues, rather than the epithelium, appear to be the site of primary replication.[135,178a] Dogs given virus intragastrically had milder disease than animals exposed to a similar dose orally, suggesting that initial replication in the orophartyngeal lymphoid tissues may be an important step in the pathogenesis of CPV-2 disease.[168]

Virus apparently generalizes from its initial sites of replication by way of the blood stream.[110,135,178a] A primary viremia was detected on the first or second day after oral exposure in 12 of 24 experimental infections reported.[110,135,166] By the 3rd and 4th or 5th day, there was a marked viremia and virus was recovered from all tissues examined.[110,135,166,178a] Viral titers in lymphoid organs and the intestine exceeded those of the serum, and fluorescent antibody studies confirmed widespread viral multiplication in these organs (Figures 7 to 9). Viral antigen was not detected in parenchymal cells of the lungs, liver, myocardium, and adrenals. It is most likely that the modest amounts of virus isolated from these organs during the viremic phase are derived from the blood rather than from viral replication in the organs themselves.

Further evidence for the proposed sequence of events — that is, of initial localized replication followed by dissemination in the blood and widespread secondary replication — comes from the comparison of dogs infected orally with those inoculated parenterally. The onset of clinical signs, virus shed, and antibody production all occur 1 to 2 days early in dogs inoculated parenterally,[110,135] although parenteral infections are similar in all other respects to those of dogs challenged orally. This suggests that parenteral inoculation bypasses a step in the pathogenesis of oral infections, possibly a period of initial replication preceeding systemic dissemination.

Active excretion of CPV-2 usually begins on the third day after oronasal exposure or the second day after parenteral infection.[75,96,110,135,166,167] The amount of infectious virus in the feces increases rapidly until the fifth or sixth day, at which time there may be more than 10^9 $TCID_{50}$ of CPV-2 per gram of stool (Figure 6)[96,110] and up to 10,240 HA units.[75] During this period, which corresponds to the period of illness in clinically affected dogs, virus is detected readily in the stool by a variety of methods including electron microscopy,[5,37] fecal HA,[75,103,152a] counterimmunoelectrophoresis,[184] viral isolation,[3,11,13,15,75] and ELISA tests.[101] After the seventh or eighth day, the amount of virus that can be recovered from stool specimens drops sharply.[75,96,110] Virus can be isolated only rarely from the feces of dogs more than 12 days postinfection (d.p.i.).[96,110,167] Local intestinal antibody may be important in the termination of fecal virus excretion; in one study, fecal HA titers were shown to be inversely proportional to the amount of coproantibody.[185]

The development of histologic lesions parallels the replication of virus.[135,178a] Lesions are not detectable on the first day after oral exposure. On the second day there is

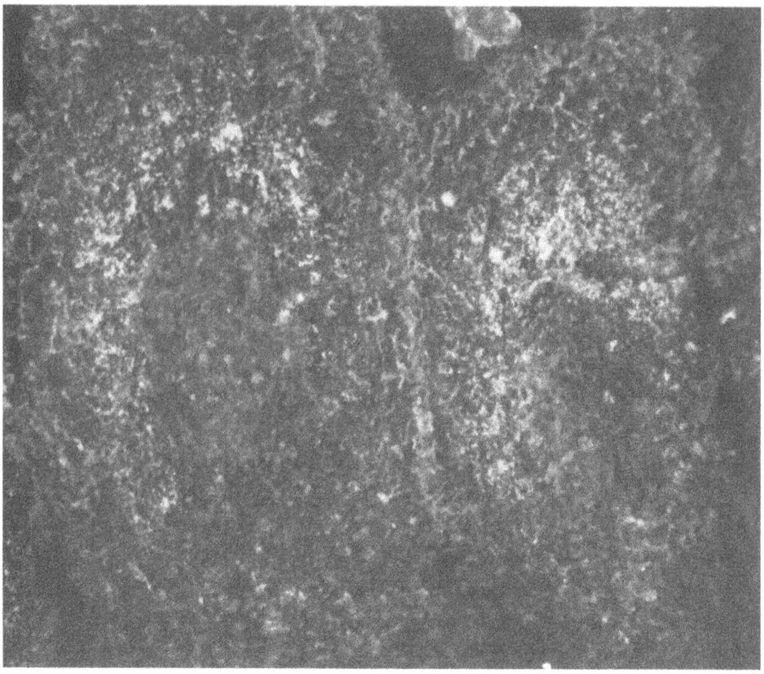

FIGURE 7. Fluorescent antibody (FA) stained section of the mesenteric lymph node of a 12-week old pup 4 days after experimental infection. Extensive viral replication is evident in the germinal centers.

histologic evidence of necrosis within the germinal centers of the tonsil and the retropharyngeal and mesenteric lymph nodes. On the third day, lymphoid necrosis is more widespread; cell loss is evident in the thymus of some dogs. A few infected cells can be detected in the lymphoid follicles of the intestinal tract. Nevertheless, intestinal lesions or infection of the epithelium are not observed until the fourth day after oral exposure.[135]

Virtually all dogs have a viremia on the fourth day after oral exposure to CPV-2.[110,135] Titers of 10^4 to $10^{6.5}$ $TCID_{50}$ of CPV-2 per m*l* of blood are present. Titers of virus in the serum are two to four orders of magnitude greater than those of washed white blood cells.[110,135] The source of the viremia is not known with certainty, but it probably results from virus-mediated lympholysis, since extensive necrosis is seen in the thymus and gut-associated lymphoid tissues.[56,69,70,133,151,160]

The effect of CPV-2 infection on the immune system has not been adequately studied. Krakowka et al.[178] reported that two of five dogs developed vaccinal distemper encephalitis when they were infected with virulent CPV-2 3 days after receiving attenuated canine distemper virus. Krakowka hypothesized that CPV-2 had an "immunomodulating" effect which accounted for the increased virulence of the distemper vaccine strain. Studies in progress[147,185a] have demonstrated a transient reduction in lymphocyte responsiveness to T cell mitogens following infection with virulent CPV-2. Similar findings have been reported in cats infected with FPV.[183] Antibody production appears unimpaired, however; infected dogs develop detectable levels of humoral antibody within 3 to 4 days of infection (see below). The significance of CPV-2-induced changes in immune responsiveness are unknown.

Infection of the intestinal epithelium is first evident on day 4 or 5 after oral infection.[135,178a] Infected cells are first noted adjacent to areas of infected lymphoid tissue

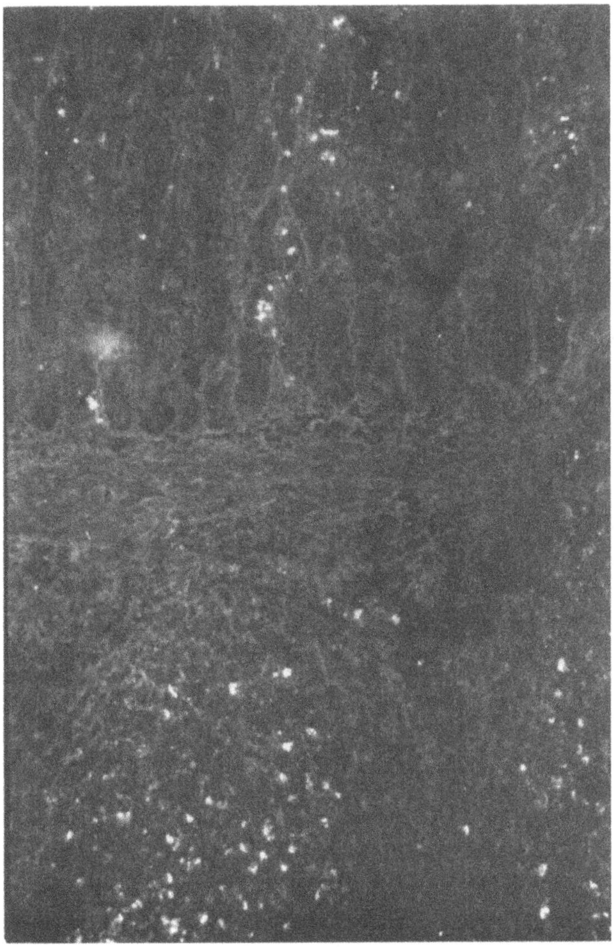

FIGURE 8. FA stained section of jejunum of 12-week-old pup 4 days after experimental infection. Note the staining of infected epithelial cells overlying infected lymphoid aggregates.

(Figure 8), suggesting that infection may spread to the epithelium from infected lymphocytes. Later, areas of the mucosa are involved that are not adjacent to lymphoid aggregations; thus spread through the epithelium and/or hematogenous infection also probably occur.

Serum antibody first becomes detectable on the 4th day after oral exposure.[110,135] Titers rise rapidly to maximal levels by the 7th day after infection (Figure 10). The rapidity of antibody production may be an important determinant of the outcome of infection. Meunier[135] showed that dogs with early and vigorous antibody production tended to have asymptomatic infections. Dogs with delayed humoral responses had a greater and more prolonged plasma viremia. They also had more severe clinical signs and intestinal lesions, and shed more virus in their feces. Meunier[135] concluded that the onset and rate of antibody production were the principal determinants of the clinical course of infection.

Lymphoid lesions continue to increase in severity until the 7th to 9th day after exposure, after which time regenerative activity predominates.[135,178a] Large numbers of lymphoblastoid cells are seen in the germinal centers of the lymph nodes, splenic nodes, and thymus. There is marked hyperplasia of the intestinal crypts and partial restoration

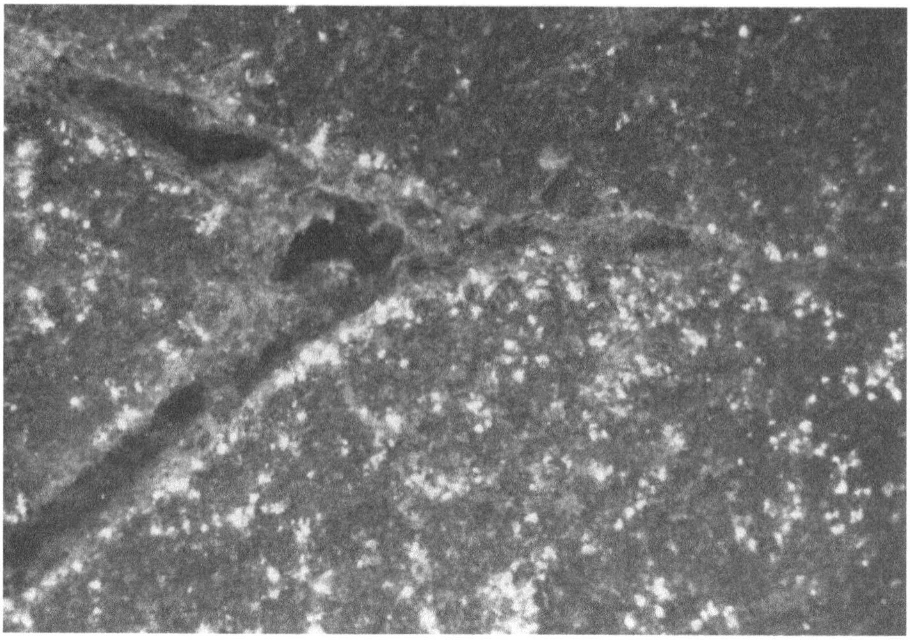

FIGURE 9. FA stained section of the thymus of a 12-week-old pup 4 days after experimental CPV-2 infection. In some lobules, as many as 50% of the lymphocytes contain viral antigen.

of the normal villous architecture. A few infected epithelial cells still can be detected by FA staining.

By 14 days after exposure, the intestinal architecture of recovered dogs is largely restored.[135] Repopulation of lymphoid organs is well underway.

B. Myocardial Form

Studies of the pathogenesis of the myocardial form of CPV-2 infection are very limited. The disease has been reproduced experimentally in one study by *in utero* inoculation of pups 5 days before birth,[175] and in a second by infection of seronegative pups at 5 days of age.[135] Attempts to reproduce the disease in pups 4 weeks and older, even using virus recovered from the myocardium, produced only the enteric form of the disease.[28,160] Epidemiologic studies also suggest that myocarditis occurs only in pups infected near the time of birth.[57,161] Thus, age at the time of infection appears to be the critical factor that determines which form of the disease will be manifest.

It has been speculated that the rapid proliferation of myocardial cells after birth favors CPV-2 replication, since parvoviruses are largely dependent on host cell DNA synthesis for their replication. In the dog, cardiac myocyte replication is rapid only during the first 2 weeks after birth.[186] Thereafter, growth is principally by hypertrophy, although DNA synthesis and nuclear kinesis continue until at least 8 weeks of age.[187] It is believed that only very young pups have a sufficient rate of myocardial proliferation to support CPV-2 replication in the heart. Thus, the shift in the clinical manifestation of infection from myocarditis to enteritis is thought to reflect the decreasing rate of myocardial cell proliferation and increasing rate of intestinal epithelial turnover that occurs in the first few weeks after birth.[135,158] Alternative hypotheses, such as a change in cell receptors with age or stage of cellular differentiation have not been excluded, however.

There is conflicting evidence regarding the effect of CPV-2 infection on reproductive performance. Meunier et al.,[151] in a study of commercial kennel of approximately 1200

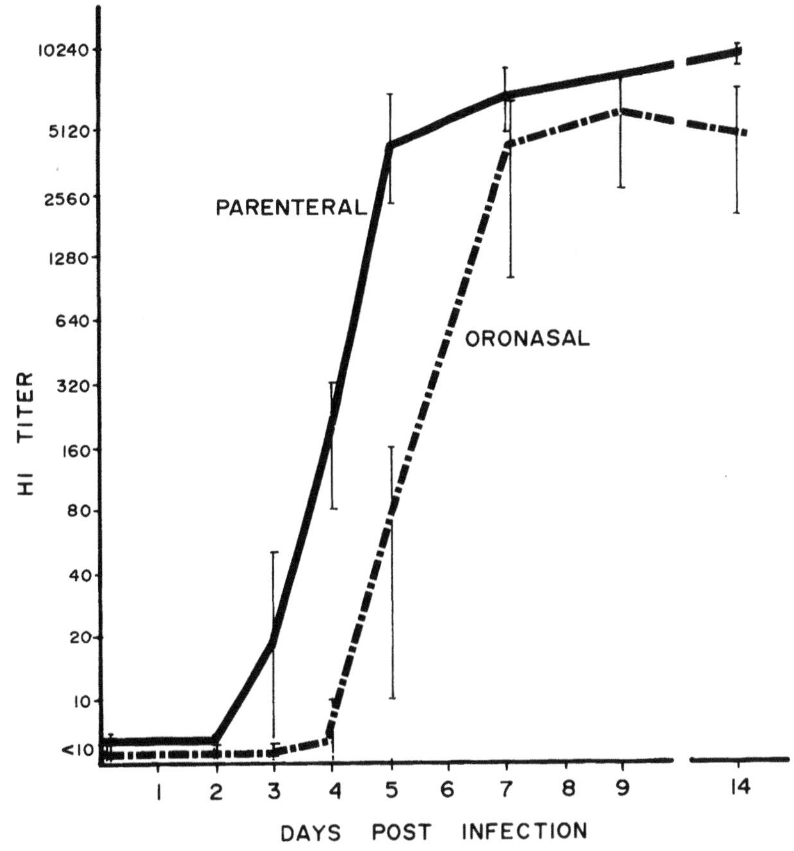

FIGURE 10. Mean and standard deviations of serum HI antibody titers in dogs after experimental CPV-2 infection by the oronasal (n = 17) or parenteral (n = 8) routes. (From Pollock, R. V. H., *Cornell Vet.*, 72, 103, 1982. With permission.)

breeding bitches, concluded that CPV-2 infection had no effect on canine reproduction. They found no difference in the average litter size, average number of pups weaned, number of successful matings or the number of stillbirths before and after the appearence of CPV-2 in the colony. Gooding and Robinson,[189a] however, in a study of a kennel of approximately 30 breeders, found significant decreases in the average number of pups per litter, the number of pups that survived to 5 weeks, and the number of bitches that produced live pups after 1978. While management changes or diseases other than CPV-2 might account for the latter observations, their concurrence with the time of the emergence of CPV-2 warrants further investigation.

V. IMMUNITY

A. Natural Infection

Recovery from CPV-2 infection confers long-lived immunity. Dogs were completely immune to reinfection when challenged orally with CPV-2 up to 20 months after initial infection.[88] They had no clinical signs of illness, their antibody titers did not increase significantly, and CPV-2 was not detected in their feces. Virus could not be recovered from any of the tissues examined from one dog killed on the 4th day after challenge.[88]

The mediator of immunity to CPV-2 has not been rigorously established, but humoral immunity is believed to be the principal determinant of resistance. In general, there is a direct correlation between serum antibody titer and resistance to infection.[88,99a,188,189] Passively acquired antibody, both maternally derived and from immune serum, is completely protective.[188,189,189a] Similar phenomena have been reported with respect to immunity to FPV infection in cats.[155,190-192]

The precise mechanism(s) by which serum antibody prevents infection are not known. Parvoviruses spread primarily by cell lysis and extracellular release.[181] Presumably, sufficient levels of humoral antibody interrupt the spread of CPV-2 from infected to uninfected cells. Very high titers appear to block the initiation of infection completely. In the presence of low levels of humoral antibody, localized infection can still occur, but viremia and generalized illness do not.[88]

The importance of secretory antibody in the intestinal tract is not known; only one preliminary study has been reported. Rice et al.[185] examined levels of coproantibody in naturally infected dogs. They found that levels of coproantibody were inversely related to the amount of virus in stool (as measured by HA), and that fecal HA titers were lower in dogs that survived infection than in those that succumbed. There was a significant correlation between increased likelihood of survival and greater amounts of coproantibody, but not serum antibody, during clinical illness. The authors concluded that "these data suggest that local intestinal immunity is more important than humoral immunity in developing immunological resistance to CPV-2 gastroenteritis".[185] This conclusion is unwarranted, however, since immunity to CPV-2 infection was not tested. The data do agree with those of Meunier,[135] *viz.*, that early and rapid onset of antibody production is correlated with reduced severity of disease and improved chance of survival. Nara et al.[192a] demonstrated a vigorous secretory IgM and IgG response to CPV-2 infection and speculated that secretory antibody may have a critical role in the elimination of virus from the host.

The role of local immunity in the prevention of disease has not been tested directly, however. It may be of minor importance; the intestinal epithelium appears to infected only secondarily,[135] passively transferred antibody can prevent infection,[188] and detectable levels of coproantibody persisted only 2 weeks after infection.[192a] Also, inactivated virus vaccines, which typically stimulate little secretory antibody, are protective (see below).

Serum antibody titers remain high in recovered dogs for at least 20 months[89] (Figure 11), suggesting continued antigenic stimulation. Other parvoviruses are known to persist for long periods in their respective hosts. In rodents, the maintainance of high antibody titers appears to result from the persistence of virus; H-1 was recovered from the livers of infected hamsters as long as 3 years after initial infection.[193] Feline panleukopenia virus was recovered from the kidneys of neonatally infected kittens 40 days after infection.[180] Although latent infections in dogs have been postulated by several authors, and serologic responses suggest that they occur, there is no direct evidence of persistent infection in dogs at this time.

B. Immunoprophylaxis

The optimal strategy for immunoprophylaxis against parvoviral infection has been an issue of considerable debate. The controversy has been fueled by the commercial implications of certain findings. Confounding factors include the difficulty of obtaining sufficient numbers of experimental animals, and of demonstrating the efficacy of vaccines against an agent which frequently produces mild or inapparent infections. Field trials are difficult to interpret because of the prevalence of prior infection, subclinical disease, and frequent re-exposure. Comparisons of vaccines of different antigenic strengths, faulty experimental designs, and the use of differing criteria for evidence of "immunity" have added to the confusion.

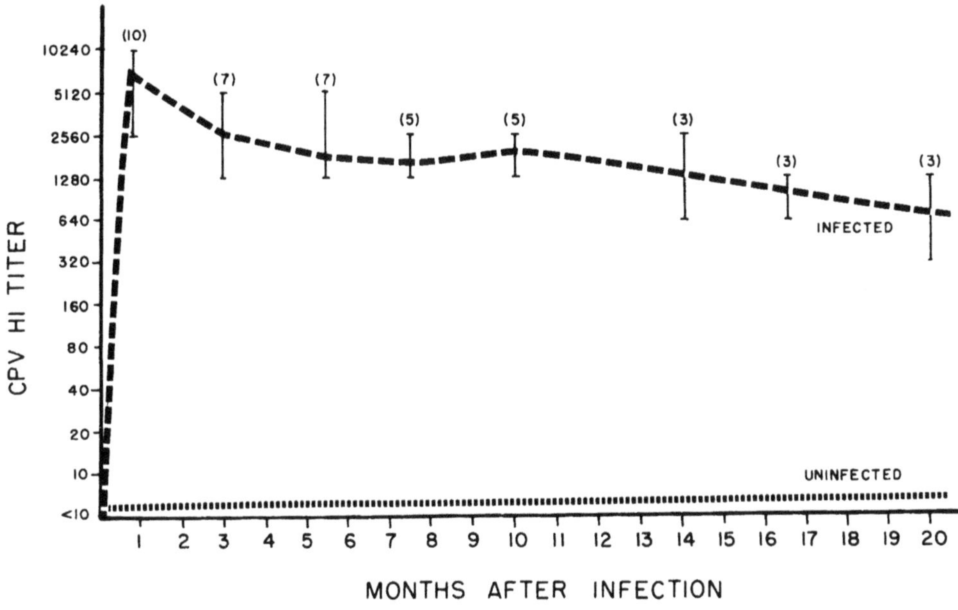

FIGURE 11. Range and mean antibody titers of dogs 20 months after infection. Numbers in parentheses indicate number tested at each time. Dogs were housed in contact with uninfected sentinel animals to exclude the possibility of undetected re-exposure. (From Pollock, R. V. H. and Carmichael, L. E., *Am. J. Vet. Res.*, 44, 169, 1983. With permission.)

1. Inactivated Vaccines

Inactivated FPV, MEV, and CPV-2 vaccines of sufficient antigenic mass immunize dogs against CPV-2 challenge.[3,6,84,86,88,90,91,193a,194,195] If "immunity" is taken to mean resistance to clinical disease, then inactivated vaccines appear to provide protection for at least 6 months and probably longer.[88] Adjuvants may extend the duration of protection significantly (see below). Definitive statements about the duration of immunity cannot be made with surety because the number of adequately controlled studies of challenge at intervals longer than 2 to 4 weeks after vaccination is exceedingly small. Studies were considered "adequately controlled" only if they entailed simultaneous challenge of similarly aged vaccinated and unvaccinated dogs at various postvaccinal intervals, if there was evidence that vaccinated dogs were seronegative prior to vaccination, and if sentinel animals were monitored to ensure that unplanned exposure to CPV-2 had not occurred between vaccination and challenge. Very few studies meet these criteria.

The magnitude of the serologic response to inactivated vaccines depends both upon the amount of inactivated virus they contain and the interval between vaccinations.[88,91,195a] Thus, comparisons of immunogenicity of FPV, CPV-2, and MEV vaccines are probably meaningless unless all contain similar amounts of antigen. The reported failure of an inactivated MEV vaccine to protect dogs in one study[196] may have been the result of inadequate antigen rather than of major antigenic differences as hypothesized.

If "immunity" is defined as resistance to infection, then the immunity engendered by inactivated vaccines is shortlived. Eugster,[84] based on fecal HA findings, suggested that subclinical infection might occur as little as 2 weeks after inactivated CPV-2 vaccination. Wierup et al.[91] reached the same conclusion, based on the appearance of atypical lymphocytes and antibody titers indicative of infection after challenge of dogs that had been vaccinated with inactivated FPV. Pollock and Carmichael[88] recovered

CPV-2 from the mesenteric lymph nodes, intestine, and feces of dogs challenged 4 and 6 months after vaccination with a commercial inactivated FPV vaccine. Clinical signs were absent in all cases, however. Viremia, lymphopenia, and thymic infection were observed in the unvaccinated but not in the vaccinated littermates.[88]

Thus, inactivated vaccines, provided they contain sufficient antigen, appear to provide protection against clinical disease for at least several months. They may, however, provide only short-lived protection against infection and shed of virus.

The effect of adjuvants has not been thoroughly studied. Evidence suggests that they may increase the efficacy of killed vaccines, but the choice of adjuvant appears important. The addition of aluminum hydroxide gel to inactivated CPV-2 or FPV vaccines had little effect on either the magnitude or duration of the serologic response to vaccination.[88,100,194] In one study,[194] addition of both aluminum hydroxide gel and a proprietary adjuvant appeared to increase the initial response to vaccination and the persistence of serum antibody. None of 10 dogs challenged 52 or 64 weeks after vaccination became ill, and 6 were completely immune to infection.[193a] Thus, it appears that the duration of protection induced by inactivated vaccine can be significantly increased with appropriate adjuvants, but further studies are needed.

2. Attenuated FPV Vaccines

Modified live feline panleukopenia virus also immunizes dogs against CPV-2 challenge.[3,87,89,197] As with inactivated vaccines, results of published studies are difficult to compare because of differences in viral strain, dose, experimental design, and the criteria for protection used.

Feline panleukopenia virus replicates to a limited extent in the dog. On the 4th day after inoculation, FPV was recovered only from the thymus and/or mesenteric lymph nodes of vaccinated dogs.[89,99a] A few animals were viremic on postvaccinal days 1 or 2. This is in contrast to the marked viremia and generalized distribution of CPV-2 in dogs on the 4th day after infection.[110,135] Feline panleukopenia virus was recovered from the intestine in one instance, and a pooled and concentrated fecal sample was also shown to contain small amounts of FPV.[89] The virus did not spread by direct contact to other dogs or cats however.

Response of dogs to attenuated FPV is extremely variable (Figure 12).[87,89] Limited data suggest that the number of dogs immunized, but not the magnitude of the individual response, is related to the amount of living virus in the vaccinal dose.[89] This probably reflects the relatively limited ability of FPV to establish infection in the dog. It was not possible to select a subpopulation of FPV with increased virulence for the dog; virus recovered from vaccinated dogs failed to establish infection when inoculated into a second group of dogs.[99a]

Dog responses to FPV can be broadly divided into two groups: those typical of response to inactivated vaccines, and those that suggest response to a replicating agent. The former are characterized by modest antibody titers that increase following a second immunization and which then decay in the absence of repeated exposure. Responses of dogs in the latter group are more like those seen following CPV-2 infection: that is, high titers which are not increased by a second vaccination and which persist in the absence of repeated exposure. It has been hypothesized that FPV replication occurs only in the latter group of dogs,[89] but direct evidence is lacking.

3. Attenuated CPV Vaccines

Attenuated canine parvovirus vaccines appear to provide the most durable protection against reinfection.[96,100,177,182] Several strains have been developed as commercial vaccines, but there are published studies on only a few; differences may exist.

The best characterized strain is CPV 916-LP developed by Carmichael and co-work-

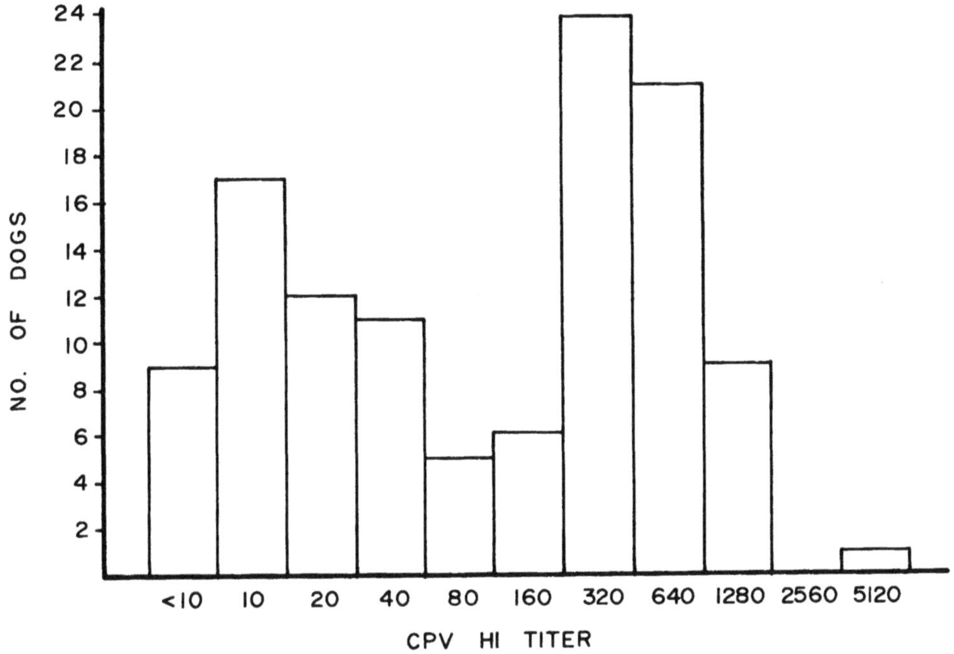

FIGURE 12. Histogram of serologic responses of 114 dogs to vaccination with $10^{5.5}$ $TCID_{50}$ of modified live feline panleukopenia virus. (From Pollock, R. V. H. and Carmichael, L. E., *Am. J. Vet. Res.*, 44, 169, 1983. With permission.)

ers.[96,182] This strain was selected by repeated passages at suboptimal temperatures and is characterized by large plaque size in A-72 cells.[96] It has an alteration of the Hinf-1 restriction map of the wild-type virus.[99,198]

The immune response to attenuated virus parallels that to wild-type virus. There is viremia and systemic distribution of the attenuated strain on the 2nd day after parenteral inoculation.[147] Viral shedding from the intestinal tract occurs on postinoculation days 3 to 7, but in amounts two to four orders of magnitude lower than from dogs that received wild-type virus.[96] Susceptible dogs are immunized by virus shed by vaccinates, a phenomenon that appears to occur with all attenuated strains studied.[199] The genetic alteration of the attenuated strain appears stable; the virus retained its properties of decreased virulence and increased plaque size through five sequential back passages in dogs.[96]

A slight reduction in the number of circulating lymphocytes is observed on the 3rd to 5th days after vaccination,[146,147] but there is no direct evidence of immunosuppression. Limited studies have failed to detect changes in lymphocyte responsiveness to phytomitogens or sheep erythrocytes in the 2 weeks following vaccination.[200] Antibody titers to other viral agents administered simultaneously in combined vaccines have been similar to those obtained with monovalent antigens.[147,177,182]

Humoral immune responses to the attenuated strains studies have paralled those to wild-type virus infection.[182] Antibody is usually detectable on the 3rd day after inoculation, although the onset may be delayed in pups with low levels of maternal antibody.[199] Maximal titers are achieved in 2 weeks and then persist at high levels for at least 2 years, even in the absence of re-exposure.[182] Five dogs challenged 24 months after vaccination were completely refractory to reinfection.[182]

4. *Effect of Maternally Derived Antibody*

The most important cause of vaccination "failure" in puppies appears to be suppres-

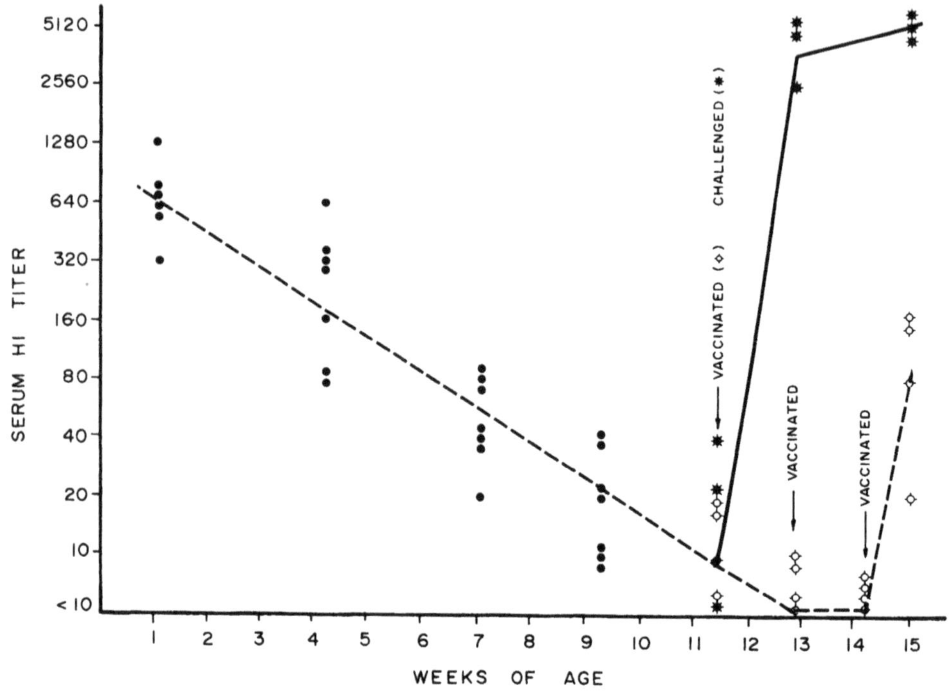

FIGURE 13. Maternal antibody decline in a litter of 7 pups and its effect on response to challenge or vaccination. At $11^1/_2$ weeks of age, 3 pups were challenged orally with CPV-2 and 4 were vaccinated with a commercial inactivated vaccine. All 3 challenged dogs became infected, but a response to vaccination was not detected in any of the vaccinated dogs until after a 3rd vaccination 19 days later. (From Pollock, R. V. H. and Carmichael, L. E., *J. Am. Vet. Med. Assoc.*, 180, 37, 1982. With permission.)

sion of an active immune response by maternally derived antibodies to CPV-2.[182,188,189] This phenomenon was first described in dogs 25 years ago with respect to canine distemper immunization.[201] The surprising finding for CPV-2 is the length of time for which these antibodies persist and are capable of interfering with active immunization. Complete suppression of response to vaccination has been documented in pups up to 18 weeks of age.[182,188,189] Inactivated vaccines appear to be particularly susceptible to maternal antibody interference, although the addition of adjuvant may partially overcome this.[193a,203]

Maternally derived antibody titers in pups decline as a log-linear function over time. The halflife is approximately 9 days.[57,188,189] A small amount of antibody crosses the placenta, but about 90% of the total maternal antibody in pups is absorbed from the colostrum.[188] In the dog, absorption of colostral antibodies is virtually complete by 72 hr after birth.[202] Titers in pups at this time are approximately equal to that of their dam.[188] Thus, given the CPV-2 antibody titer of the bitch, it should be possible to predict the age at which pups will respond to vaccination. Indeed, a nomograph relating these parameters has been published,[182] but it has not been tested prospectively.

Initially, pups are refractory to both infection[151,188] and immunization.[188,189] Pups with HI antibody titers ≥80 resisted oral challenge; they did not become ill, infectious virus was not detected in their feces, and they exhibited no antibody response to challenge.[188] Maternal antibody titers continued to decline uninterrupted in those pups and when challenged 10 weeks later, they were completely susceptible.[188]

Low levels of maternal antibody (HI titers 10 to 80) suppress responses to vaccination, especially to inactivated vaccines or live FPV, but they no longer provide protection against infection (Figure 13).[188,189,189a] Thus, all pups born to immune bitches ex-

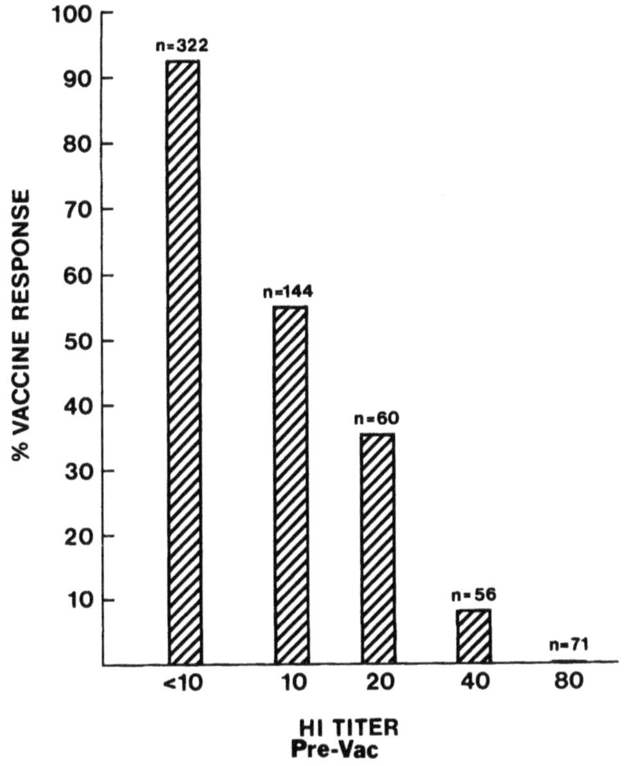

FIGURE 14. Relationship between the prevaccination HI antibody titer of pups and the percent response to attenuated canine parvoviral vaccine. A total of 724 paired sera are presented. The number (n) of paired (pre- and postvaccination sera) at each titer value is indicated. (From Carmichael, L. E., Joubert, J. C., and Pollock, R. V. H., *Cornell Vet.*, 73, 13, 1983. With permission.)

perience a period during which they are suceptible to infection, but refractory to immunization. As a result, CPV-2 disease among puppies has proved extremely difficult to control in large, heavily contaminated breeding colonies.[189]

Maternally derived antibody in sufficient amounts can also suppress response to attenuated live CPV-2 vaccines.[182,199] None of 182 pups with HI antibody titers ≥ 80 at the time of vaccination mounted an active immune response.[182] Immunization rates were inversely correlated with the amount of maternal antibody at the time of vaccination (Figure 14). Rates greater than 90% were observed only in pups that were seronegative by the HI test. Limited data suggest that attenuated CPV-2 vaccines may be slightly less suceptible to interference by maternal antibody than inactivated virus or attenuated FPV vaccines.[188] In some kennels, attenuated CPV-2 vaccines have been more effective than other types in reducing puppy mortality.[182,199,204]

VI. SUMMARY AND SUGGESTIONS FOR FUTURE STUDIES

Canine parvovirus type-2 is a new canine pathogen. Its origin is unknown, but it is closely related serologically and biochemically to known parvoviruses, particularly to feline panleukopenia and mink enteritis virus. Although the DNA sequences and antigenic determinants of these viruses are similar, their host ranges both in vitro and in vivo are distinct. They are therefore excellent models to elucidate the molecular basis of host range.

The pathogenesis of CPV-2 disease is still largely unexplored. Infection of susceptible dogs results in systemic distribution of the virus in all, but clinical disease in only a few. Factors that determine the outcome of infection are unknown. In older dogs, clinical signs reflect viral destruction of intestinal epithelium, although direct epithelial infection does not appear to be the primary event in the pathogenesis of CPV-2 enteritis. In very young puppies, a progressive myocarditis may be initiated, the pathobiology of which is not known, but which may prove to be a valuable model for viral-induced cardiomyopathy.

Recovery from infection confers long-lived immunity, although again the precise mechanisms responsible are unknown. Whether persistent infection occurs is unresolved, as are the short- and long-term effects of infection on the immune system. Immunization is now widely practiced, but the performance of all vaccines needs further critical study. Maternally derived antibody supresses response to vaccination in young dogs and is an important barrier to control. Alternate or more effective methods of immunization of pups are needed.

REFERENCES

1. Appel, M. J. G., Cooper, B. J., Greisen, H., and Carmichael, L. E., Status report: canine viral enteritis, *J. Am. Vet. Med. Assoc.*, 173, 1516, 1978.
2. Appel, M. J. G., Cooper, B. J., Greisen, H., Scott, F., and Carmichael, L. E., Canine viral enteritis. I. Status report on corona- and parvolike viral enteritides, *Cornell Vet.*, 69, 123, 1979.
3. Appel, M. J. G., Scott, F. W., and Carmichael, L. E., Isolation and immunisation studies of a canine parvo-like virus from dogs with haemorrhagic enteritis, *Vet. Rec.*, 105, 156, 1979.
4. Black, J. W., Holscher, M. A., Powell, H. S., and Byerly, C. S., Parvoviral enteritis and panleukopenia in dogs, *Vet. Med. Small Anim. Clin.*, 74, 47, 1979.
5. Burtonboy, G., Coignoul, F., Delferriere, N., and Pastoret, P.-P., Canine hemorrhagic enteritis: detection of viral particles by electron microscopy, *Arch. Virol.*, 61, 1, 1979.
6. Carmichael, L. E. and Binn, L. N., New enteric viruses in the dog, in *Advances in Veterinary Science and Comparative Medicine*, Vol. 25, Cornelius, C. E. and Simpson, C. F., Eds., Academic Press, New York, 1, 1981.
7. Eugster, A. K. and Nairn, C., Diarrhea in puppies: parvovirus-like particles demonstrated in their feces, *Southw. Vet.*, 30, 50, 1977.
8. Eugster, A. K., Bendele, R. A., and Jones, L. P., Parvovirus infection in dogs, *J. Am. Vet. Med. Assoc.*, 173, 1340, 1978.
9. Fritz, T. E., Canine enteritis caused by a parvovirus — Illinois, *J. Am. Vet. Med. Assoc.*, 174, 5, 1979.
10. Gagnon, A. N. and Povey, R. C., A possible parvovirus associated with an epidemic gastroenteritis of dogs in Canada, *Vet. Rec.*, 104, 263, 1979.
11. Hitchcock, L. M. and Scarnell, J., Canine parvovirus isolated in UK, *Vet. Rec.*, 105, 172, 1979.
12. Johnson, R. and Smith, J. R., Parvovirus enteritis in dogs, *Aust. Vet. Pract.*, 9, 197, 1979.
13. Johnson, R. H. and Spradbrow, P. B., Isolation from dogs with severe enteritis of a parvovirus related to feline panleukopaenia virus, *Aust. Vet. J.*, 55, 151, 1979.
14. Kelly, W. R., An enteric disease of dogs resembling feline panleukopaenia, *Aust. Vet. J.*, 54, 593, 1978.
15. McCandlish, I. A. P., Thompson, H., Cornwell, H. J. C., Laird, H., and Wright, N. C., Isolation of a parvovirus from dogs in Britain, *Vet. Rec.*, 105, 167, 1979.
16. Moraillon, A., Moraillon, R., DeLisle, F., Cotard, J. P., Pouchelon, J. L., and Clerc, B., Identification en France et etude generale de la gastroenterite canine parvovirus, *Le Point Vet.*, 9(44), 13, 1979.
17. Thomson, G. W. and Gagnon, A. N., Canine gastroenteritis associated with a parvovirus-like agent, *Can. Vet. J.*, 19, 346, 1978.
18. Touratier, L., La parvovirus canine en France et dans le monde, *Bull. Soc. Vet. Prat. Fr.*, 64, 263, 1980.

19. Huxtable, C. R., Howell, J. M., Robinson, W. R., Wilcox, G. E., and Pass, D. A., Sudden death in puppies associated with suspected viral myocarditis, *Aust. Vet. J.*, 55, 37, 1979.
20. Kelly, W. R. and Atwell, R. B., Diffuse subacute myocarditis of possible viral aetiology. A cause of sudden death in pups, *Aust. Vet. J.*, 55, 36, 1979.
21. Robinson, W. F., Wilcox, G. E., and Flower, R. L. P., Evidence for a parvovirus as the aetiologic agent in myocarditis of puppies, *Aust. Vet. J.*, 55, 294, 1979.
22. Robinson, W. F., Huxtable, C. R., Pass, D. A., and Howell, J., Clinical and electrocardiographic findings in suspected viral myocarditis of pups, *Aust. Vet. J.*, 55, 351, 1979.
23. Smith, J. R., Farmer, T. S., and Johnson, R. H., Serological observations on the epidemiology of parvovirus enteritis of dogs, *Aust. Vet. J.*, 56, 149, 1980.
24. Walker, S. T., Feilen, C. P., and Love, D. N., Canine parvovirus infection, *Aust. Vet. Pract.*, 9(3), 151, 1979.
25. Burtonboy, G., Coignoul, G., and Pastoret, P.-P., L'enterite a parvovirus du chien, *Ann. Med. Vet.*, 123, 123, 1979.
26. Schwers, A., Pastoret, P.-P., Burtonboy, G., and Thiry, E., Frequence en Belgique de l'infection a *Parvovirus* chez le chien, avant et apres l'observation des premiers cas cliniques, *Ann. Med. Vet.*, 123, 561, 1979.
27. Hayes, M. A., Russell, R. G., Mueller, R. W., and Lewis, R. J., Myocarditis in young dogs associated with a parvovirus-like agent, *Can. Vet. J.*, 20, 126, 1979.
28. Hayes, M. A., Russell, R. G., and Babiuk, L. A., Sudden death in young dogs with myocarditis caused by parvovirus, *J. Am. Vet. Med. Assoc.*, 174, 1197, 1979.
29. Bucci, T. J., Botros, and El Molla, M., Canine parvovirus infection: a brief review and report of first cases in Egypt, *J. Egyp. Vet. Med. Assoc.*, 42, 21, 1982.
30. Jalanka, H., Possible introduction of canine parvovirus enteritis into Finland, *Suomen Elanilaakar-ilent.*, 86, 15, 1980.
31. Neuvonen, E., Veijalainen, P., and Kangas, J., Canine parvovirus infection in housed raccoon dogs and foxes in Finland, *Vet. Rec.*, 110, 448, 1982.
32. Lescure, F., Guelfi, J. F., and Regnier, A., La parvovirose du chien, *Rev. Med. Vet.*, 131, 7, 1980.
33. Touratier, L., A propos d'une epizootie canine de gastro-enterite a parvovirus, *Bull. Acad. Vet. Fr.*, 52, 605, 1979.
34. Touratier, L., Epizootic parvovirus gastroenteritis and myocarditis in dogs, *Tijdschr. Diergeneesk.*, 105, 369, 1980.
35. Becker, G. and Becker, C.-H., Erste Beobachtungen uber eine Parvovirus-Enteritis bei Hunden in der DDR, *Mh. Vet. Med.*, 35, 891, 1980.
36. Bergmann, V., Baumann, G., and Fuchs, H.-W., Zur postmortalen Diagnostik der Parvovirus-Enteritis des Hundes, *Mh. Vet. Med.*, 35, 894, 1980.
37. Arens, M. and Krauss, H., Zum Nachweis von Parvovirus bei infektiosen Gastroenteritiden des Hundes mittels Immunelektronmikroscopie, *Berl. Muench. Tieraerztl. Wschr.*, 93, 156, 1980.
38. Bohm, H. O., Parvovirus bedingte Enteritis and Myocarditis beim Hund, *Tieraerztl. Umsch.*, 35, 229, 1980.
39. Hoffman, R., Frese, K., Reinacher, M., and Krauss, H., Parvovirusinfektion bei akuten Magen- und Darmerkrankungen des Hundes, *Berl. Muench. Tieraerztl. Wschr.*, 93, 121, 1980.
40. Kraft, W., Graf, R., Schwarz, H., Gerbig, T., Benary, F., Geyer, S., and Krebs, C., Parvovirus-Enteritis des Hundes — Clinik, Diagnose, Differentialdiagnose, Therapie, *Kleintierpraxis*, 25, 81, 1980.
41. Niemand, H. G., Niemand, S., and Wendel, E., Parvovirusinfektion von Hunden im grossraum Mannheim, *Berl. Muench. Tieraerztl. Wschr.*, 93, 211, 1980.
42. Thiel, W., Myocarditis bei Hundewelpen, *Berl. Muench. Tieraerztl. Wschr.*, 93, 271, 1980.
43. Boros, G. and Burtha, A., Occurrence of acute enteritis of parvovirus origin in dogs in Hungary, *Magy. Allatorv. Lapja*, 36, 247, 1981.
44. Sheahan, B. J. and Grimes, T. D., Sudden death in puppies, *Ir. Vet. J.*, 33, 94, 1979.
45. McNulty, M. S., Curran, N. L., McFerran, J. B., and Collins, D. S., Viruses and diarrhoea in dogs, *Vet. Rec.*, 108, 350, 1980.
46. Neill, S. D., McNulty, M. S., Bryson, D. G., and Ellis, W. A., Microbiological findings in dogs with diarrhoea, *Vet. Rec.*, 109, 538, 1981.
47. Perl, S., Jacobson, B., Klopfer, U., and Kuttin, E. S., First report of canine parvovirus infection in Israel-Histopathological findings, *Refu. Vet.*, 37, 110, 1980.
48. Cammarata, G., Finazzi, M., Cammarata, M. P., and Mandelli, G., Miocardite da parvovirus nel cucciolo, *Riv. Zootec. Vet.*, 8, 149, 1980.
49. Azetaka, M., Kirasawa, T., Konishi, S., and Ogata, M., Studies on canine parvovirus isolation, experimental infection and serologic survey, *Jpn. J. Vet. Sci.*, 43, 243, 1981.
50. Mohri, S., Handa, S., Wada, T., and Tokiyoshi, S., Sero-epidemiologic survey on canine parvovirus infection, *Jpn. J. Vet. Sci.*, 44, 543, 1982.

51. Ingh, Th. S. G. A. M. van den, Linde-Sipman, J. S. van den, and Webster, P. W., Parvo-virus-like particles in myocarditis in pups, *J. Small Anim. Pract.*, 21, 81, 1980.
52. Osterhaus, A. D. M. E., Drost, G. A., Wirahadiredja, R. M. S., and Ingh, Th. S. G. A. M. van den, Canine viral enteritis: Prevalence of parvo-, corona- and rotavirus infections in dogs in The Netherlands, *Vet. Q.*, 2, 181, 1980.
53. Osterhaus, A. D. M. E., Steenis, G. van, and deKreek, P., Isolation of a virus closely related to panleukopenia virus from dogs with diarrhea, *Zbl. Vet. Med. B*, 27, 11, 1980.
54. Gumbrell, R. C., Parvovirus infection in dogs, *N.Z. Vet. J.*, 27, 113, 1979.
55. Horner, G. W., Hunter R., and Chisholm, E. G., Isolation of a parvovirus from dogs with enteritis, *N.Z. Vet. J.*, 27, 280, 1979.
56. Parrish, C. R., Oliver, R. E., Julian, A. F., Smith, B. F., and Kyle, B. H., Pathological and virological observations on canine parvoviral enteritis and myocarditis in the Wellington region, *N.Z. Vet. J.*, 28, 238, 1980.
57. Parrish, C. R., Oliver, R. E., and McNiven, R., Canine parvovirus infections in a colony of dogs, *Vet. Microbiol.*, 7, 317, 1982.
58. Krohn, B. and Blakstad, E., Gastroenteritis of dogs due to parvovirus. Five case reports, *Nor. Vet. Tidsskr.*, 92, 249, 1980.
59. Bastianello, S. S., Canine parvovirus myocarditis: clinical signs and pathological lesions encountered in natural cases, *J. S. Afr. Vet. Assoc.*, 52, 105, 1981.
60. Rensburg, I. B. J. van, Botha, W. S., Lange, A. L., and Williams, M. C., Parvovirus as a cause of enteritis and myocarditis in puppies, *J. S. Afr. Vet. Assoc.*, 50, 249, 1979.
61. Klingeborn, B. and Moreno-Lopez, J., Diagnostic experience from an epidemic of canine parvoviral enteritis, *Zbl. Vet. Med. B*, 27, 483, 1980.
62. Olson, P., Hedhammer, A., and Klingeborn, B., Erfarenheter av parvovirus som orsak till akuta gastroenteriter hos hund, *Sven. Vet.*, 32, 189, 1980.
63. Bestetti, G., Hani, H., Dudan, F., Meister, V., Waber, S., and Luqinbuhl, H., Panleukopenieahnliche Enteritis und plotzliche Todesfalle bei Welpen infolge Myocarditis, wahrscheinlich verursacht durch Parvoviren, *Schweiz. Arch. Tierheilkd.* 121, 663, 1979.
64. Tingpalapong, M., Whitmire, R. E., Watts, D. M., Burke, D. S., Binn, L. N., Tesaprateep, T., Laungtongkum, S., and Marchwicki, R. H., Epizootic of viral enteritis in dogs in Thailand, *Am. J. Vet. Res.*, 43, 1687, 1982.
65. Else, R. W., Fatal haemorrhagic enteritis in a puppy associated with a parvovirus infection, *Vet. Rec.*, 106, 14, 1980.
66. Jefferies, A. R. and Blakemore, W. F., Myocarditis and enteritis in puppies associated with parvovirus, *Vet. Rec.*, 104, 221, 1979.
67. Thompson, H., McCandlish, I. A. P., Cornwell, H. J. C., Wright, N. G., and Rogerson, P., Myocarditis in puppies, *Vet. Rec.*, 104, 107, 1979.
68. Carpenter, J. L., Roberts, R. M., Harpster, N. K., and King, N. W., Intestinal and cardiopulmonary forms of parvovirus infection in a litter of pups, *J. Am. Vet. Med. Assoc.*, 176, 1269, 1980.
69. Nelson, D. T., Eustis, S. L., McAdaragh, J. P., and Stotz, I., Lesions of spontaneous canine viral enteritis, *Vet. Pathol.*, 16, 680, 1979.
70. Pletcher, J. M., Toft, J. D., Frey, R. M., and Casey, H. W., Histopathologic evidence for parvovirus infection in dogs, *Am. J. Vet. Med. Assoc.*, 175, 825, 1979.
71. Pollock, R. V. H. and Carmichael, L. E., Canine viral enteritis: recent developments, *Mod. Vet. Pract.*, 60, 375, 1979.
72. Blackburn, N. K. and LeBlank Smith, P. M., Serological diagnosis of canine parvovirus infection, *Zimbabwe Vet. J.*, 12, 64, 1981.
73. Robinson, W. F., Huxtable, C. R., and Pass, D. A., Canine parvoviral myocarditis: a morphologic description of the natural disease, *Vet. Pathol.*, 15, 282, 1980.
74. Binn, L. N., Lazar, E. C., Eddy, G. A., and Kajima, M., Recovery and characterization of a minute virus of canines, *Infect. Immun.*, 1, 503, 1970.
75. Carmichael, L. E., Joubert, J. C., and Pollock, R. V. H., Hemagglutination by canine parvovirus: serologic studies and diagnostic applications, *Am. J. Vet. Res.*, 41, 784, 1980.
76. Walker, S. T., Feilen, C. P., Sabine, M., Love, D. N., and Jones, R. F., A serological survey of canine parvovirus infection in New South Wales, Australia, *Vet. Rec.*, 106, 324, 1980.
76a. Jones, B. R., Robinson, A. J., Fray, L. M., and Lee, E. A., A longitudinal serological survey of parvovirus infection in dogs, *N.Z. Vet. J.*, 30, 19, 1982.
77. Burtonboy, G., Bazin, H., and Delferriere, N., Rat hybridoma antibodies against canine parvovirus, *Arch. Virol.*, 71, 291, 1982.
78. Flower, R. L. P., Wilcox, G. E., and Robinson, W. F., Antigenic differences between canine parvovirus and feline panleukopenia virus, *Vet. Rec.*, 107, 254, 1980.
79. Lenghaus, C. and Studdert, M. J., Relationships of canine panleukopaenia (enteritis) and myocarditis parvoviruses to feline panleukopaenia virus, *Aust. Vet. J.*, 56, 152, 1980.

80. Parrish, C. R., Carmichael, L. E., and Antczak, D. F., Antigenic relationships between canine parvovirus type 2, feline panleukopenia virus and mink enteritis virus using conventional antisera and monoclonal antibodies, *Arch. Virol.*, 72, 267, 1982.
81. Tratschin, J.-D., McMaster, G. K., Kronauer, G., and Siegl, G., Canine parvovirus: relationship to wild-type and vaccine strains of feline panleukopenia virus and mink enteritis virus, *J. Gen. Virol.*, 61, 33, 1982.
82. Appel, M. J. G., and Parrish, C. R., Raccoons are not susceptible to canine parvovirus, *J. Am. Vet. Med. Assoc.*, 181, 489, 1982.
83. Binn, L. N., Marchwicki, R. H., Eckermann, E. H., and Fritz, T. E., Viral antibody studies of laboratory dogs with diarrheal disease, *Am. J. Vet. Res.*, 42, 1665, 1981.
84. Eugster, A. K., Studies on canine parvovirus infections: development of an inactivated vaccine, *Am. J. Vet. Res.*, 41, 2020, 1980.
85. Mengeling, W. L., Bunn, T. O., and Paul, P. S., Antigenic relationship between porcine and canine parvoviruses, *Am. J. Vet. Res.*, 44, 865, 1983.
86. Chapek, M. L., McClaughry, L. E., and Wilkins, L. M., Efficacy and safety of an inactivated feline parvovirus vaccine against canine parvovirus infection, *Mod. Vet. Pract.*, 60, 261, 1980.
87. Chappuis, G. and Duret, C., Innocuite et antigenicite pour le chien d'une vaccin a virus vivant modife de la panleucopenie feline, *Le Point Vet.*, 10, 77, 1980.
88. Pollock, R. V. H. and Carmichael, L. E., Dog Response to inactivated canine parvovirus and feline panleukopenia virus vaccines, *Cornell Vet.*, 72, 16, 1982.
89. Pollock, R. V. H. and Carmichael, L. E., Use of modified live feline panleukopenia virus vaccine to immunize dogs against canine parvovirus, *Am. J. Vet. Res.*, 44, 169, 1983.
90. Wierup, M., Hok, K., Klingeborn, B., Rivera, F., Morein, B., Karlsson, K.-A., Hedhammer, A., Ekman, L., and Olson, P., Svenskt vaccin utprovat som skydd mot parvovirusinfektion hos hund, *Sven. Vet.*, 32, 195, 1980.
91. Wierup, M., Olson, P., Hedhammer, A., Klingeborn, B., and Karlsson, K.-A., Evaluation of a killed feline panleukopenia virus vaccine against canine parvoviral enteritis in dogs, *Am. J. Vet. Res.*, 43, 2183, 1982.
92. Parrish, C. R. and Carmichael, L. E., Antigenic structure and variation of canine parvovirus type-2, feline panleukopenia virus and mink enteritis virus, *Virology*, 129, 401, 1983.
93. McMaster, G., Tratschin, J.-D., and Siegl, G., Comparison of canine parvovirus with mink enteritis virus by restriction site mapping, *J. Virol.*, 38, 368, 1981.
94. Paradiso, P. R., Rhode, S. L. III, and Singer, I. I., Canine parvovirus: a biochemical and ultrastructural characterization, *J. Gen. Virol.*, 62, 113, 1982.
95. Burtonboy, G., Identification of canine parvovirus CPV 2 by means of rat monoclonal antibodies, in *Proc. Conf. Int. Virol. Comp.*, Banff, Vol. 78, 1982.
96. Carmichael, L. E., Joubert, J. C., and Pollock, R. V. H., A modified live canine parvovirus strain with novel plaque characteristics. I. Viral attenuation and dog response, *Cornell Vet.*, 71, 408, 1981.
97. Crandell, R. A., Fabricant, C. G., and Nelson-Rees, W. A., Development, characterization and viral susceptibility of a feline *Felis catus* renal cell (CRFK), *In Vitro*, 9, 176, 1973.
98. Binn, L. N., Marchwicki, R. H., and Stephenson, E.H., Establishment of a canine cell line: derivation, characterization, and viral spectrum, *Am. J. Vet. Res.*, 41, 855, 1980.
99. Parrish, C. R., unpublished data, 1983.
99a. Parrish, C. R., Canine Parvovirus and Feline Panleukopenia Virus: Structure and Function, Ph.D. thesis, Cornell University, Ithaca, N.Y., 1984.
100. Pollock, R. V. H., Canine Parvovirus: Host Response and Immunoprophylaxis, Ph.D. thesis, Cornell University, Ithaca, N.Y., 1981.
101. Mathys, A., Mueller, R., Pedersen, N. C., and Theilen, G. H., Comparison of hemagglutination and competitive enzyme-linked immunosorbent assay procedures for detecting canine parvovirus in feces, *Am. J. Vet. Res.*, 44, 152, 1983.
102. Fastier, L. B., A single radial haemolysis test for measuring canine parvovirus antibody, *Vet. Rec.*, 108, 299, 1981.
103. Mathys, A., Mueller, R., Pedersen, N. C., and Theilen, G. H., Hemagglutination with formalin-fixed erythrocytes for detection of canine parvovirus, *Am. J. Vet. Res.*, 44, 150, 1983.
104. Bachmann, P. A., Hoggan, M. D., Melnick, J. L., Pereira, H. G., and Vago, C., Parvoviridae, *Intervirology*, 5, 83, 1975.
105. Williams, F. P., Astrovirus-like, coronavirus-like, and parvovirus-like particles detected in the diarrheal stools of beagle pups, *Arch. Virol.*, 66, 215, 1980.
105a. Carmen, P. S. and Povey, R. C., Comparison of the viral proteins of canine parvovirus-2, mink enteritis virus and feline panleukopenia, *Vet. Microbiol.*, 8, 423, 1983.
106. Paradiso, P. R., Analysis of protein-protein interaction in the parvovirus H-1 capsid, *J. Virol.*, 46, 94, 1983.
107. Tattersall, P., Shatkin, A. J., and Ward, D. C., Sequence homology between the structural polypeptides of minute virus of mice, *J. Mol. Biol.*, 111, 375, 1977.

108. Peterson, J. L., Dale, R. M. K., Karess, R., Leonard, D., and Ward, D. C., Comparison of parvovirus structural proteins: evidence for post translational modification, in *Replication of Mammalian Parvoviruses,* Ward, D. C. and Tattersall, P., Eds., Cold Spring Harbor Laboratory, Cold Spring Harbor, N.Y., 1978, 431.
109. Molitor T. W., Joo, H. S., and Collett, M. S., Porcine parvovirus: virus purification and structural and antigenic properties of virion polypeptides, *J. Virol.,* 66, 261, 1983.
110. Pollock, R. V. H., Experimental canine parvovirus infection in dogs, *Cornell Vet.,* 72, 103, 1982.
111. Rhode, S. L. III, Replication process of the parvovirus H-1. IX. Physical mapping studies of the H-1 genome, *J. Virol.,* 22, 446, 1977.
112. Berns, K. I. and Hauswirth, W. W., Parvovirus DNA structure and replication, in *Replication of Mammalian Parvoviruses,* Ward, D. C. and Tattersall, P. C., Eds., Cold Spring Harbor Laboratory, Cold Spring Harbor, N.Y., 1978, 13.
113. Astell, C. R., Smith, M., Chow, M. B., and Ward, D. C., Structure of the 3′ hairpin terminus of four rodent parvovirus genomes: nucleotide sequence homology at origin of DNA replication, *Cell,* 17, 691, 1979.
114. Astell, C. R., Thomson, M., Merchlinsky, M., and Ward, D. C., The complete sequence of minute virus of mice, an autonomous parvovirus, *Nuc. Acids Res.,* 11, 999, 1983.
115. Pintel, D., Dadachanji, D., Astell, C. R., and Ward, D. C., The genome of minute virus of mice, an autonomous parvovirus, encodes two overlapping transcription units, *Nucl. Acids Res.,* 11, 1019, 1983.
116. Rhode, S. L. III and Paradiso, P. R., Parvovirus genome: nucleotide sequence of H-1 and mapping of its genes by hybrid-arrested translation, *J. Virol.,* 45, 173, 1983.
117. Rhode, S. L. III, personal communication, 1983.
118. Green, M. R., Lebovitz, R. M., and Roeder, R. G., Expression of the autonomous parvovirus H-1 genome: evidence for a single transcriptional unit and multiple spliced polyadenylated transcripts, *Cell,* 17, 967, 1979.
119. Faust, E. A. and Ward, D. C., Incomplete genomes of the parvovirus minute virus of mice, selective conservation of genome termini, including the origin of DNA replication, *J. Virol.,* 32, 276, 1979.
120. Rhode, S. L. III, Defective interfering particles of parvovirus H-1, *J. Virol.,* 27, 347, 1978.
121. Linser, P. and Armentrout, R. W., Binding of minute virus of mice to cells in culture, in *Replication of Mammalian Parvoviruses,* Ward, D.C. and Tattersall, P., Eds., Cold Spring Harbor Laboratory, Cold Spring Harbor, N.Y., 1978.
122. Linser, P., Bruning, H., and Armentrout, R. W., Specific binding sites for a parvovirus, minute virus of mice, on cultured mouse cells, *J. Virol.,* 24, 211, 1977.
123. Tennant, R. W. and Hand, R. E., Requirement of cellular DNA synthesis for Kilham rat virus replication, *Virology,* 42, 1054, 1970.
124. Tattersall, P., Replication of the parvovirus MVM. I. Dependence of virus multiplication and plaque formation on cell growth, *J. Virol.,* 10, 586, 1972.
125. Kollek, R. and Goulian, M., Synthesis of parvovirus H-1 replicative form viral DNA by DNA polymerase γ, *Proc. Nat. Acad. Sci. U.S.A.* 78, 6206, 1981.
125a. Astell, C. R., Thomson, M., Chow, M. B., and Ward, D. C., Structure and replication of minute virus of mice DNA, *Cold Spring Harbor Symp. Quant. Biol.,* 47, 751, 1982.
126. Kollek, R., Tseng, B. T., and Goulian, M., DNA polymerase requirements for parvovirus H-1 replication in vitro, *J. Virol.,* 41, 982, 1982.
127. Pritchard, C., Stout, E. R., and Bates, R. C., Replication of parvoviral DNA: characterization of a nuclear lysate system, *J. Virol.,* 37, 352, 1981.
128. Tseng, B. T., Grufstrom, R. H., Revie, D., Oersel, W., and Goulian, M., Studies on the early intermediates in the synthesis of DNA in animal cells, *Cold Spring Harbor Symp.,* 43, 263, 1979.
129. Revie, D., Tseng, B. T., Grafstrom, R. H., and Goulian, M., Covalent association of protein with replicative form DNA of parvovirus H-1, *Proc. Nat. Acad. Sci. U.S.A.,* 76, 5539, 1979.
130. Carlson, J. H. and Scott, F. W., Feline panleukopenia. II. The relationship of intestinal mucosal cell proliferation rates to viral infection and development of lesions, *Vet. Pathol.,* 14, 173, 1977.
131. Csiza, C. K., Scott, F. W., deLahunta, A., and Gillespie, J. H., Pathogenesis of feline panleukopenia virus in susceptible newborn kittens. I. Clinical signs, hematology, serology, and virology, *Infect. Immun.,* 3, 833, 1971.
132. Csiza, C. K., deLahunta, A., Scott, F. W., and Gillespie, J. H., Pathogenesis of feline panleukopenia virus in susceptible newborn kittens. II. Pathology and immunofluorescence, *Infect. Immun.,* 3, 838, 1971.
133. Cooper, B. J., Carmichael, L. E., Appel, M. J. G., and Greisen, H., Canine viral enteritis. II. Morphologic lesions in naturally occurring parvovirus infection, *Cornell Vet.,* 69, 134, 1979.
134. McCandlish, I. A. P., Thompson, H., Cornwell, H. J., and Macartney, L., Canine parvovirus infection, *Vet. Rec.,* 107, 204, 1980.
135. Meunier, P. C., The Pathogenesis of Canine Parvovirus Infection, Ph.D. thesis, Cornell University, Ithaca, N.Y., 1983.

136. Tattersall, P., Susceptibility to minute virus of mice as a funtion of host cell differentiation, in *Replication of Mammalian Parvoviruses*, Ward, D. C. and Tattersall, P., Eds., Cold Spring Harbor Laboratory, Cold Spring Harbor, N.Y., 1978, 131.
137. Miller, R. A., Ward, D. C., and Ruddle, F. H., Embryonal carcinoma cells (and their somatic cell hybrids) are resistant to infection by the murine parvovirus MVM, which does infect other tetracarcinoma-derived cell lines, *J. Cell Physiol.*, 91, 393, 1977.
138. Lipton, H. L. and Johnson, R. T., The pathogenesis of rat virus infections in the newborn hamster, *Lab. Invest.*, 27, 508, 1972.
139. Spalholz, B. A. and Tattersall, P., Interactions of minute virus of mice with differentiated cells: strain-dependent target cell specificity is mediated by intracellular factors, *J. Virol.*, 46, 937, 943, 1983.
140. Tattersall, P. and Bratton, J., Reciprocal productive and restrictive virus-cell interactions of immunosuppressive and prototype strains of minute virus of mice, *J. Virol.*, 46, 944, 1983.
141. Mitra, S., Snyder, C. E., Bates, R. C., and Banerjee, P. T., Comparative physicochemical and biological properties of two strains of Kilham rat virus, a non-defective parvovirus, *J. Gen. Virol.*, 61, 43, 1983.
142. Mann, P. C., Buch, M., Appel, M. J. G., Beehler, B. A., and Montali, R. J., Canine parvovirus infection in South American canids, *J. Am. Vet. Med. Assoc.*, 177, 779, 1980.
143. Evermann, J. F., Foreyt, W., Maag-Miller, L., Leathers, C. W., McKeirnan, A. J., and LeaMaster, B., Acute hemorrhagic enteritis associated with canine coronavirus and parvovirus infections in a captive coyote population, *J. Am. Vet. Med. Assoc.*, 177, 784, 1980.
144. Fletcher, K.C., Eugster, A. K., Schmidt, R. E., and Hubbard, G. B., Parvovirus infection in maned wolves, *J. Am. Vet. Med. Assoc.*, 175, 897, 1979.
145. Bieniek, H. J., Encke, W., Gandras, R., and Vogt, P., Parvovirus-Infektion beim Mahnenwolf, *Kleintierpraxis*, 26, 291, 1981.
146. Pollock, R. V. H., unpublished data, 1981.
146a. Toma, B., Chappuis, G., and Elliott, M., Recherche des anticorps du parvovirus et du corovirus canins chez l'homme, *Rec. Med. Vet.*, 158, 607, 1982.
147. Carmichael, L. E., unpublished data, 1982.
148. Nettles, V. F., Pearson, J. E., Gustafson, G. A., and Blue, J. L., Parvovirus infection in translocated raccoons, *J. Am. Vet. Med. Assoc.*, 177, 787, 1980.
149. Moraillon, A., Moraillon, R., Person, J. M., and Parodi, A. L., Parvovirose canine: l'ingestion d'organes de vison atteint d'enterite a virus declenche chez le chien une maladie indentique a la maladie spontanee, *Rec. Med. Vet.*, 156, 539, 1980.
150. Helfer-Baker, C., Evermann, J. F., McKeirnan, A. J., Morrison, W. B., Slack, R. L., and Miller, C. W., Serological studies on the incidence of canine enteritis viruses, *Canine Pract.*, 7(3), 37, 1980.
151. Meunier, P. C., Glickman, L. T., Appel, M. J. G., and Shin, S. J., Canine parvovirus in a commercial kennel: Epidemiologic and pathologic findings, *Cornell Vet.*, 71, 96, 1981.
152. Hirasawa, T., Azetaka, M., and Konishi, S., Prevalence and conversion of canine parvovirus antibody in various dog colonies in Japan, *Jpn. J. Vet. Sci.*, 44, 997, 1982.
152a. Studdert, M. J., Oda, C., Riegl, C. A., and Roston, R. P., Aspects of the diagnosis, pathogenesis and epidemiology of canine parvovirus, *Aust. Vet. J.*, 60, 197, 1983.
153. Rohovsky, M. W. and Griesemer, R., Experimental feline infectious enteritis in the germfree cat, *Pathol. Vet.*, 4, 391, 1967.
154. Carlson, J. H., Scott, F. W., and Duncan, J. R., Feline panleukopenia. I. Pathogenesis in germfree and specific pathogen-free cats, *Vet. Pathol.*, 14, 79, 1977.
155. King, D. A. and Croghan, D. L., Immunofluorescence of feline panleucopenia virus in cell culture: determination of immunological status of felines by serum neutralization, *Can. J. Comp. Med. Vet. Sci.*, 29, 85, 1965.
156. Carpenter, J. L., Feline panleukopenia: clinical signs and differential diagnosis, *J. Am. Vet. Med. Assoc.*, 158, 857, 1971.
157. Horvath, Z., Papp, L., and Bartha, A., Epizootiological and clinical study of feline panleucopenia, *Acta. Vet. Acad. Sci. Hung.*, 24, 7, 1974.
158. McCandlish, I. A. P., Thompson, H., Fisher, E. W., Cornwell, H. J. C., Macartney, J., and Walton, I. A., Canine parvovirus infection, *In Practice*, 3, 5, 1981.
159. Sabine, M., Herbert, L., and Love, D. N., Canine parvovirus in Australia during 1980, *Vet. Rec.*, 110, 551, 1982.
160. Robinson, W. F., Wilcox, G. E., and Flower, R. L. P., Canine parvoviral disease: experimental reproduction of the enteric form with a parvovirus isolated from a case of myocarditis, *Vet. Pathol.*, 17, 589, 1980.
161. Mulvey, J. J., Bech-Nielsen, S., Haskins, M. E., Jesyk, P. F., Taylor, H. W., and Eugster, A. K., Myocarditis induced by parvoviral infection in weanling pups in the United States, *J. Am. Vet. Med. Assoc.*, 177, 695, 1980.

161a. Atwell, R. B. and Kelly, W. R., Canine parvovirus: a cause of chronic myocardial fibrosis and adolescent congestive heart failure, *J. Small Anim. Pract.*, 21, 609, 1980.
162. Harcourt, R. A., Spurling, N. W., and Pick, C. R., Parvovirus infection in a beagle colony, *J. Small Anim. Pract.*, 21, 293, 1980.
162a. Lenghaus, C. and Studdert, M. J., Generalized parvovirus disease in neonatal pups, *J. Am. Vet. Med. Assoc.*, 181, 41, 1982.
162b. Johnson, B. J. and Castro, A. E., Isolation of canine parvovirus from a dog brain with severe necrotizing vasculitis and encephalomalacia, *J. Am. Vet. Med. Assoc.*, 184, 1398, 1984.
163. Kramer, J. M., Meunier, P. C., and Pollock, R. V. H., Canine parvovirus: update, *Vet. Med. Small Anim. Clin.*, 75, 1541, 1980.
164. Moreau, P. M., Canine viral enteritis, *Compendium Cont. Ed.*, 2, 540, 1980.
165. Jacobs, R. M., Weiser, M. G., Hall, R. L., and Kowalski, J. J., Clinicopathologic features of canine parvoviral enteritis, *J. Am. Anim. Hosp. Assoc.*, 16, 809, 1980.
166. McArdaragh, J. P., Eustis, S. L., Nelson, D. T., Stotz, I., and Kenefick, K., Experimental infection of conventional dogs with canine parvovirus, *Am. J. Vet. Res.*, 43, 693, 1982.
167. Carman, S. and Povey, C., Successful experimental challenge of dogs with canine parvovirus-2, *Can. J. Comp. Med.*, 46, 33, 1982.
168. Potgieter, L. N. D., Jones, J. B., Patton, C. S., and Webb-Martin, T. A., Experimental parvovirus infection in dogs, *Can. J. Comp. Med.*, 45, 212, 1981.
169. Woods, C. B., Pollock, R. V. H., and Carmichael, L. E., Canine parvoviral enteritis, *J. Am. Anim. Hosp. Assoc.*, 16, 171, 1980.
170. Boosinger, T. R., Rebar, A. H., DeNicola, D. B., and Boon, G. D., Bone marrow alterations associated with canine parvoviral enteritis, *Vet. Pathol.*, 19, 558, 1982.
171. Jesyk, P. F., Haskins, M. E., and Jones, C. L., Myocarditis of probable viral origin in pups of weaning age, *J. Am. Vet. Med. Assoc.*, 174, 1204, 1979.
172. Hezel, B., Thornburg, L. P., and Kinter, L. D., Inclusion body myocarditis: cause of acute death in puppies, *Vet. Med. Small Anim. Clin.*, 174, 1627, 1979.
173. Cimprich, R. E., Robertson, J. L., Kutz, S. A., Struve, P. S., Detweiler, D. K., DeBaeke, P. J., and Streett, C. S., Degenerative cardiomyopathy in experimental beagles following parvovirus exposure, *Toxicol. Pathol.*, 9, 19, 1981.
174. Ilgen, B. E. and Conroy, J. D., Fatal cardiomyopathy in an adult dog resembling parvovirus-induced myocarditis: a case report, *J. Am. Anim. Hosp. Assoc.*, 18, 613, 1982.
175. Lenghaus, C., Studdert, M. J., and Finnie, J. W., Acute and chronic canine parvovirus myocarditis following intrauterine inoculation, *Aust. Vet. J.*, 56, 465, 1980.
176. Burren, V. S., Mason, K. V., and Mason, S. M., Sudden death in pups due to myocarditis, possibly of viral origin, *Aust. Vet. Pract.*, 9, 25, 1979.
177. Bass, E. P., Gill, M. A., and Beckenhauer, W. H., Development of a modified live, canine origin parvovirus vaccine, *J. Am. Vet. Med. Assoc.*, 181, 909, 1982.
178. Krakowka, S., Olsen, R. G., Axthelm, M. K., Rice, J., and Winters, K., Canine parvovirus infection potentiates canine distemper encephalitis attributable to modified live-virus vaccine, *J. Am. Vet. Med. Assoc.*, 180, 137, 1982.
178a. O'Sullivan, G., Durham, P. J. K., Smith, J. R., and Campbell, R. S. F., Experimentally induced severe parvoviral enteritis, *Aust. Vet. J.*, 61, 1, 1984.
179. Merickel, B. S., Hahn, F. F., Hanika-Rebar, C., Muggenburg, B. A., Brownstein, D. A., Rebar, H., and DeNicola, D., Acute parvoviral enteritis in a closed beagle dog colony, *Lab. Anim. Sci.*, 30, 874, 1980.
180. Csiza, C. K., Scott, F. W., deLahunta, A., and Gillespie, J. H., Immune carrier state of feline panleukopenia virus-infected cats, *Am. J. Vet. Res.*, 32, 419, 1971.
181. Siegl, G., Biology and pathogenicity of autonomous parvoviruses, in *The Parvoviruses*, Berns, K., Ed., Plenum Press, New York, 1984.
182. Carmichael, L. E., Joubert, J. C., and Pollock, R. V. H., A modified live canine parvovirus vaccine. II. Immune response, *Cornell Vet.*, 73, 13, 1983.
183. Schultz, R. D., Mendel, H., and Scott, F. W., Effect of feline panleukopenia virus infection on development of humoral and cellular immunity, *Cornell Vet.*, 66, 324, 1976.
184. Schwers, A., Pastoret, P.-P., Dagenais, L., and Aguilar-Setien, A., Utilisation d'une technique de contre-immunoelectro-osmophorese (CIEOP) pour la detection des anticorps envers le parvovirus canin et des antigenes dans les matieres fecales, *Ann. Med. Vet.*, 124, 255, 1980.
185. Rice, J. B., Winters, K. A., Krakowka, S., and Olsen, R. G., Comparison of systemic and local immunity in dogs with canine parvovirus gastroenteritis, *Infect. Immun.*, 38, 1003, 1982.
185a. Olsen, R. G. and Krakowka, S., Immune dysfunctions associated with viral infections, *Comp. Cont. Educ. Vet. Pract.*, 6, 422, 1984.
186. Bishop, S. P. and Hine, P., Cardiac muscle cytoplasmic and nuclear development during canine neonatal growth, in *Recent Advances in Studies on Cardiac Structure and Metabolism*, Roy, P. E. and Harris, P., Eds., University Park Press, Baltimore, 1975, 77.

187. Bishop, S. P., Effect of aortic stenosis on myocardial cell growth, hyperplasia and ultra structure in neonatal dogs, in *Recent Advances in Studies on Cardiac Structure and Metabolism*, Dhalle, N. S., Ed., University Park Press, Baltimore, 1972, 637.
188. Pollock, R. V. H. and Carmichael, L. E., Maternally derived antibody to canine parvovirus infection: transfer, decline and interference with vaccination, *J. Am. Vet. Med. Assoc.*, 180, 37, 1982.
189. Moraillon, A., Moraillon, R., and Marjollet, S., Sur les echecs de la vaccination du chiot contre la parvovirose en milieu contamine, *Rec. Med. Vet.*, 158, 205, 1982.
189a. Gooding, G. E. and Robinson, W. F., Maternal antibody, vaccination and reproductive failure in dogs with parvovirus infection, *Aust. Vet. J.*, 59, 170, 1982.
190. Fastier, L. B., Feline panleukopenia — a serological study, *Vet. Rec.*, 83, 653, 1968.
191. Scott, F. W., Csiza, C. K., and Gillespie, J. H., Maternally derived immunity to feline panleukopenia, *J. Am. Vet. Med. Assoc.*, 156, 439, 1970.
192. Johnson, R. H., Serologic procedures for the study of feline panleukopenia, *J. Am. Vet. Med. Assoc.*, 158, 876, 1971.
192a. Nara, P. L., Winters, K., Rice, J. B., Olsen, R. G., and Krakowka, S., Systemic and local intestinal antibody in dogs given both infective and inactivated canine parvovirus, *Am. J. Vet. Res.*, 44, 1989, 1983.
193. Toolan, H. W., The picodna viruses: H, RV, and AAV, *Int. Rev. Exp. Pathol.*, 6, 135, 1968.
193a. Povey, R. C., Carmen, P. S., and Ewert, E., The duration of immunity to an inactivated canine parvovirus vaccine. A 52- and 64-week post vaccination challenge study, *Can. Vet. J.*, 24, 245, 1983.
194. Povey, C., Development of a vaccine incorporating killed virus of canine origin for the prevention of canine parvovirus infection, *Can. Vet. J.*, 23, 15, 1982.
195. Smith, J. R., Johnson, R. H., and Farmer, T. S., Canine parvovirus vaccine, *Aust. Vet. J.*, 56, 611, 1980.
195a. Murisier, N., Pfister, R., Ohder, H., and Kihm, U., Die serologishe immunatwort nach vakzination mit inaktvierten parvovirus — impfstoffen beim hund, *Schweizer Arch. Tierheilkund*, 125, 851, 1983.
196. Carmen, S. and Povey, C., The failure of an inactivated mink enteritis virus vaccine in four preparations to provide protection to dogs against challenge with canine parvovirus-2, *Can. J. Comp. Med.*, 46, 47, 1982.
197. Moraillon, A., Canine parvovirus: safety and efficacy of attenuated feline panleucopenia virus, *Vet. Rec.*, 107, 512, 1980.
198. Siegl, G., personal communication, 1982.
199. Carmichael, L. E., Pollock, R. V. H., and Joubert, J. C., Response of puppies to canine-origin parvovirus vaccines, *Mod. Vet. Pract.*, 65, 99, 1984.
200. Montgomery, S., personal communication, 1983.
201. Baker, J. A., Robson, D. S., Gillespie, J. H., Burgher, J. A., and Doughty, M. F., A nomograph that predicts the age to vaccinate puppies against distemper, *Cornell Vet.*, 49, 158, 1959.
202. Carmichael, L. E., Robson, D. S., and Barnes, F. D., Transfer and decline of maternal infectious canine hepatitis antibody in puppies, *Proc. Soc. Exp. Biol. Med.*, 109, 677, 1962.
203. Wilson, J. H. G. and Hermann-Dekkers, W. M., Experiments with a homologous, inactivated canine parvovirus vaccine in vaccination programmes for dogs, *Vet. Q.*, 4(3), 108, 1982.
204. Glickman, L. T. and Appel, M. J. G., A controlled trial of an attenuated canine origin parvovirus vaccine, *Comp. Cont. Educ.*, 4, 888, 1982.

Chapter 11

PATHOGENESIS OF PORCINE POLIOENCEPHALOMYELITIS

John F. Long

TABLE OF CONTENTS

I. INTRODUCTION

Surveys have shown porcine enteroviruses to be ubiquitous in the swine population.[8] While a majority of infections are asymptomatic, there is a broad spectrum of pathogenicity among various porcine enteroviruses, ranging from no clinically detectable disease to severe paralysis.

In addition to neurological disease, porcine enteroviruses have been associated with other clinical syndromes, including diarrhea,[1,8,35] pneumonia,[39,48] pericarditis and myocarditis,[33,35,48] exanthems,[33] and antenatal and perinatal disease of pigs.[57] From the standpoint of comparative pathology, it is notable that human enteroviruses have been associated with syndromes comparable with all of those described in porcine enterovirus infections and additionally with syndromes involving skeletal muscle, the eye, and occasionally others.[14]

This report will be limited to studies of porcine enteroviruses producing neurological disease, expecially focusing on pathogenetic aspects. Wherever appropriate, comparisons with human polioviruses will be made. The only known natural host for porcine enteroviruses is the pig.[8]

II. EPIDEMIOLOGIC PERSPECTIVES OF PORCINE POLIOENCEPHALOMYELITIS

The initial description of disease produced by what is now known to be the porcine enterovirus group was by Trefny in Czechoslovakia in 1929.[55] The lesions histologically were those of a polioencephalomyelitis, and in the subsequent year the disease was found to be experimentally transmissible.[28] This disease, known as Teschen disease, has continued to recur sporadically, mainly in central Europe but also in Africa. In subsequent years additional polioencephalomyelitides of swine were identified in western Europe,[42] North America,[30,45] Australia,[12] and other countries (reviewed by Mills and Neilsen[42]).

The Porcine enteroviruses have now been serologically classified.[10,56] The virulent strains associated with classical Teschen disease appear to be restricted to those geographical areas in which clinical disease occurs, and they have not been isolated in North America.[8] Less virulent viruses appear to be ubiquitous. This is illustrated by the observation that exclusion of the infection from specific pathogen-free herds over a prolonged period is difficult or impossible to achieve.[8] Endemic infection with several serogroups of porcine enteroviruses are routinely found in conventional herds. Singh and Bohl[50] demonstrated waves of infection with six different serotypes over a period of 26 months in a long-term study of enterovirus infection in a single herd.

Transmission of porcine enterovirus infection occurs by ingestion of the virus.[8] Infection is usually acquired by young pigs shortly after weaning when the level of maternally derived antibodies has declined and pigs from several litters are mixed together. It persists for at least several weeks. Adults rarely excrete virus but have high antibody levels.[8] Pigs of any age are, however, fully susceptible to infection with a porcine enterovirus belonging to a serogroup to which they have not been previously exposed,[8] although they may not necessarily develop clinical signs of disease.

III. PORCINE ENTEROVIRUSES

Porcine enteroviruses have been studied by many authors and have been reviewed and summarized by Derbyshire in 1981.[8] Porcine enteroviruses resemble the enteroviruses of other species in their basic properties and are classified within the family Picornaviradae.[38] The virions are spherical, 25 to 31 nm in diameter and nonenveloped. They contain a core of single-stranded ribonucleic acid (RNA) which is surrounded by a capsid of a polypeptide composition similar to that of other enteroviruses.[51]

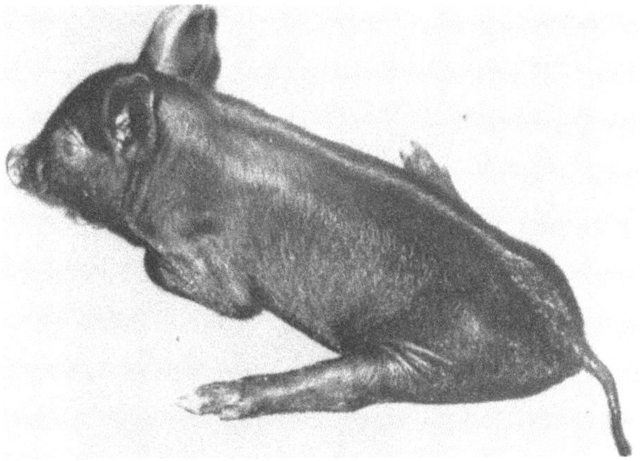

FIGURE 1. Experimental germfree pig with rigid spastic paralysis of both rear limbs. (From Long, J. F., Koestner, A., and Kasza, L., *Am. J. Vet. Res.*, 27, 274, 1966. With permission.)

Porcine enteroviruses are readily grown in cell cultures of porcine origin. Certain strains can be cultivated in monkey kidney or baby hamster kidney cell lines.[8] They have also been shown to multiply in porcine intestinal organ cultures.[9] The different strains produce one of two kinds of cytopathic effect (CPE) in porcine kidney cultures. Both types are rapidly cytolytic. Strains belonging to serogroups 1 to 7 produce changes in porcine kidney (PK) cells similar to those produced by other enteroviruses in kidney cell cultures.[8] The changes are characterized by the development of foci of rounded refractile cells which rapidly detach from the glass surface (type 1 CPE), while serogroup 8 strains produce a more unusual CPE characterized by granular cells with peripheral cytoplasmic protrusions. Teschen disease virus (Serogroup 1), as well as the Ohio viruses used in the following report, all produce type 1 CPE.[8]

IV. CLINICAL FEATURES ASSOCIATED WITH NEUROTROPIC STRAINS OF PORCINE ENTEROVIRUSES (OHIO ISOLATES)

In order to assess the clinical signs produced by a neurotropic enterovirus alone (without being obscured by effects of secondary microbial invaders) Long et al.[34,35] used germfree pigs as experimental subjects. The inoculum was prepared from isolates of pathogenic enteroviruses obtained from natural outbreaks in Ohio. The oral route of administration was used so as to simulate natural exposure.

Although there were minor variations in time of onset of various clinical manifestations between different strains of pathogenic porcine enteroviruses, a typical sequence was recorded as follows[35] (with porcine enteroviruso 02b): no clinical signs were detected until the 7th day. At this time, a sudden rise in body temperature occurred in all experimentally exposed pigs. Body temperatures rose to an average of 103.9°F. Diarrhea occurred 24 hr after the onset of fever, persisted 24 to 48 hr, and then rapidly subsided. Paralysis and paresis developed in several pigs per litter on the 10th to 11th day following inoculation. The fever persisted to a varying degree until about the 14th day, after which it returned to the level of the controls.

The paralysis associated with enteroviral infection usually took the form of flaccid paralysis although the spastic form was sometimes seen (Figure 1).[34] By using germfree pigs and by maintaining nursing care, the pigs could be kept alive for indefinite periods and thus exhibit the full spectrum of the clinical signs. It would seem reasonable to

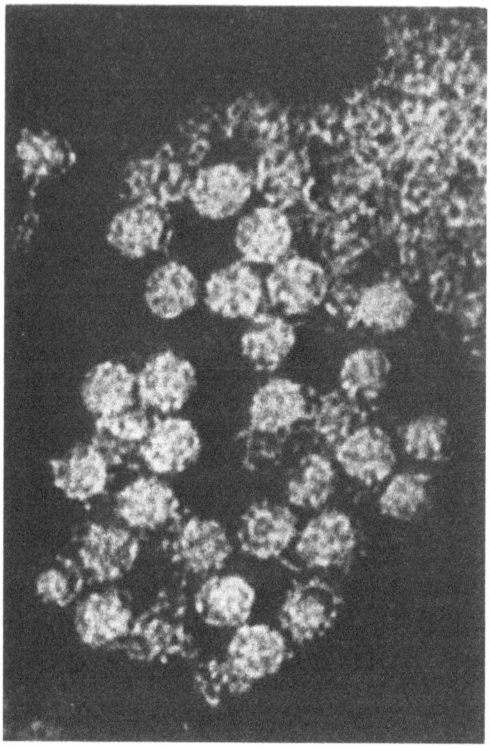

FIGURE 2. Negatively stained purified particles. The individual capsomeres can be recognized. The viral particles in this preparation measure 27 nm; magnification × 260,000. (From Koestner, A., Kasza, L., and Kindig, O., *Am. J. Pathol.*, 48, 129, 1966. With permission.)

believe that under field conditions, secondary microbial invasion would often result in mortality of the litter before paralysis could develop.

V. PATHOGENESIS — NEUROTROPIC STRAINS OF PORCINE ENTEROVIRUSES (OHIO ISOLATES)

In order to eliminate as many variables as possible, a study (of pathogens of porcine polioencephalomyelitis in pigs) was undertaken in which a single strain of porcine enterovirus (03b) (Figures 2 and 3)[31] was used as the inoculum. This virus was obtained from the brain of a pig with the naturally occurring disease.[26,30,34] The experimental animals consisted entirely of germfree pigs, thus avoiding colostral immunity, as well as secondary microbial infection. The germfree isolators provided a uniform environment. Thus, the development of lesions could be attributed exclusively to the experimentally administered virus.

The virus was given by the oral route. The findings in experimentally infected pigs were compared with littermate controls at sequential intervals throughout the course of the disease, utilizing techniques of virus isolation as well as light and electron microscopy.

A. Isolation of Virus at Sequential Intervals During Infection[27]

By 2 days postinoculation the virus was isolated from the oronasal cavity (turbinates,

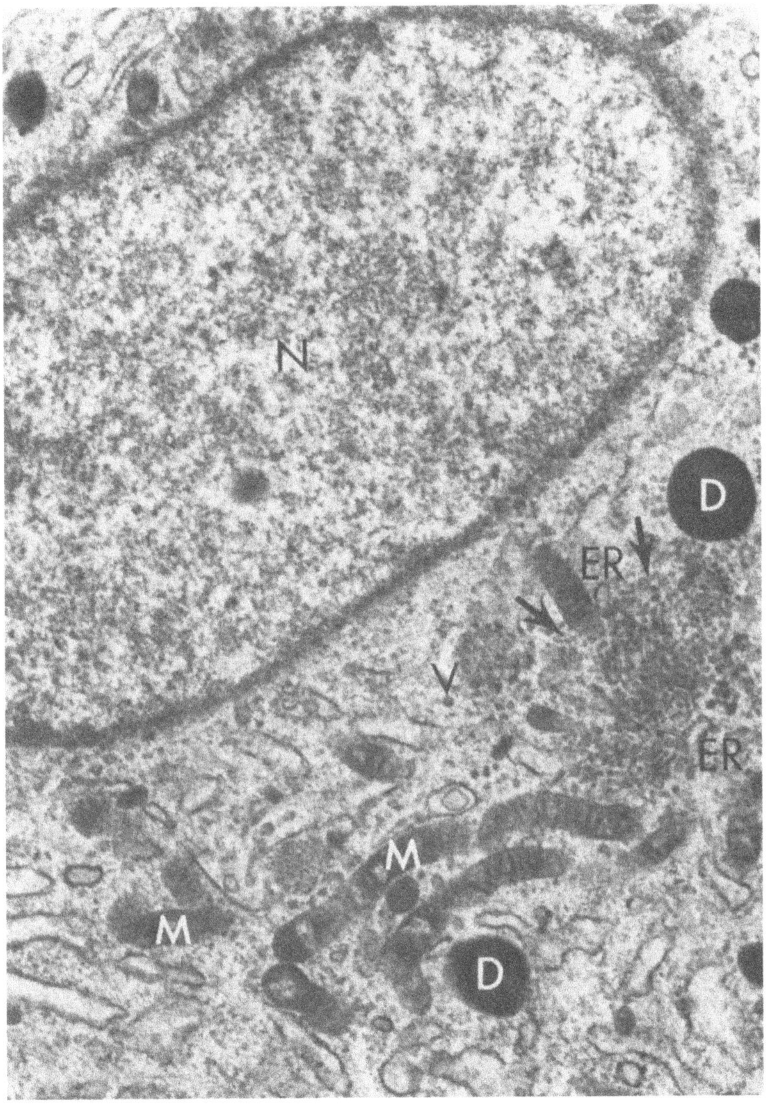

FIGURE 3. Juxta-nuclear location of a replication site (arrows) illustrating immature and mature viral particles (V). The nucleus (N) exhibits margination of chromatin. There is an association of ergastoplasm (ER) to the replication site. Mitochondria (M); dense bodies (D); magnification × 23,000. (From Koestner, A., Kasza, L., and Kindig, O., *Am. J. Pathol.*, 48, 129, 1966. With permission.)

tonsils) and intestinal tract (ileum, spiral colon, and mesenteric lymph nodes). Viremia was detected on the 4th day postinoculation (p.i.). By this time and up to the 10th day p.i., virus could be isolated from a variety of parenchymatous organs (thymus, salivary glands, heart, lung, diaphragm, liver, spleen, kidney). Virus continued to be isolated from tonsils and lung until the 12th day p.i. and from the lower alimentary canal until the study was terminated at 28 days postinoculation.[27]

Following the viremia at the 4th day p.i., virus was recovered from the central nervous system at sites including the olfactory bulb, cerebral cortex, cerebellum, and lumbar cord during 4 to 7 days postinoculation. It was isolated from these plus medulla (but not cerebrum) consistently from the 8th to 10th day p.i. Virus was still demon-

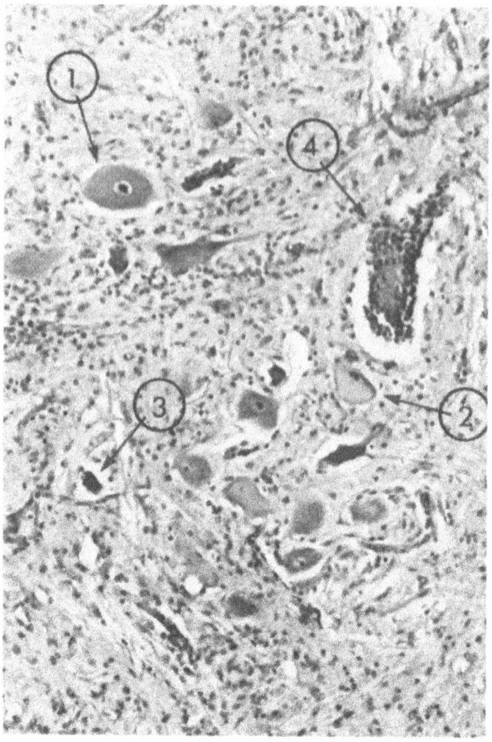

FIGURE 4. Spinal cord of experimental germfree pig. Notice various stages of neuronal degeneration including (1) swelling, (2) central chromatolysis, and (3) shrinkage; (4) perivascular cuffing is also found. H & E stain; magnification × 42. (From Long, J. F., Koestner, A., and Kasza, L., *Am. J. Vet. Res.*, 27, 274, 1966. With permission.)

strated in the medulla and lumbar spinal cord at 12 days p.i. Virus was not recovered from the CNS after the 12th day p.i.

These results were generally comparable with those of others[2] working with different strains of porcine enteroviruses. They found an initial and rapid multiplication in the intestinal tract and mesenteric lymph nodes, a viremia extending through about the 5th day, p.i. localization in the CNS with the highest viral titer occurring on about the 6th day, and presence of the virus only in the intestine after about the 11th day.

The absence of neutralizing antibodies in serum from control pigs and pigs during the initial phase of infection indicates that the progressive increases in titer found during the later stages represent a specific response to the viral infection.[27]

B. Histopathogenesis of CNS Lesions (Light Microscopy)

The course of the disease in 23 germfree pigs was studied utilizing techniques of light microscopy, including the use of silver carbonate impregnation.[36] The histopathologic changes during the course of the disease were described as follows:

Preparalytic period (0 to 8 days p.i.) — No morphologic change was observed in neurons during the first 5 days p.i. On the 6th day, small groups of motor neurons in the ventral horns were swollen as indicated by an increase in size and a rounded contour. The Nissl bodies were normal in appearance at this time.

On the 7th to 8th day, diffuse chromatolysis was observed in small groups of motor neurons (Figure 4). The nerve cells had rounded contours with loss of Nissl bodies.

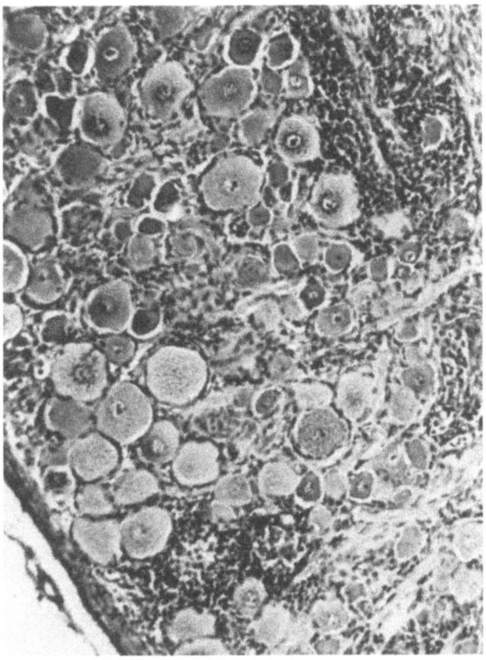

FIGURE 5. Dorsal root ganglion of experimental germfree pig. Notice degeneration of neurons and proliferation of satellite cells. H & E stain; magnification × 42. (From Long, J. F., Koestner, A., and Kasza, L., *Am. J. Vet. Res.*, 27, 274, 1966. With permission.)

The cytoplasm contained diffusely distributed fine granules. In other neurons, central chromatolysis was observed. The degeneration of neurons was associated with mild perivascular cuffing and proliferation of adventitial cells and microglia. Endothelial swelling also occurred. Activation of oligodendrocytes was indicated by swelling, with the nucleus remaining in a central location. The axons leading from affected groups of motor neurons appeared normal. There was degeneration of sensory neurons in the dorsal root ganglia, as indicated by swelling and chromatolysis, and proliferation of satellite cells (Figure 5).

Onset of paralysis (8 to 12 days p.i.) — The histopathologic changes in nerve cells were of the same general nature as in the preparalytic period but much more advanced. The lesions at this time were distributed widely throughout the spinal cord. By the 10th p.i. day, nearly all neurons in the ventral horns were altered. The lesions ranged from swelling with chromatolysis to necrosis and neuronophagia.

Various degrees of chromatolysis were seen, ranging in degree from mild changes (similar to Figure 6) to complete disintegration of chromatin (Figure 7). The amount of phagocytic reaction around necrotic neurons was variable. While most had microglial accumulations (Figure 8), others had little or no glial response. In some, vacuolization of the cytoplasm was a predominant feature. In some areas the neurons had disappeared, their former locations being marked by glial nodules (Figure 9). Large cytoplasmic vacuoles remaining from the destroyed neuron were frequently seen outside the nodule. Neurofibrils were often intact in cell remnants in spite of chromatolysis and karyorrhexis (Figure 10). The ventral horns were more severely affected than the dorsal horns.

By the 10th day the destruction of neurons and the formation of sheets of lipid

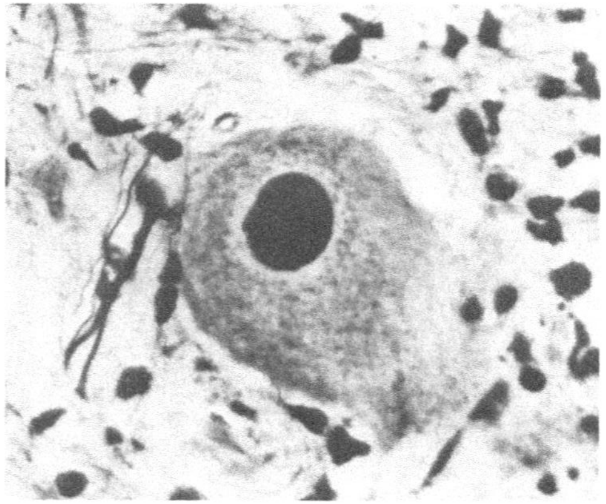

FIGURE 6. Ganglion cell with acute swelling and diffuse chromatolysis (7 days postinoculation). Magnification × 500. (From Long, J. F., Koestner, A., and Liss, L., *Pathol. Vet.*, 4, 186, 1967, S. Karger AG, Basel. With permission.)

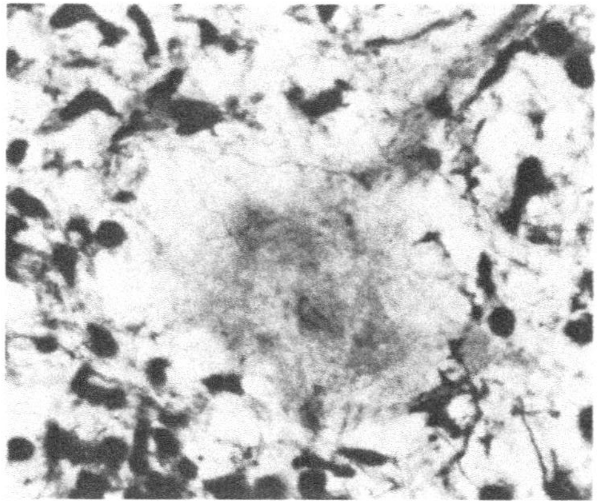

FIGURE 7. Ganglion cell with advanced chromatolysis. The nucleus has undergone dissolution (10 days postinoculation). Magnification × 500. (From Long, J. F., Koestner, A., and Liss, L., *Pathol. Vet.*, 4, 186, 1967, S. Karger AG, Basel. With permission.)

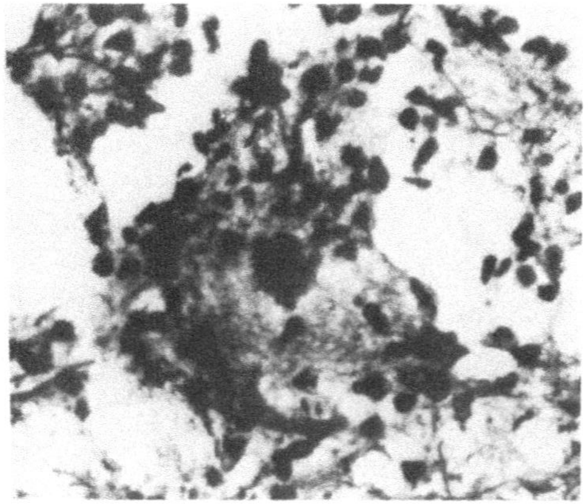

FIGURE 8. Large numbers of microglia surrounding a ganglion cell in an area of malacia (10 days postinoculation). Magnification × 500. (From Long, J. F., Koestner, A., and Liss, L., *Pathol. Vet.*, 4, 186, 1967, S. Karger AG, Basel. With permission.)

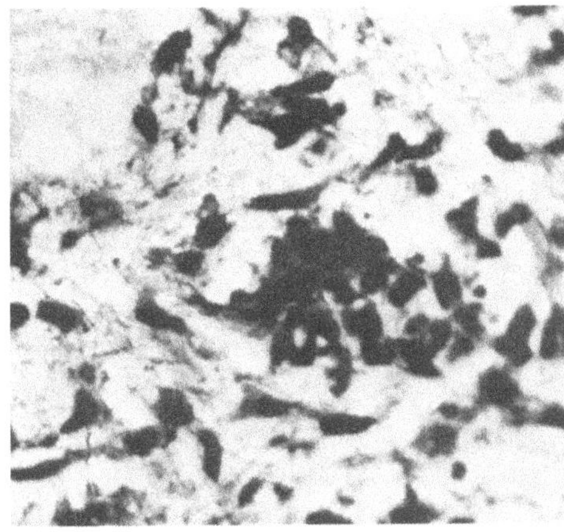

FIGURE 9. Ganglion cell undergoing neuronophagia (10 days postinoculation). Magnification × 500. (From Long, J. F., Koestner, A., and Liss, L., *Pathol. Vet.*, 4, 186, 1967, S. Karger AG, Basel. With permission.)

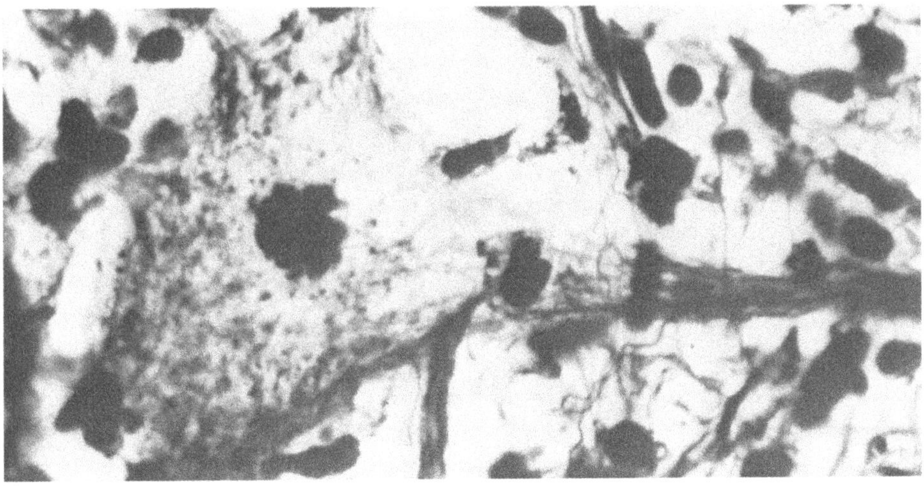

FIGURE 10. Ganglion cell with chromatolysis and karyorrhexis. Neurofibrils in the processes are still intact (10 days postinoculation). Magnification × 1250. (From Long, J. F., Koestner, A., and Liss, L., *Pathol. Vet.*, 4, 186, 1967, S. Karger AG, Basel. With permission.)

phagocytes were often seen on such a massive scale that cavitation of the ventral horns was grossly visible on the stained sections.

Axonal degeneration occurred in tracts originating from necrotic motor neurons (Figure 11). In some tracts almost all the axons were affected while in others only a few were involved, indicating the degree of destruction of the individual ganglion cells. The degeneration of the axons in the dorsal roots was a prominent feature at this stage and reflected the intensity of the destruction of the dorsal root ganglia.

The oligodendroglial response was manifested by satellitosis of affected neurons as well as swelling of interfascicular oligodendrocytes along nerve tracts.

The principal astrocytic changes were represented by clasmatodendrosis and the formation of plump astrocytes (gemistocytes).

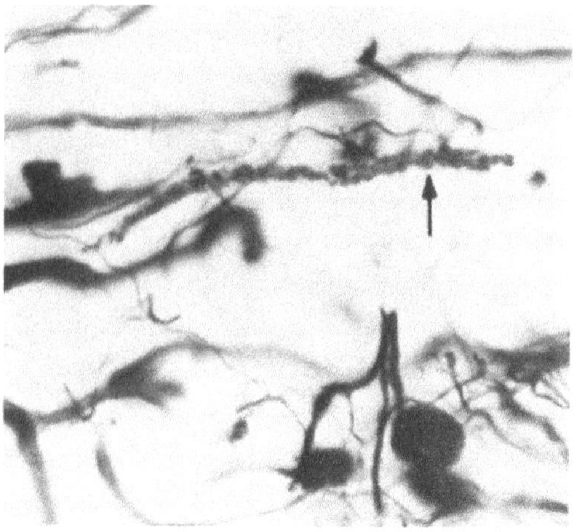

FIGURE 11. Arrow points to axon undergoing degeneration (10 days postinoculation). Magnification × 1250. (From Long, J. F., Koestner, A., and Liss, L., *Pathol. Vet.*, 4, 186, 1967, S. Karger AG, Basel. With permission.)

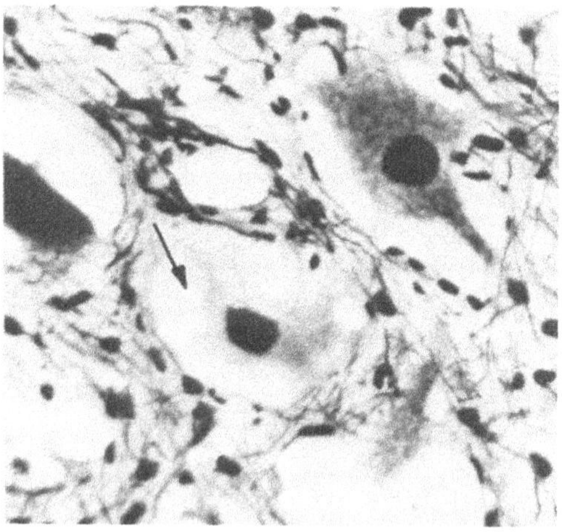

FIGURE 12. Ganglion cell (arrow) with numerous vacuoles in the cytoplasm (16 days postinoculation). Magnification × 315. (From Long, J. F., Koestner, A., and Liss, L., *Pathol. Vet.*, 4, 186, 1967, S. Karger AG, Basel. With permission.)

Destruction of individual ependymal cells sometimes occurred. This was indicated by cytoplasmic granularity and degeneration of processes.

Postparalytic period (13 to 17 days p.i.) — Active neuronal destruction appeared to cease about 2 weeks p.i. There were large foci containing phagocytic cells and there was some evidence of completed neuronophagia. The surviving neurons were generally rounded. Many ganglion cells had large vacuoles in their cytoplasm (Figure 12). Nissl bodies reappeared in surviving neurons. The Nissl granules were either distributed throughout the cell body or were clustered in the perinuclear region.

Axonal degeneration in both the dorsal roots and in tracts originating from the ventral horns was characterized by coarse argyrophilic granules.

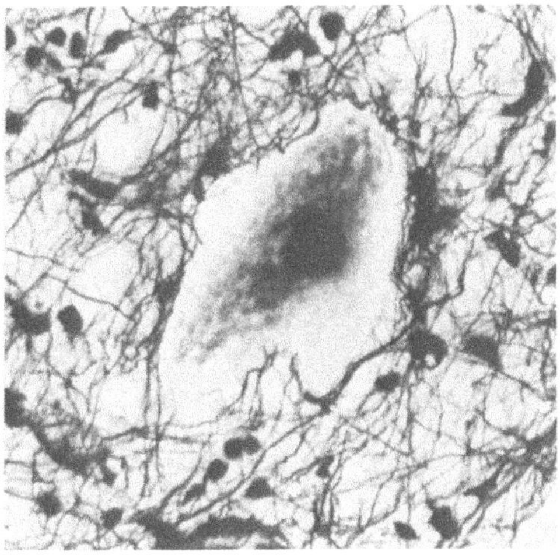

FIGURE 13. Dense mesh of astrocytic processes around degenerating ganglion cell (28 days postinoculation). Magnification × 500. (From Long, J. F., Koestner, A., and Liss, L., *Pathol. Vet.*, 4, 186, 1967, S. Karger AG, Basel. With permission.)

Microglial cells were less numerous than in the paralytic period and were mostly in focal aggregates throughout the gray matter.

Oligodendroglial involvement in the form of satellitosis and activation along nerve tracts was frequently seen.

The astrocytic reaction was characterized by hyperplasia and hypertrophy with proliferation of processes. It was most noticeable in the ventral horns especially in areas deprived of neurons.

Convalescent period (after 17 days p.i.) — Ganglion cells were absent from large areas in the ventral horns. The few surviving neurons had a slightly rounded contour, thicker processes, and, occasionally, had vacuolated cytoplasm which sometimes contained neutral fat. They occupied small areas limited by the encroaching mesh of astrocytic fibers (Figure 13).

Microglial cells were fewer in number and were mainly concentrated in the ventral horns in areas formerly occupied by ganglion cells.

The predominant lesion was the proliferative astrocytosis (Figure 14). Massive concentrations of astrocytic fibers were consistently found in the ventral horns. The cord was often grossly distorted due to the loss of neurons and their replacement by astrocytic scars.

C. Electron Microscopic Evaluation of CNS at Sequential Intervals During Infection

CNS changes were studied by electron microscopy in germfree pigs given the virus orally, after, 7, 10, 14, and 19 days p.i.[29] Electron microscopy of the replication of porcine polioencephalomyelitis virus in tissue culture was available for comparison (Figure 3).[31]

The most frequent lesion in the spinal cord, encountered on the 7th day p.i., but also present at later stages, was a diffuse chromatolysis of motor neurons (Figure 15). The endoplasmic reticulum was deprived of ribosomes and appeared slightly and irregularly dilated. Clusters of ribosomes were still present and more or less concentrated in areas of Nissl bodies. Some concentration of ribosomes was observed along the nuclear membrane. At the 7th and particularly at the 10th day p.i. an increase in extracellular space was apparent. This was seen in the perivascular area as well as in the

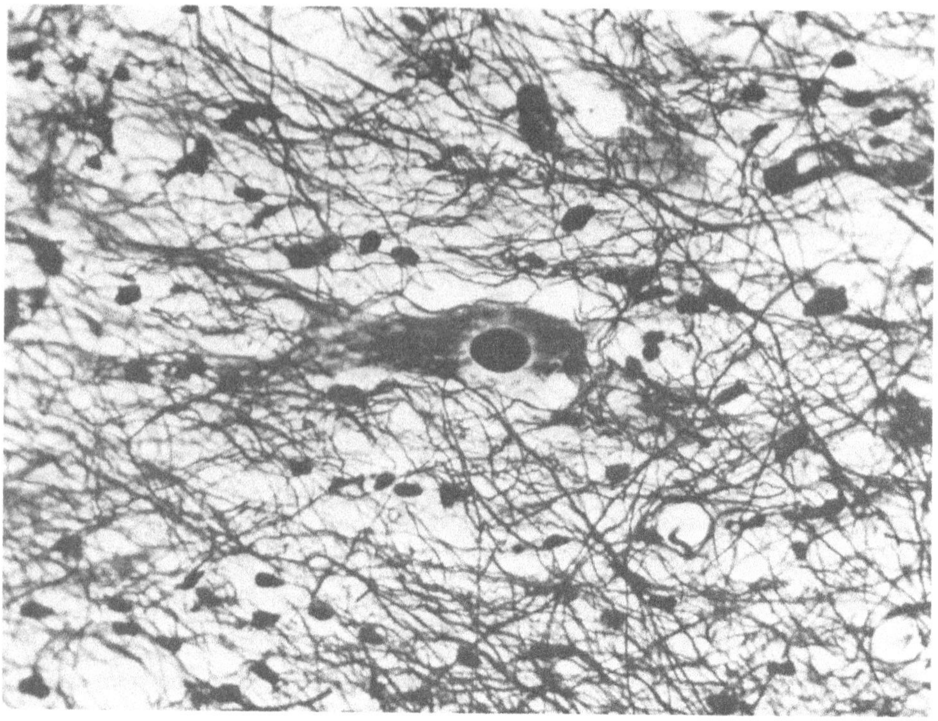

FIGURE 14. Dense astrocytic scar surrounding recovering ganglion cell (28 days postinoculation). Magnification × 500. (From Long, J. F., Koestner, A., and Liss, L., *Pathol. Vet.*, 4, 186, 1967, S. Karger AG, Basel. With permission.)

pericellular region. The synaptic boutons, normally densely lining the neuronal membranes, were separated by wide spaces (Figure 16). In some capillaries, endothelial cells contained large pinocytotic vesicles (Figure 17). Occasionally, individual leukocytes had migrated through the vascular wall and were observed within a dilated perivascular space. At the 14th day p.i. crystalline arrays of particles measuring 26 nm in diameter were found in capillary endothelial cells, astrocyte footpads, and in glial processes in proximity to ganglion cell borders.

At this stage varying degrees of ultrastructural changes were seen in motor neurons. The most striking lesion consisted of a dilatation of endoplasmic reticular cisternae. The mitochondria were well preserved even though there was swelling and loss of cristae. In other neurons there was an increase in cytolysosomes and lipid droplets. The most severe lesions consisted of a complete vesiculation of ganglion cells with dense mitochondria compressed between the vesicles.

By the 19th day p.i. ganglion cells were replaced by proliferating glial cells in the form of glial nodules. The vessels at this time were in close apposition to the surrounding brain and no perivascular space was discernible. This was also true for vessels which were surrounded by a cuff of leukocytes, proliferating glial, and perithelial cells.

VI. DISCUSSION OF PATHOGENESIS

The oral route of infection is considered to be the principal natural route in porcine polioencephalomyelitis. Enteroviruses are resistant to stomach acidity and are adapted to replicate within the gastrointestinal tract. It is well established that the mucosal-associated immune system of the gut plays a key role in poliovirus infection in man[46]

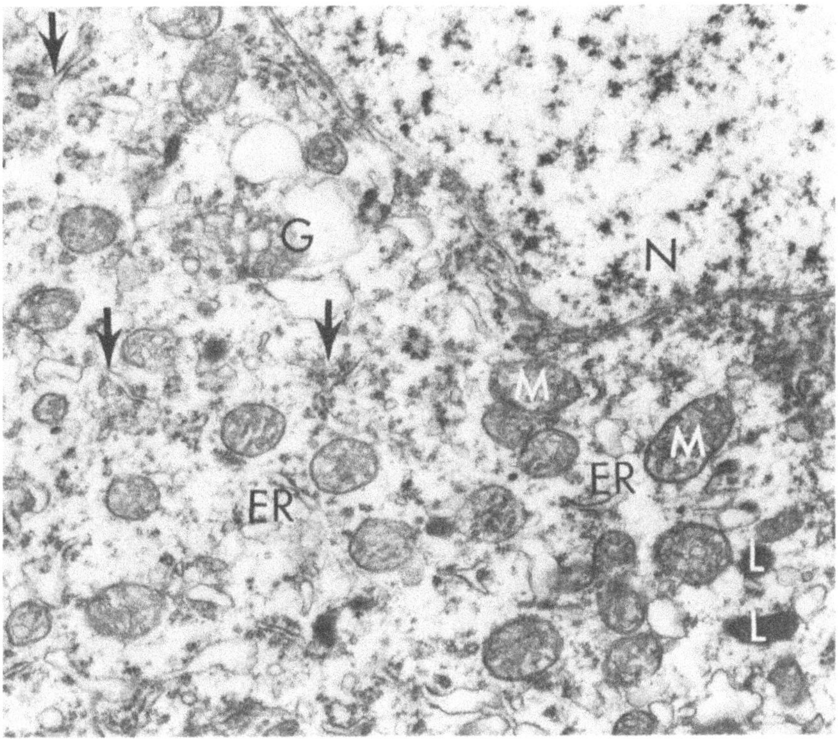

FIGURE 15. Motor neuron from the ventral horn of the cervical cord in a germfree pig infected with porcine polioencephalomyelitis virus. Nissl bodies (arrows) show depletion of ribosomes (chromatolysis). There is loss of ribosomal clusters but some concentration along the nuclear membrane. Nucleus (N); Golgi apparatus (G); endoplasmic reticulum (ER); lysosomes (L); mitochondria (M). Magnification × 31,600. (From Koestner, A., Kasza, L., and Holman, J. E., *Am. J. Pathol.*, 49, 325, 1966. With permission.)

and in porcine polioencephalomyelitis in the pig.[16] In both animal species parenteral administration of the inactivated vaccine failed to elicit gut immunity.[16,44] The principal immune mechanism in mucosal immune systems, associated with protection from disease, is the secretion of IgA antibody. Hazlett and Derbyshire found that the predominant gastrointestinal virus-neutralizing component in piglets was IgA. However, some of the neutralizing copro-antibody was IgG and IgM.[17] These authors found that inactivated porcine poliovirus vaccine elicited IgG and IgM serum antibody but little IgA.

The role of cell-mediated immunity (CMI) in the protection of porcine polio virus is not clear. Brundage[6] found that orally infected piglets yield a weak cell-mediated response in blood lymphocytes and lamina propria lymphocytes as measured by the macrophage-inhibition test.[6]

Further studies are needed to ascertain the role of CMI in viral intestinal infections. The influence of poliovirus on the immune response or the trafficking of lymphocytes from the gut to mucosal surfaces may be of prime importance. It is known that type I human poliovirus vaccine depresses the delayed-type hypersensitivity response in patients.[3] How this virus effect on the host alters the host's capability to restrain disease is unknown.

Lesions of a specific nature have not been described in the intestine with porcine enteroviral infections.[8] The porcine enteroviral infections do not appear to cause villous atrophy.[8]

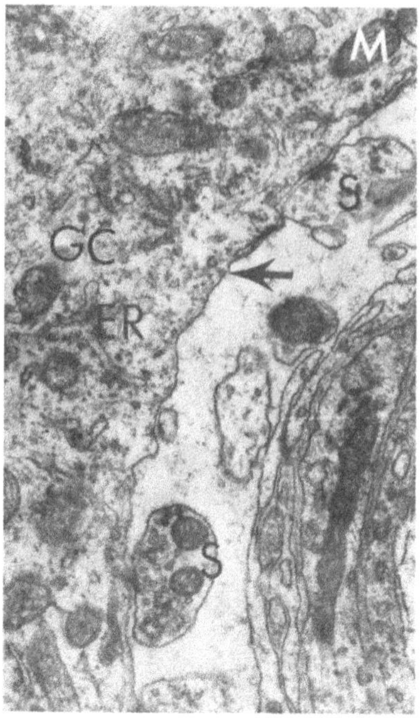

FIGURE 16. An increased extracellular space is apparent along a neuronal membrane (arrow). The synaptic boutons (S) are widely separated. There is no indication that glial processes occupy this space. The section is from the ventral horn of the cervical cord of a pig 10 days after inoculation. Ganglion cell (GC); endoplasmic reticulum (ER); mitochondria (M). Magnification × 23,700. (From Koestner, A., Kasza, L., and Holman, J. E., *Am. J. Pathol.*, 49, 325, 1966. With permission.)

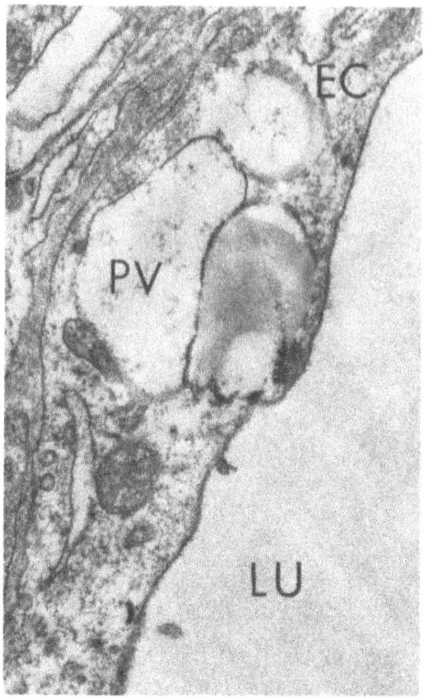

FIGURE 17. Large pinocytotic vesicles (PV) appearing in a capillary endothelial cell (EC) of an infected pig 10 days postinoculation. Capillary lumen (LU). Magnification × 25,000. (From Koestner, A., Kasza, L., and Holman, J. E., *Am. J. Pathol.*, 49, 325, 1966. With permission.)

Porcine enteroviruses characteristically replicate readily in porcine intestine. To identify sites of greatest replication, Derbyshire and Collins,[9] using pig intestinal organ cultures, found that the Talfan strain of pig enterovirus multiplied to a high titer in ileum and even higher in colon. In human poliovirus infections in monkeys the enterovirus multiplied first in the peritonsilar lymphatic cells and in Peyer's patches. Infection of reticuloendothelial cells of the lamina propria of the intestine and vascular endothelial cells have also been shown by immunofluorescence studies.[25] Penetration from gut lumen to the lymphoid cells may be mediated by the microfold or M cells which overlie Peyer's patches and transport particulate antigens.[19] Mild gastrointestinal symptoms may precede paralytic human poliomyelitis. Following replication in lymph nodes, the virus is spread to the bloodstream.

Viremia follows regularly in virulent serogroup 1 viruses of porcine polioencephalomyelitis.[8] It has also been found in other strains capable of producing CNS disease through about the 5th day postinoculation.[2,27] The virus evidently reaches the CNS during the viremia.

Some viruses may infect the vascular endothelial cells of the CNS in the course of the infection of adjacent glia and neurons. This mechanism has been suggested for poliovirus infections in monkeys[4] and porcine polioencephalomyelitis infections[29] and is based in part on the presence of crystalline arrays of particles (suggestive of virus) in the cerebroendothelial cells. This concept is supported by studies of Watanabe et al.[58] They exposed pigs to Teschen virus by the intranasal route and found the capillaries in all brain sections to show fluorescence at the 8th day. Although Purkinje cells showed fluorescence at this time it was not seen in neurons of the cerebrum or thalamus at this time. Fluorescence which subsequently developed in neurons of the cerebrum and thalamus was granular and located in the cytoplasm.

In some virus infections, virus may be carried across the endothelial cells in infected leukocytes.[52] Passage of virus through areas of permeability has also been postulated, but in experimental studies no virus has been shown to cause infection which was preferentially related to the anatomical sites of permeability.[19]

The glia have long been suspected of forming a channel for the transport of material from blood to neurons. Inert ferratin particles, after crossing the cerebral capillaries, are found within the astrocytic footplates and subsequently in neurons.[5] Koestner et al.[29] found crystalline arrays (suggestive of virus) in astrocytic processes during infection with porcine polioencephalomyelitis.

With regard to the question in enterovirus infections as to whether nerve cell damage is due primarily to virus multiplication in the neurons or secondarily to the effects of virus multiplication in the supporting tissue of the CNS, several studies were compared by Hashimoto and Hagiwara.[15] In a study of enterovirus 71 infection in monkeys, specific immunofluorescence was found in degenerating or necrotic neurons but not in glial cells, vascular endothelial cells, or mononuclear cells.[15] Likewise, Jubelt et al.[23] in immunofluorescent studies, found poliovirus antigen only in the cytoplasm of the motor neurons of mice infected with Lansing type 2 strain. Kovacs et al.[32] and Kanamitsu et al.[25] detected specific immunofluorescence in the vascular endothelial and mononuclear inflammatory cells as well as in the motoneurons of the CNS in monkeys infected with poliovirus. In contrast, Simon[49] (cited in Reference 15) found poliovirus antigen in the vascular endothelial and mononuclear inflammatory cells but not in any of the neurons.

Blinzinger et al.[4] studied the lumbar spinal cord of a monkey infected with a highly virulent strain of type 3 poliovirus utilizing both immunohistochemistry and electron microscopy. By immunohistochemistry they demonstrated the presence of poliovirus antigen within the walls of intraspinal blood vessels, as well as within the cytoplasm of many mononuclear cells. By electron microscopy they found aggregates of dense particles, often arranged in ordered lattices which they felt resembled poliovirus crystals. These were seen in the cytoplasm of many endothelial cells, monocytes, histiocytes, and macrophages. Several authors[13,15,53] mention the fact that crystalline inclusion bodies have been found by electron microscopy in a wide variety of cells where no evidence of infection was known to exist so their nature and significance is often undetermined. Wolinsky et al.[60] believe that there has been, as yet, no convincing fine structural demonstration of viral particles in poliovirus infections since the virus does not appear to form paracrystalline arrays in neural tissues. Thus, in their study with poliovirus, they consider that confirmation that the particulate structures seen in neurons and the phagolysosomes of microglial rod cells are truly viral must await electron microscopic immunocytochemical delineation.

Recent studies of human poliovirus infection in mice have indicated axonal spread within the CNS, based on the observation that cordectomy prevents rostral or caudal spread of virus in the spinal cord despite patent CSF pathways. Furthermore, rates of transport increase with age, corresponding to acceleration of fast transport with age.[24]

With porcine polioencephalomyelitis infections, many strains of virus result in clinical signs of disease only in pigs under a certain age. The mechanisms of this age-dependent susceptibility or resistance has, in the case of some enteroviruses, been shown to depend on factors such as maturation of the immune system, route of virus inoculation, and age-related intrinsic resistance to viral replication.[54]

With regard to the vulnerability of the diverse cell populations within the CNS, there is variation of susceptibility to infection with different viruses. As summarized by Johnson,[20] viruses, having no inherent mobility, are moved passively in extracellular spaces and encounter cell membranes at random. To be susceptible to infection by the virus, the host cell must have specific receptor sites accessible on its cytoplasmic membrane. In view of the extreme specialization of the cytoplasmic membranes of neural cells, great variation in their susceptibility to different viruses would be anticipated.[20] Furthermore, this selective vulnerability may vary between species or at different ages within a given species.[20] Motor nerve cells of the spinal cord seem to be selectively susceptible to poliomyelitis viruses, and this selectivity explains the clinical presentations of flaccid paralysis.[19] A similar predilection of the virus for motor nerve cells of the spinal cord has similarly been consistently observed in porcine polioencephalomyelitis.

The distribution of the lesions in porcine polioencephalomyelitis is based in part on the locations of the susceptible cells as well as the virulence of the virus. Generally the lesions involved the entire cerebrospinal axis with the principal changes in the gray matter. The frequency and intensity of lesions increased caudally from the olfactory bulbs, corpus striatum, hypothalamus and thalamus with greatest severity in the midbrain, pons, and medulla oblongata. Limited cellular accumulations were noted in the leptomeninges adjacent to superficial parenchymal lesions. Cerebellar and cerebral cortexes were occasionally affected. The ventral horns of the spinal cord were consistently involved throughout its entire length. Whereas there is a general similarity in the distribution of lesions in the porcine CNS with the various strains of virulent neurotropic enteroviruses, variations do nevertheless occur[11,42] and in some there is a greater tendency to involve the cerebral and cerebellar cortexes. The dorsal root ganglia at all levels of the spinal cord are quite consistently involved. The Gasserian ganglia are involved with at least some strains.[18] The nature of the lesion in neurons, that of chromatolysis and necrosis is followed by reactive changes of neuronophagia. The typical vascular response consists of perivascular cuffing with proliferating adventitial cells, lymphocytes, plasmacytes, and microglia with eventual astrocytic scar formations.

Ogra and Karzon[43] reviewed the course of poliovirus infection in humans and primate experimental models. They divided the course of infection into (1) implantation and replication of virus at the primary portal of entry — the nasopharynx and lower alimentary tract; (2) spread of virus to tonsils, deep cervical nodes, Peyer's patches, and mesenteric lymph nodes; (3) spread via the blood stream (minor viremia) to other susceptible extraneural tissues (liver, spleen, bone marrow, peripheral lymphoid tissue, etc.); (4) heavy sustained shedding into the blood stream (major viremia); and (5) establishment of infection in the central nervous system. Thus, the pathogenesis of porcine polioencephalomyelitis appears to be comparable in most major respects to that of poliovirus infections in humans.

Recently there has been an upsurge of interest in the possible role of persistent enteroviral infections of the nervous system in human disease.[40] In addition to reports in patients with identifiable immunodeficiencies,[7,59] there is also speculation regarding the possibility of enteroviral persistence having a role in certain degenerative diseases such as amyotrophic lateral sclerosis.[21,41]

Reports indicate neurologic impairment as possible long-term sequelae of nonpolio enterovirus infection in children, especially in those who incurred the infection during the first year of life.[47] The assessment of the role of the virus in causing injury during acute infection with subsequent impairment of cerebral ontogenesis, and the possible role of injury due to viral persistence need to be determined.

In a study by Miller[41] it was found that the mouse-adapted Lansing strain of poliovirus type II could be recovered by isolation techniques from the brain of asymptomatic mice as long as 77 days after intracerebral inoculation and without a neutralizing antibody response to the virus.

The regulation of viral replication during in vivo infection is still not well understood. The production of neutralizing antibody has been thought to be paramount in the control of enteroviral infections. Its importance for prevention is accepted, but its role in eradicating infections is less apparent. The maintenance of a prolonged asymptomatic infection in the poliovirus-inoculated mice in the study by Miller[41] clearly indicates that there are other important factors in the control of viral replication in this case.

The slowly progressive enterovirus CNS infections in patients with immune deficiencies have a similarity to the intracerebral infection in mice with the Lansing strain virus studied by Miller,[41] in that the infection progresses at an indolent pace in the absence of viral neutralizing antibodies.[41]

Although successful cure of infection in one of these patients by the administration of antiviral antibodies has been reported,[37] studies with a mouse enterovirus (Theiler's encephalomyelitis virus) have indicated that the presence of neutralizing antibodies is not sufficient to eradicate virus even when antibodies are present in high titer.

Infection with viruses has been shown to be modulated by factors including abnormalities of viral replication, defective interfering particles, interferon, and immune responses. It is probable that there are numerous mechanisms which could channel infection into a chronic form. These mechanisms might act individually, or in sequence, or in unison.[22]

In view of the basic similarities in pathogenesis between porcine polioencephalomyelitis and poliovirus infections of humans, the pig would provide a suitable model for investigation of the role of the immune system in the development of persistent infection of the nervous system caused by an enterovirus in its natural host.

REFERENCES

1. Anderson, A., Picornaviruses of animals: clinical observations and diagnosis, in *Comparative Diagnosis of Viral Diseases*, Kurstak, E. and Kurstak, C., Eds., Academic Press, New York, 1981, 3.
2. Baba, S. P., Bohl, E. H., and Meyer, R. C., Infection of germfree pigs with a porcine enterovirus, *Cornell Vet.*, 56, 386, 1966.
3. Berkovich, S. and Starr, S., Effects of live type I poliovirus vaccine and other viruses on the tuberculin test, *N. Engl. J. Med.*, 274, 67, 1966.
4. Blinzinger, K., Simon, J., Magrath, D., and Boulger, L., Poliovirus crystals within the endoplasmic reticulum of endothelial and mononuclear cells in the monkey spinal cord, *Science*, 163, 1336, 1969.
5. Bondareff, W., Distribution of ferratin in the cerebral cortex of the mouse revealed by electron microscopy, *Exp. Neurol.*, 10, 377, 1964.
6. Brundage, L. J., Derbyshire, J. B., and Wilkie, B. N., Cell-mediated responses in a porcine enterovirus infection in piglets, *Can. J. Comp. Med.*, 44, 61, 1980.
7. Davis, L. E., Bodian, D., Price, D., et al., Chronic progressive poliomyelitis secondary to vaccination of an immunodeficient child, *N. Engl. J. Med.*, 297, 241, 1977.
8. Derbyshire, J. B., Porcine enterovirus infections, in *Diseases of Swine*, 5th ed., Leman, A. D., Glock, R. D., Mengeling, W. L., Penny, R. H. C., Scholl, E., and Straw, B., Eds., Iowa State University Press, Ames, 1981, 265.
9. Derbyshire, J. B. and Collins, A. P., The multiplication of Talfan virus in pig intestinal organ cultures, *Res. Vet. Sci.*, 12, 367, 1971.
10. Dunne, H. W., Wang, J. T., and Ammerman, E. H., Classification of North American porcine enteroviruses: a comparison with European and Japanese strains, *Infect. Immun.*, 4, 619, 1971.
11. Edington, N., Christofinis, G. J., and Betts, A. O., Pathogenicity of Talfan and Konratice strains of Teschen virus in gnotobiotic pigs, *J. Comp. Pathol.*, 82, 393, 1972.
12. Gardiner, M. R., Polio-encephalomyelitis of pigs in Western Australia, *Aust. Vet. J.*, 38, 24, 1962.
13. Grimley, P. M. and Schaff, Z., Significance of tubuloreticular inclusions in the pathobiology of human disease, *Pathobiol. Annu.*, 6, 221, 1976.

14. Grist, N. R., Bell, E. J., and Assaad, F., Enteroviruses in human disease, in *Prog. Med. Virol.*, S. Karger, Basel, 1978, p. 114.
15. Hashimoto, I. and Hagiwara, A., Studies on the pathogenesis of and propagation of enterovirus 71 in poliomyelitis-like disease in monkeys, *Acta Neuropathol.*, 58, 125, 1982.
16. Hazlett, D. T. G. and Derbyshire, J. B., Neutralizing activity in the gastrointestinal contents in piglets vaccinated with a live or formaldehyde-inactivated porcine enterovirus, *Can. J. Comp. Med.*, 40, 370, 1976.
17. Hazlett, D. T. G. and Derbyshire, J. B., Characterization of the local and systemic virus-neutralizing activity in swine vaccinated with a porcine enterovirus, *Can. J. Comp. Med.*, 41, 257, 1977.
18. Holman, J. E., Koestner, A., and Kasza, L., Histopathogenesis of porcine polioencephalomyelitis in the germfree pig, *Pathol. Vet.*, 3, 633, 1966.
19. Johnson, R. T., Pathogenesis of CNS infections, in *Viral Infection of the Nervous System*, Raven press, New York, 1982, 37.
20. Johnson, R. T., Selective vulnerability of neural cells to viral infections, *Brain*, 103, 447, 1980.
21. Johnson, R. T., Degenerative diseases, in *Viral Infection of the Nervous System*. Raven Press, New York, 1982, 288.
22. Johnson, R. T., Lazzarini, R. A., and Waksman, B. H., Mechanisms of virus persistence, *Ann. Neurol.*, 9, 616, 1981.
23. Jubelt, B., Gallez-Hawkins, G., Narayan, O., and Johnson, R. T., Pathogenesis of human poliovirus infection in mice. I. Clinical and pathological studies, *J. Neuropathol. Exp. Neurol.*, 39, 138, 1980.
24. Jubelt, B., Narayan, O., and Johnson, R. T., Pathogenesis of human poliovirus infection in mice. II. Age-dependency of paralysis, *J. Neuropathol. Exp. Neurol.*, 39, 149, 1980.
25. Kanamitsu, M., Kasamaki, A., Ogawa, M., Kasahara, S., and Inamura, M., Immunofluorescent study on the pathogenesis of oral infection of poliovirus in monkey, *Jpn. J. Med. Sci. Biol.*, 20, 175, 1967.
26. Kasza, L., Swine polioencephalomyelitis viruses isolated from the brains and intestines of pigs, *Am. J. Vet. Res.*, 26, 131, 1965.
27. Kasza, L., Holman, J., and Koestner, A., Swine polioencephalomyelitis virus in germfree pigs: viral isolation, immunoreaction, and serum electrophoresis, *Am. J. Vet. Res.*, 28, 461, 1967.
28. Klobouk, A., cited from Kaplan, M. M. and Meranze, D. R., Porcine virus encephalomyelitis and its possible biological relationship to human poliomyelitis, *Vet. Med.*, 43, 330, 1948.
29. Koestner, A., Kasza, L., and Holman, J. E., Electron microscopic evaluation of the pathogenesis of porcine polioencephalomyelitis, *Am. J. Pathol.*, 49, 325, 1966.
30. Koestner, A., Long, J. F., and Kasza, L., Occurrence of viral polioencephalomyelitis in suckling pigs in Ohio, *J. Am. Vet. Med. Assoc.*, 140, 811, 1962.
31. Koestner, A., Kasza, L., and Kindig, O., Electron microscopy of tissue cultures infected with porcine polioencephalomyelitis virus, *Am. J. Pathol.*, 48, 129, 1966.
32. Kovacs, E., Baratawidjaja, R. K., and Labzoffsky, N. A., Visualization of poliovirus type III in paraffin sections of monkeys spinal cord by indirect immunofluorescence, *Nature (London)*, 200, 497, 1963.
33. Lai, S. S., McKercher, P. D., Moore, D. M., and Gillespie, J. H., Pathogenesis of swine vesicular disease in pigs, *Am. J. Vet. Res.*, 40, 463, 1979.
34. Long, J. F., Koestner, A., and Kasza, L., Infectivity of three porcine polioencephalomyelitis viruses for germfree and pathogen-free pigs, *Am. J. Vet. Res.*, 27, 274, 1966.
35. Long, J. F., Kasza, L., and Koestner, A., Pericarditis and myocarditis in germfree and colostrum-deprived pigs experimentally infected with a porcine polioencephalomyelitis virus, *J. Infect. Dis.*, 120, 245, 1969.
36. Long, J. F., Koestner, A., and Liss, L., Experimental porcine polioencephalomyelitis in germfree pigs. A silver carbonate study of neuronal degeneration and glial response, *Pathol. Vet.*, 4, 186, 1967.
37. Mease, P. J., Ochs, H. D., and Wedgwood, R. J., Successful treatment of echovirus meningoencephalitis and myositis-fasciitis with intravenous immune globulin therapy in a patient with X-linked agammaglobulinemia, *N. Engl. J. Med.*, 304, 1278, 1981.
38. Melnick, J. L., et al., Picornaviridae, *Intervirology*, 4, 303, 1974.
39. Meyer, R. C., Woods, G. T., and Simon, J., Pneumonitis in an enterovirus infection in swine, *J. Comp. Pathol.*, 76, 397, 1966.
40. Miller, J. R., Prolonged intracerebral infection with poliovirus in asymptomatic mice, *Ann. Neurol.*, 9, 590, 1980.
41. Miller, J. R., Persistent infection by poliovirus: experimental studies, in *Human Motor Neuron Diseases*, Rowland, L. P., Ed. Raven Press, New York, 1982, 311.
42. Mills, J. H. L., and Neilsen, S. W., Porcine polioencephalomyelitides, in *Adv. Vet. Sci.*, 12, 33, 1968.
43. Ogra, P. L. and Karzon, D. T., Formation and function of poliovirus antibody in different tissues, *Prog. Med. Virol.*, 13, 156, 1971.

44. Ogra, P. L., Karzon, D. T., Righthand, F., and MacGillivray, M., Immunoglobulin response in serum and secretions after immunization with live and inactivated polio vaccine and natural infection, *N. Engl. J. Med.*, 279, 893, 1968.
45. Richards, W. P. C. and Savan, M., Viral encephalomyelitis of pigs. A preliminary report on the transmissibility and pathology of a disease observed in Ontario, *Cornell Vet.*, L, 132, 1960.
46. Sabin, A. B., Alvarez-Amezquita, R. M., et al., Live orally-given polio vaccine, *JAMA*,173, 1521, 1960.
47. Sells, C. J., Carpenter, R. L., and Ray, C. G., Sequela of central-nervous-system enterovirus infections, *N. Engl. J. Med.*, 293, 1, 1975.
48. Sibalin, M. and Lannek, N., An enteric porcine virus producing encephalomyelitis and pneumonitis in baby pigs, *Arch. Ges. Gesamte Virusforschung*, 10, 31, 1960.
49. Simon, J., Magrath, D. I., and Boulger, L. R., The role of the defensive mechanism in experimental poliomyelitis, *Prog. Immunobiol. Scand.*, 4, 643, 1970.
50. Singh, K. V., and Bohl, E. H., The pattern of enteroviral infections in a herd of swine, *Can. J. Comp. Med.*, 36, 243, 1972.
51. Sulochana, S. and Derbyshire, J. B., Sturctural proteins of porcine enteroviruses, *Vet. Microbiol.*, 3, 23, 1978.
52. Summers, B. A., Greisen, H. A., and Appel, M. J. G., Possible initiation of viral encephalomyelitis in dogs by migrating lymphocytes infected with distemper virus, *Lancet*, 1, 187, 1978.
53. Tajima, M. and Kudow, S., Morphology of the Warthin-Finkeldey giant cells in monkeys with experimentally induced measles, *Acta Pathol. Jpn.*, 26, 367, 1976.
54. Tardieu, M., Powers, M. L., and Weiner, H. L., Age dependent susceptibility to Reovirus type 3 encephalitis: role of viral and host factors, *Ann. Neurol.*, 13, 602, 1983.
55. Trefny, L., Massive illness of swine in Teschen area, *Zverol. Obzor.*, 23, 235, 1930; cited in *Jahresber. Vet. Med.*, 50, 395, 1930.
56. Wang, J. T. and Dunne, H. W., Comparison of porcine picornaviruses isolated in North America and their identification with SMEDI viruses, *Am. J. Vet. Res.*, 30, 1677, 1969.
57. Wang, J. T., Dunne, H. W., Griel, L. C., Hokanson, J. F., and Murphy, D. M., Mortality, antibody development, and viral persistence in porcine fetuses incubated *in utero* with SMEDI (Entero-) virus, *Am. J. Vet. Res.*, 34, 785, 1973.
58. Watanabe, H., Popisil, Z., and Mensik, J., Study on the pigs infected with virulent Teschen disease virus (KNM strain) with special reference to immunofluorescence, *Jpn. J. Vet. Res.*, 19, 87, 1971.
59. Wilfert, C. M., Buckley, R. H., Mohanakumar, T., et al., Persistent and fatal central-nervous-system echovirus infections in patients with agammaglobulinemia, *N. Engl. J. Med.*, 296, 1486, 1977.
60. Wolinsky, J. S., Jubelt, B., Burke, S., and Narayan, O., Hematogenous origin of the inflammatory response in acute poliomyelitis, *Ann. Neurol.*, 11, 59, 1982.

Chapter 12

HEPATITIS B VIRUS

J. J. Alexander and S. Aspinall

TABLE OF CONTENTS

I. INTRODUCTION

The acute clinical manifestations of human disease caused by hepatitis B virus (HBV) infections have been known for centuries, but the nature of the infectious entity causing transmissible long incubation or serum hepatitis was unknown until comparatively recently. In 1964 Blumberg et al.[1] described the Australia antigen, an antigen initially found in the blood of an Australian aborigine and shown to occur too in the blood of patients with acute long incubation hepatitis. Australia antigen subsequently proved to be the surface antigen (HBsAg) of the complete infectious HBV which was identified in 1970.[2] Progress in HBV research has been remarkable over the last decade especially considering the restrictions which limit conventional laboratory investigations of this virus. The natural host is man with the chimpanzee the only other susceptible animal. HBV cannot be grown in tissue culture although it is isolated from the liver and blood of infected persons. The patient may show clinical symptoms of hepatitis of varying severity or may be a clinically inapparent persistent carrier of the virus and therefore serve as a reservoir for new infections. About 200 million people worldwide are estimated to be HBV carriers.[3] There is persuasive evidence too that HBV infection is associated with the subsequent development of cirrhosis and primary liver cancer (PLC), one of the most common and invariably fatal malignancies known.[3,4]

In 1978[5] a virus morphologically similar to HBV was isolated from woodchucks and named the woodchuck hepatitis B virus. In 1980 a similar, but antigenically distinct virus was found in ground squirrels[6] and another HBV-like virus was demonstrated in Pekin ducks, the duck hepatitis B virus (DHBV).[7] Thus, a number of HBV-like viruses exist and Robinson[8] has suggested that this group of morphologically and biochemically similar small DNA-containing viruses, which multiply primarily in the liver, should be classified collectively into a separate group, the Hepadna viruses.

II. STRUCTURE OF HEPATITIS B VIRUS

The virion (Figure 1) is a 42-nm particle surrounded by a lipoprotein coat which constitutes the HBsAg. Two other morphological particles are found in large excess in infected blood. These are 22-nm spheres and tubules of 22-nm diameter with lengths ranging from 50 to 250 nm. Both are noninfectious and are composed HBsAg only. The virion, or Dane particle, has a nucleocapsid within the HBsAg envelope comprising a core antigen (HBsAg) within which is a circular partially double-stranded DNA molecule of 3.2 kilo base pairs.[9] Within the HBcAg are a DNA polymerase,[10] a protein kinase,[11] and a protein covalently linked to the DNA,[12] the function of which is unknown. Associated with the virion is hepatitis B e antigen (HBeAg) which also occurs free in infectious blood. HBeAg is a breakdown product of HBcAg[13,14] although it has antigenic and immunogenic properties separate from HBsAg.

The two components of HBsAg are a 22-kdalton polypeptide and a 22-kdalton glycoprotein; the latter being the 22-kdalton polypeptide containing a carboyhdrate moeity.[15] Thus, a single gene codes for HBsAg. All HBs antigens contain a common antigenic specificity "a"[16] together with a number of mutually exclusive antigenic subtypes "d" or "y" and "w" or "r". The prevalence of the subtypes differs globally, ad being more common in Africa and ay more common is Asia.[17] HBsAg consists of one major 19-k polypeptide[18] and since HBeAg is derived from HBcAg one viral gene will code for both antigens.

The entire sequence of HBV DNA has been determined[19,20,21] and both surface (S) and core (C) genes have been located (Figure 1). while it is not certain if the other virally associated proteins, DNA polymerase, protein kinase, and the DNA-linked protein, are of viral or cellular origin the former is more likely, and a large open reading frame on the HBV DNA, overlapping both C and S gene sequences, representing about

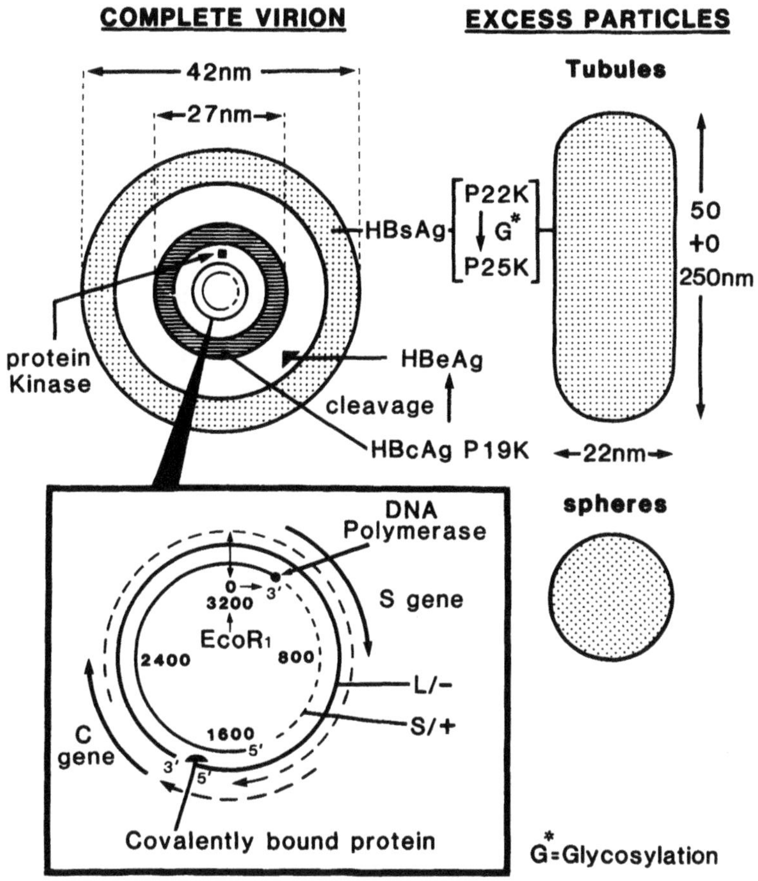

FIGURE 1. The structure of HBV particles and the nature of the genome.[22]

75% of the genome, has been identified. This sequence may represent the viral DNA polymerase gene.[22] Likewise another smaller reading frame which would specify a basic protein may represent the gene for the DNA associated protein. The gene coding for protein kinase, which preferentially phosphorylates serine residues on the HBcAg, has not been located but may represent a portion of the large open reading frame particularly if the products of the region have multiple functions.[23]

The circular nature of virion-associated HBV DNA is maintained by the short overlapping sequence of nucleotides around the 1.6 kilo base region between the long or minus strand and the short or plus strand, since the minus strand is not a covalently closed circle. The plus strand in infectious virions is incomplete, the gap varying from 15% to 40% of the genome although the 5′ end of the short strand is always located at the same fixed position.[24]

Virions isolated from infected blood have endogenous DNA polymerase activity and if incubated together with nucleotide triphosphates, synthesis of the plus strand will continue. Thus, diagnostically it is possible to determine of HBsAg positive blood contains infectious virus.

III. REPLICATION

The mechanisms by which HBV replicates are not yet comletely understood but the fact the HBV-like viruses have recently been found in other nonhuman species has provided additional sources of experimental material.

HBV was considered a strictly hepatotropic virus, but recently HBV DNA has been found in Kaposi sarcoma cells.[25] DHBV DNA and antigens have also been demonstrated in the kidney, pancreas, and liver of infected Pekin ducks.[26]

Both HBsAg and HBcAg have been located in infected human hepatocytes by immunofluorescent and immunoperoxidase techniques. HBcAg occurs both in the nucleus and the cytoplasm while HBsAg is intracytoplasmic. Burrell et al.[27] studied human liver biopsy and autopsy material to locate both HBcAg and HBsAg by immunofluorescent techniques. In the same tissue sections viral DNA was located by *in situ* hybridization with cloned radiolabeled HBV DNA probes. Large amounts of cytoplasmic single-stranded HBV DNA were found in foci of 5 to 50 cells which showed signs of cell damage. This was a significant finding since no other small DNA-containing viruses replicate in the cytoplasm.

Summers and Mason[28] have carried out more extensive studies on the replication of DHBV (which is closely related to HBV), and have shown that the mode of replication of DHBV takes place by a previously unknown mechanism — via an RNA intermediate. The complete DHBV DNA minus strand is transcribed into an RNA plus strand which has been termed the pregenome. The pregenome serves as a template for the synthesis of progeny DNA minus strands and this transcription is mediated by viral DNA polymerase. In the process of pregenome is degraded. Viral polymerase also later transcribes the complementary plus DNA strand. The early transcription of pregenomic RNA to DNA minus strands takes place within structures which resemble immature core particles. Complementary DNA plus strand synthesis may continue in mature core particles and final polymerization to complete the plus strand may be accomplished by the progeny virion during the next round of infection.

The replication strategy of DHBV and probably all HBV-like viruses therefore resembles that used by retroviruses in that a reverse transcription step is included. A replicating RNA intermediate — the pregenome — is required to produce DNA for progeny virions in HBV-like viruses while retroviruses require a DNA intermediate to produce the RNA-containing virions.

Other similarities too, exist between HBV-like viruses and the retroviruses. The overlap at the 5′ ends of the plus and minus strands would resemble the long terminal repeats found in retroviruses if the HBV DNA was linear. Only one strand of the DNA contains the base sequences for gene products and all genes are read off in the same direction. The configuration of HBV virion-associated DNA resembles the DNA intermediate postulated for the retrovirus group.[29,30]

IV. TRANSMISSION

HBV is transmitted parenterally by whole blood, blood products, and body secretions such as semen and saliva. Once routine testing of blood for HBsAg was introduced, most cases of HBV-associated post-transfusion hepatitis were eliminated. Those at high risk of becoming infected are male homosexuals, drug addicts sharing syringes, and people receiving tattos. Other groups such as surgeons, dentists, and personnel in medical pathology laboratories are also exposed to infection. Children born to HBsAg and HBeAg positive mothers too, have a high risk of infection and frequently become carriers themselves.[31] Although the exact mode of transmission in these cases is not known, recent findings suggest that *in utero* infections may occur although most infections probably take place in the perinatal period.

V. PATHOGENESIS

HBV multiplies in hepatocytes. Clinical symptoms may be absent or mild or severe or fulminant. Factors determining each individual response are unknown, although it is generally believed that manifestations of disease are governed by the host's immune system, notably the cell mediated immune response.[32] Only recently has there been any evidence for direct virus-induced cytolysis.[27] Viral DNA is integrated into host cell chromosomes in many long-term carriers,[33] thus, the nature of the association of HBV within infected cells too may influence the short- or long-term pathology. Healthy carriers may eventually develop cirrhosis or PLC.

VI. HEPATITIS B VIRUS AND PRIMARY LIVER CANCER

On a global basis, populations with a high HBV carrier rate have a high incidence of PLC. In western societies where the carrier rate is less than 0.2%, PLC is rarely encountered but in some regions of Africa and Asia with carrier rates of 10 to 20%, PLC is the commonest malignancy.[4,34,35] An extensive prospective survey in Taiwan has shown that among HBsAg positive carriers the risk of developing PLC is 490-fold greater than that measured among HBsAg negative people.[36,37] Not all PLC cases develop in HBsAg positive individuals, however, but HBV DNA has been demonstrated in the hepatocytes of PLC patients who showed no other serological markers of ongoing or past infection with HBV.[38] While the presence of viral DNA within PLC cells may constitute the definitive marker to establish a causative association, a number of cell lines established from PLC do not contain HBV DNA although they show liver-specific markers, such as the production of alpha-fetoprotein, and have other characteristics associated with malignancy, for example, growth in soft agar and tumor formation in nude mice.[39-41]

Among animal species harboring HBV-like viruses only the woodchuck (*Marmota monax*) has a high rate of PLC associated with chronic woodchuck HBV infection.[42] In an overall perspective the evidence points to a close association between HBV infection followed many years later by the development of PLC in some individuals. The nature of the initial interaction between virus and host, the immune response options mounted by the host, and the resulting effects of long term virus persistence remain speculative. Undoubtedly, the combined approach of molecular biology and immunopathology will identify some of these strategies. However the technology is available at present to undertake a vaccination campaign among PLC-susceptible populations which would demonstrate within 2 to 3 decades whether or not HBV is the dominant cause of PLC in these areas.

VII. VACCINES

The successful eradication or control of many viral diseases has been achieved by the development of vaccines. HBV vaccines cannot be produced by conventional methods but some of the unique features of this virus have led to the recent development of effective, but expensive vaccines. Currently the production of cheaper vaccines is a distinct possibility.

In 1970 the first HBV vaccine was prepared by boiling HBsAg positive serum.[43] It was noninfectious and induced protective immunity.[44] Subsequent vaccines were prepared from serum containing mainly 22-nm particles which were semipurified and formalin treated.[45,46] Later refinements of the earlier purification methods led to the production of licensed vaccines, which are prepared from 22-nm particles obtained from blood. Vaccine lots are safety tested by inoculation into chimpanzees. An extensive trial was conducted in male homosexual volunteers and three doses of 20-μg amounts

Table 1
IN VITRO SYSTEMS FOR HBV STUDIES

Persistently infected cell lines derived from human PLC		Recombinant HBV-DNA systems	
Designation	Ref.	Species	Ref.
PLC/PRF/5	50, 51, 52	Vaccinia virus	49
Hep3B	40	Bacteria	19, 20, 21, 55, 56
Delsh-5	53	Yeast	57, 58
HuH-1	54	Mouse cells	59
HuH-4	54	Rat cells	60

induced protective antibodies to HBsAg in over 95% of vaccinees. The vaccine was also effective in postexposure prophylaxis.[47] Babies of carrier mothers who are vaccinated shortly after birth also develop antibodies,[48] so it is possible to eradicate a potential reservoir of human carriers.

Advances in recombinant DNA techniques have provided the means of producing specific subunit vaccines in a number of alternative biological systems. Recently, vaccinia virus recombinants containing the HBsAg gene segment have been constructed.[49] The S gene sequence, under control of vaccinia virus early promotors, produced and excreted into the tissue culture medium 1.7 μg HBsAg per 5×10^6 infected tissue culture cells within 24 hr. Rabbits inoculated intradermally with this vaccinia virus recombinant developed high serum titers of anti-HBsAg within a month.

The combination of recombinant techniques and the finding that HBV and PLC appear to be causally related has led to the development of a number of in vitro experimental systems.

VIII. IN VITRO STUDIES

While HBV cannot be grown in vitro, viral gene expression has been investigated in a number of persistently infected cell lines derived from human PLC. Viral DNA too has been introduced into prokaryotic and eukaryotic cells by recombinant techniques (Table 1).

The HBsAg-producing cell lines established PLC synthesize only HBsAg although the entire HBV DNA is present, at least in both PLC/PRF/5 and Hep3B lines.[40,61] No infectious virions are produced but the HBsAg is morphologically,[52] biochemically,[62] and immunologically similar to the 22-nm particles extracted from human serum[63] and the amounts produced are in the region of 0.1 to 1 μg/10^6 cells per 24 hr.

None of the bacterial recombinant HBV DNA systems satisfactorily synthesized HBsAg; the molecules, when produced, were degraded and had a deleterious effect on the procaryotic host cells. However HBcAg was produced when the C gene coding region was introduced into bacteria.[20] Plasmids containing S gene coding sequences expressed HBsAg in yeast cells. The antigen was extracted from cell lystates in amounts ranging from 2 to 5 μg/200 mℓ culture[57] and from 1 to 3.4 μg/mℓ of cell lysate, or an estimated 500,000 molecules per cells.[58]

Mammalian cells lacking thymidine kinase (TK) were cotransfected with plasmids containing tandem head to tail multiple copies of HBV DNA and plasmids containing herpes simplex virus TK^+ genes as the selectable marker.[59,60] The amounts of antigen produced by TK^+ and HBsAg positive cell clones in 24 hr ranged from 2 to 4×10^4 particles or 2 to 4×10^6 HBsAg polypeptides per cell[59] to 1 to 1.5 μg HBsAg per 10 cm^2 confluent culture dish.[60] One HBsAg positive rat cell line also produces HBcAg and HBeAg and contains up to 20 complete HBV genome copies integrated into cellular

Table 2
HBV ANTIGENS EXPRESSED IN IN VITRO SYSTEMS

PLC cell lines	HBV-DNA present	HBsAg	HBcAg
PLC/PRF/5	4—6 genome copies	+	−
Hep3B	2—3 genome copies	+	−
Delsh-5	NT[a]	+	NT
HuH-1	NT	+	NT
HuH-4	NT	+	NT
Transformants			
Vaccinia virus	S gene	+	−
Bacteria	S gene	±	−
Bacteria	C gene	−	+
Yeast	S gene	+	−
Mouse cells	2—3 genome copies	+	−
Rat cells	18—20 genome copies	+	+

[a] NT = not tested.

DNA.[60] Mammaliam cell lines transformed in vitro synthesize HBsAg in amounts similar to that produced by cell lines derived from human PLC which contain one to six HBV DNA gene copies. Messenger RNA synthesis required for the production of HBcAg may require a precursor greater than the length of the entire single genome[60] although DNA in the PLC/PRF/5 cell line contains C gene sequences which are highly methylated while the S gene regions have few methylated bases which could also explain nonexpression of HBcAg.[64] Prolonged treatment of PLC/PRF/5 cells with the demethylating agent 5′ azacytidine however did not lead to expression of HBcAg as measured by radioimmunoassay or immunofluorescence.[65]

The presence of viral DNA and gene expression in both naturally derived human cell lines and in prokaryotic and eukaryotic systems transformed in vitro is summarized in Table 2.

A number of these in vitro experimental systems have been used in various other studies which have shown that HBsAg production is not affected by exogenously added interferon,[66,67] and that the cells cannot be cured of HBsAg production by growth in medium containing antibody to HBsAg.[68] PLC/PRF/5, Hep3B and mouse cells transfected with HBV DNA form rapidly growing tumors in nude mice and the mice become serologically positive for HBsAg.[67,69]

PLC/PRF/5 cultures have been used as target cells to measure the effect of peripheral blood mononuclear cells from patients with HBV infections, but the in vitro cytotoxicity measured did not reflect the in vivo pathological differences found between asymptomatic HBV carriers and patients with chronic liver disease associated with HBV.[70]

IX. SUMMARY

Viruses and the diseases they cause have been studied traditionally by developing experimental systems, either eggs, laboratory animals, or tissue cultures, in which the virus replicates. HBV is an exception. It is also exceptional in that infection may cause a variety of pathological conditions ranging from acute disease to persistent infection with or without clinical manifestations, or lead to the development of PLC many years later. The reasons why remain unknown. Knowledge of the molecular structure of this small DNA-containing virus is almost complete. Studies into the probable mode of replication indicate a novel mechanism, via an RNA intermediate, and comparisons with RNA-containing tumor-causing viruses are apparent.

A number of experimental systems exist which contain some or all of the HBV genome. Some have been derived from human PLC, some have been produced by recombinant DNA techniques. Studies on in vitro experimental systems have shed little light on the immunopathology of HBV infection. There are other nonlaboratory animal species which harbor HBV-like viruses, and one of these has a high incidence of PLC. Effective vaccines have been produced from the blood of infected individuals and a variety of other potential vaccine sources have been constructed.

Current and future studies on HBV and related viruses both at the molecular level and at that of the infected patient may reveal as many surprises in the next few years as have been uncovered in the past decade.

REFERENCES

1. Blumberg, B., Alter, H., and Visnich, S., A new antigen in leukemia sera, *JAMA*, 191, 541, 1965.
2. Dane, D., Cameron, C., and Briggs, M., Virus like particles in serum of patients with Australia-antigen associated hepatitis, *Lancet*, i, 695, 1970.
3. Szmuness, W., Hepatocellular carcinoma and the hepatitis B virus: evidence for a causal association, *Prog. Med. Virol.*, 24, 40, 1978.
4. MacSween, R., Pathology of viral hepatitis and its sequelae, *Clin. Gastroenterol.*, 9, 23, 1980.
5. Summers, J., Smolec, J., and Snyder, R., A virus similar to human hepatitis B virus associated with hepatitis and hepatoma in woodchucks, *Proc. Natl. Acad. Sci. U.S.A.*, 75, 4533, 1978.
6. Marion, P., Oshiro, L., Regnery, D., Scullard, G., and Robinson, W., A virus in Beechy ground squirrels that is related to hepatitis B virus of humans, *Proc. Natl. Acad. Sci. U.S.A.*, 77, 2941, 1980.
7. Mason, W., Seal, G., and Summers, J., Virus of Pekin ducks with structural and biological relatedness to human hepatitis B virus, *J. Virol.*, 36, 829, 1980.
8. Robinson, W., Genetic variation among hepatitis B and related viruses, *Ann. N.Y. Acad. Sci.*, 354, 371, 1980.
9. Robinson, W., Clayton, D., and Greenman, R., DNA of a human hepatitis B virus candidate, *J. Virol.*, 14, 384, 1974.
10. Kaplan, P., Greenman, R., Gerin, J., Purcell, R., and Robinson, W., DNA polymerase associated with human hepatitis B antigen, *J. Virol.*, 12, 995, 1973.
11. Albin, C. and Robinson, W., Protein kinase activity in hepatitis B virus, *J. Virol.*, 34, 297, 1980.
12. Gerlich, W. and Robinson, W., Hepatitis B virus contains protein attached to the 5′ terminus of its complete DNA strand, *Cell*, 21, 801, 1980.
13. Hindman, S., Gravelle, C., Murphy, B., Bradley, D., Budge, W., and Maynard, J., "e" Antigen, Dane particles, and serum DNA polymerase activity in HB_sAg carriers, *Ann. Intern. Med.*, 85, 458, 1976.
14. MacKay, P., Lees, J., and Murray, K., The conversion of hepatitis B core antigen synthesized in *E. coli* into e antigen, *J. Med. Virol.*, 8, 237, 1981.
15. Shiraishi, H., Kohama, T., Shirachi, R., and Ishida, N., Carbohydrate composition of hepatitis B surface antigen, *J. Gen. Virol.*, 36, 207, 1977.
16. Almeida, J., Zuckerman, A., Taylor, P., and Waterson, A., Immune electron microscopy of the Australia SH (serum hepatitis) antigen, *Microbios*, 2, 117, 1969.
17. Mazzur, S., Burgert, S., and Blumberg B., Geographical distribution of antigen determinants d, y and w, *Nature (London)*, 247, 38, 1974.
18. Hrushka, J. and Robinson, W., The protein of hepatitis B Dane particle cores, *J. Med. Virol.*, 1, 119, 1977.
19. Galibert, F., Mandart, E., Fitoussi, F., Tiollais, P., and Charnay P., Nucleotide sequence of hepatitis B virus genome (subtype ayw) cloned in *E. coli*, *Nature (London)*, 281, 646, 1979.
20. Pasek, M., Goto, T., Gilbert, W., Zink, B., Schaller, H., Mackay, P., Leadbetter, G., and Murray, K., Hepatitis B virus genes and their expression in *E. coli*, *Nature (London)*, 282, 575, 1979.
21. Valenzuela, P., Gray, P., Quiroga, M., Zaldiuar, J., Goodman, H., and Rutter, W., Nucleotide sequence of the gene coding for the major protein of hepatitis B surface antigen, *Nature (London)*, 280, 815, 1979.

22. Tiollais, P., Charnay, P., and Vyas, G., Biology of hepatitis B virus, *Science*, 213, 406, 1981.
23. Gerlich, W., Goldman, U., Muller, R., Stibbe, W., and Wolff, W., Specificity and localization of the hepatitis B virus-associated protein kinase, *J. Virol.*, 42, 761, 1982.
24. Landers, T., Greenberg, H., and Robinson, W., Structure of hepatitis B Dane particle DNA polymerase reaction, *J. Virol.*, 23, 368, 1977.
25. Siddiqui, A., Hepatitis B virus DNA in Kaposi sarcoma, *Proc. Natl. Acad., Sci. U.S.A.*, 80, 4861, 1983.
26. Halpern, M., England, J., Deery, D., Petcu, D., Mason, W., and Molnar-Kimber, K., Viral nucleic acid synthesis and antigen accumulation in pancreas and kidney of Pekin ducks infected with hepatitis B virus, *Proc. Natl. Acad. Sci. U.S.A.*, 80, 4865, 1983.
27. Burrel, C., Gowans, E., Jilbert, A., Lake, J., and Marmion, B., Hepatitis B virus DNA detection by *in situ* hybridization: implications for viral replication strategy and pathogenesis of chronic hepatitis, *Hepatology*, 2, 85S, 1982.
28. Summers, J. and Mason, W., Replication of the genome of a hepatitis B-like virus by reverse transcription of an RNA intermediate, *Cell*, 29, 403, 1982.
29. Editorial, Is hepatitis B virus a retrovirus in disguise?, *Science*, 217, 1021, 1982.
30. Editorial, A growing role for reverse transcription, *Nature (London)*, 299, 204, 1982.
31. Stevens, C., Beasely, R., Tsiu, J., and Lee, W., Vertical transmission of hepatitis B virus antigen in Taiwan, *N. Engl. J. Med.*, 292, 771, 1975.
32. Budillon, G., Scala, G., D'Onofrio, C., Cassano, S., and DeRitis, F., Diminished active T rosette levels and increased spontaneous B lymphocyte blastogenesis in hepatitis B virus positive chronic active hepatitis, *Clin. Exp. Immunol.*, 52, 472, 1983.
33. Shafritz, D., Hepatitis B virus DNA molecules in the liver of HB_sAg carriers: mechanistic considerations in the pathogenesis of hepatocellular carcinoma, *Hepatology*, 2, 35S, 1982.
34. Blumberg, B. and London, W., Hepatitis B virus and the prevention of primary hepatocellular carcinoma, *N. Engl. J. Med.*, 304, 782, 1981.
35. Prince, A., Evidence suggesting hepatitis B virus is a tumor-inducing virus in man: estimate of the risk of development of hepatocellular carcinoma in chronic HG_sAg carriers and controls, in, *The Role of Viruses in Human Cancer*, Vol. 1, Giraldo, G. and Beth, E., Eds., Elsevier/North Holland, Amsterdam, 1981, 141.
36. Beasley, R., Lin, C., Hwang, L., and Chien, C., Hepatocellular carcinoma and hepatitis B virus. A prospective study of 22,707 men in Taiwan, *Lancet*, ii, 1129, 1981.
37. Beasley, R., Hepatitis B virus as the etiologic agent in hepatocellular carcinoma — epidemiology considerations, *Hepatology*, 2, 21S, 1982.
38. Brechot, C., Scotto, J., Charnay, P., Hadchouel, M., Degos, F., Trepo, C., and Tiollais, P., Detection of hepatitis B virus DNA in liver and serum: a direct appraisal of the chronic carrier state, *Lancet*, ii, 765, 1981.
39. Koshy, R., Maupas, P., Muller, R., and Hofschneider, P., Detection of hepatitis B virus-specific DNA in the genomes of human hepatocellular carcinoma and liver cirrhosis tissues, *J. Gen. Virol.*, 57, 95, 1981.
40. Aden, D., Fogel, A., Plotkin, S., Damjanov, I., and Knowles, B., Controlled synthesis of HB_sAg in a differentiated human liver carcinoma-derived cell line, *Nature (London)*, 282, 615, 1979.
41. Nakabayashi, H., Taketa, K., Miyano, K., Yamane, T., and Sato, J., Growth of human hepatoma cell lines with differentiated function in chemically defined medium, *Cancer Res.*, 42, 3858, 1982.
42. Snyder, R., Hepatomas of captive woodchucks, *Am. J. Pathol.*, 52, 32, 1968.
43. Krugman, S., Giles, J., and Hammond, J., Hepatitis virus: effect of heat on the infectivity and antigenicity of the MS-1 & MS-2 strains, *J. Inf. Dis.*, 122, 432, 1970.
44. Krugman, S. and Giles, J., Virol hepatitis, type B (MS-2-strains). Further observations on natural history and prevention, *N. Engl. J. Med.*, 288, 755, 1973.
45. Purcell, R. and Gerin, J., Hepatitis B subunit vaccine: a preliminary report of safety and efficacy tests in chimpanzees, *Am. J. Med. Sci.*, 270, 395, 1975.
46. Maupas, P., Barin, F., Chiron, J., Coursaget, P., Goudeau, A., Perrin, J., Denis, F., and Diop-Mar, I., Efficacy of hepatitis B vaccine in prevention of early HB_sAg carrier state in children — controlled trial in an endemic area (Senegal), *Lancet*, i, 289, 1981.
47. Szmuness, W., Stevens, C., Harley, E., Zang, E., Oleszko, W., William, D., Sadovsky, R., Morrison, J., and Kellner, A., Hepatitis B vaccine. Demonstration of efficacy in a controlled clinical trial in a high risk population in the United States, *N. Engl. J. Med.*, 303, 833, 1980.
48. Stevens, C., personal communication, 1982.
49. Smith, G., Mackett, M., and Moss, B., Infectious vaccinia virus recombinants that express hepatitis B virus surface antigen, *Nature (London)*, 302, 490, 1983.
50. Alexander, J., Bey, E., Geddes, E., and Lecatsas, G., Establishment of a continuously growing cell line from primary carcinoma of the liver, *S. Afr. Med. J.*, 50, 2124, 1976.

51. Macnab, G., Alexander, J., Lecatsas, G., Bey, E., and Urbanowicz, J., Hepatitis B surface antigen produced by a human hepatoma cell line, *Br. J. Cancer,* 34, 509, 1976.
52. Alexander J., Macnab, G., and Saunders R., Studies on *in vitro* production of hepatitis B surface antigen by a human hepatoma cell line, in, *Perspectives in Virology,* Vol. 10, Pollard, M., Ed., Raven Press, New York, 1978, 103.
53. Das, P., Nayak, N., Tsiquane, K., and Zuckerman, A., Establishment of a human hepatocellular carcinoma cell line releasing hepatitis B surface antigen. *Br. J. Exp. Pathol.,* 61, 648, 1980.
54. Huh, N. and Utakoji, T., Production of HB,Ag by two new human hepatoma cell lines and its enhancement by dexamethazone, *Gann,* 72, 178, 1981.
55. Burrell, C., Mackay, P., Greenaway, P., Hofschneider, P., and Murray, K, Expression in *Escherichia coli* of hepatitis B virus DNA sequences cloned in plasmid pBR322, *Nature (London),* 279, 43, 1979.
56. Charnay, P., Pourcel, C., Louise, A., Fritsch, A., and Tiollais, P., Cloning in *Escherichia coli* and physical structure of hepatitis B virion DNA, *Proc. Natl. Acad. Sci. USA,* 76, 2222, 1979.
57. Velenzuela, P., Medina, A., Rutter, W., Ammerer, G., and Hall, B., Synthesis and assembly of hepatitis B virus surface antigen particles in yeast, *Nature (London),* 298, 347, 1982.
58. Miyanohaka, A., Toh-E, A., Nozaki, C., Hamada, F., Ohtoma, N., and Matsubara, K., Expression of hepatitis B surface antigen gene in yeast, *Proc. Natl. Acad. Sci. USA,* 80, 1, 1983.
59. Dubois, M., Pourcel, C., Rousset, S., Chany, C., and Tiollais, P., Excretion of hepatitis B surface antigen particles from mouse cells transformed with cloned viral DNA, *Proc. Natl. Acad. Sci. USA,* 77, 4549, 1980.
60. Gough, N. and Murray, K., Expression of hepatitis B surface, core and e antigen genes by stable rat and mouse cell lines, *J. Mol. Biol.,* 162, 43, 1982.
61. Marion, P., Salazar, F., Alexander, J., and Robinson, W., State of hepatitis B viral DNA in a human hepatoma cell line, *J. Virol.,* 33, 795, 1980.
62. Marion, P., Salazar, F., Alexander, J., and Robinson, W., Polypeptides of hepatitis B surface antigen produced by a hepatoma cell line, *J. Virol.,* 32, 796, 1979.
63. Daemer, R., Feinstone, S., Alexander, J., Tully, J., Tully, W., London, W, Wong, W., and Purcell, R., PLC/PRF/5 (Alexander) hepatoma cell line. Studies on infectivity and synthesis of hepatitis B virus antigens, *Infect. Immun.,* 30, 607, 1980.
64. Miller, R. H. and Robinson, W. S., Integrated hepatitis B virus DNA sequences specifying the major viral core polypeptide are methylated in PLC/PRF/5 cells, *Proc. Natl. Acad. Sci. USA,* 80, 2534, 1983.
65. Aspinall, S. and Alexander, J., unpublished data, 1983.
66. Desmyter, J., De Groot, G., Ray, M., Bradburne, A., Desmet, V., De Somer, P., and Alexander, J., Tumorigenicity and interferon properties of the PLC/PRF/5 human hepatoma cell line, *Prog. Med. Virol.,* 27, 103, 1981.
67. Lemon, S. and Bancroft, W., Lack of specific effect of adenine arabinoside, human interferon and ribavirin on *in vitro* production of hepatitis B surface antigen, *J. Infect. Dis.,* 140, 798, 1979.
68. Alexander, J., McElligott, S., and Saunders, R., Antibody to hepatitis B surface antigen is not cytoxic to antigen-secreting hepatocytes, *S. Afr. Med. J.,* 54, 973, 1978.
69. Knowles, B., Howe, C., and Aden, D., Human hepatocellular carcinoma lines secrete the major plasma proteins and hepatitis B surface antigen, *Science,* 209, 497, 1980.
70. Deinstag, J. and Bhan, A., Enhanced *in vitro* cell mediated cytotoxicity in chronic hepatitis infection: absence of specificity for virus expressed antigen on target cell membranes, *J. Immunol.,* 125, 2269, 1980.

INDEX

A

B

C

D

E

F

G

H

I

K

L

O

P

R

S

T

U

V

W

Milton Keynes UK
Ingram Content Group UK Ltd.
UKHW050307111024
449327UK00043B/2275

9 780367 252212